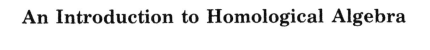

An Introduction to Homological Algebra

AN INTRODUCTION TO HOMOLOGICAL ALGEBRA

JOSEPH J. ROTMAN

Department of Mathematics
University of Illinois
Urbana, Illinois

Academic Press
San Diego New York Boston
London Sydney Tokyo Toronto

ACADEMIC PRESS, INC.
A Division of Harcourt Brace & Company
525 B Street, Suite 1900, San Diego, California 92101-4495

United Kingdom Edition published by
ACADEMIC PRESS LIMITED
24–28 Oval Road, London NW1 7DX

Library of Congress Cataloging-in-Publication Data

Rotman, Joseph J.   Date.
    An introduction to homological algebra.

    Bibliography: p.
    1.   Algebra, Homological. I. Title.
QA169.R66   512'.55   78-20001
ISBN 0-12–599250-5

AMS (MOS) 1970 Subject Classifications: 18-01,
13-01, 16-01, 20-01

To my mother, my teacher, Rose

לאמי–מורתי , רוז

Contents

Preface

In 1968, I taught a course in homological algebra at Urbana; those lectures were published in 1970 as my "Notes on Homological Algebra". Since much of the basic material there has now drifted into earlier courses, I have rewritten the notes, incorporating new items to take advantage of the present times. As it is convenient that a book be essentially self-contained, the old material remains, though redone. Therefore, students at different levels should find something of interest here.

Learning homological algebra is a two-stage affair. First, one must learn the language of Ext and Tor and what it describes. Second, one must be able to compute these things, and, often, this involves yet another language: spectral sequences. The following exercise appears on page 105 of Lang's book, "Algebra": "Take any book on homological algebra and prove all the theorems without looking at the proofs given in that book." My guess is that Lang was trying to foster the healthy attitude that, in spite of its cumbersome, perhaps forbidding, appearance, homological algebra is actually accessible to those who wish to learn it. If nothing else, this book is my attempt to make the subject lovable.

The origins of our subject are the origins of algebraic topology. At first glance, our opening sentence, "We begin with a brief discussion of line integrals in the plane", must appear strange; upon reflection, how could we begin any other way? A simple definition may mystify a reader if it seems to have been offered up by some oracle. There are no oracles here. Indeed, one role of the introductory first chapter is to give the etymology, both linguistic and mathematical, of the word *homology*.

With Chapter 1, the next three chapters form a short course in module theory. The main subject of homological algebra is the pair of bifunctors Hom and tensor. Thus, Chapter 2 studies their behavior on all modules, whereas Chapter 3 studies how they act on certain special modules: projectives, injectives, and flats. These special modules are used at once to characterize Hom and tensor (Watts' theorems, which gave rise to adjoint functor theorems in category theory). We also introduce localization for later applications to polynomial rings and to regular local rings. To make all more concrete, to see why certain earlier hypotheses are needed, and to make contact with "honest" mathematics, Chapter 4 examines the interplay between the ring of scalars and the special modules. In particular, we prove the Quillen–Suslin theorem that projective $R[x_1, \ldots, x_n]$-modules are free when R is a field.

Chapter 5 is important etymology: group extensions are considered so that one can see the familiar formulas being born, crying for a projective resolution. At last, Chapter 6 defines homology and derived functors. The next two chapters treat Ext^n, the derived functors of Hom, and Tor_n, the derived functors of tensor. We show that Ext was named for its relation to extensions, and that Tor was named for its relation to torsion.

Chapters 9 and 10 apply Ext and Tor to rings and to groups. An introduction must show why its subject is valuable, and we have chosen these two areas as the most familiar ones in algebra. Dimension theory of rings organizes much of the material in Chapter 4 and also yields new results (we describe Serre's characterization of regular local rings, and the theorem of Auslander–Buchsbaum–Nagata that these rings have unique factorization). Applications to groups include a discussion of the Schur multiplier, Maschke's theorem, the Schur–Zassenhaus theorem, and a description of the theorem of Stallings–Swan characterizing free groups as those groups of cohomological dimension one.

We have tried to keep the needs of algebraic topologists in mind, partly because one ought, and partly because the algebra is better understood when one sees its topological sources. A short discussion of Eilenberg–MacLane spaces does point out the intimate connection between topology and homological algebra; indeed, we even prove the universal coefficient theorem for group cohomology by looking at its counterpart for singular cohomology.

The final chapter, spectral sequences, is almost a text by itself. This is not unexpected, for we have already mentioned that it is a second language. (Here I have not given etymology, although Lyndon [1948] is recommended for an algebraic viewpoint.) I think this exposition differs from others in its emphasis on illustrating how spectral sequences are used. Also, exact couples seem to eliminate most of the horrors of many indices. If I have succeeded, the reader will know when a spectral sequence is likely to arise, and he will know what to do with it if it does.

And now the pleasant paragraph in which I can thank people from whom I have learned. I. Kaplansky, S. MacLane, J. Gray, and S. Gersten gave lectures that taught me much. L. McCulloh has made many valuable suggestions. E. Davis allowed me to use his account of Quillen's solution of Serre's problem. R. Treger told me of Vaserstein's proof of Suslin's solution of Serre's problem, and was kind enough to translate it for me from the Russian. An incomplete list of others who helped: P. M. Cohn, A. Fauntleroy, E. Green, P. Griffith, the late A. Learner, A. Mann, D. Robinson, M. Schacher, and A. Zaks. Of course, I am indebted to the fine books on homological algebra, especially those of Cartan–Eilenberg [1956] and MacLane [1963].

I am grateful to the Lady Davis Foundation; most of this book was written while I was a Lady Davis Fellow and Visiting Professor at the Technion, Israel Institute of Technology, Haifa, and at the Hebrew University of Jerusalem. My warm thanks to both of these institutions for their hospitality.

Finally, I thank Mrs. Esther Tuval for a superb job of typing my manuscript.

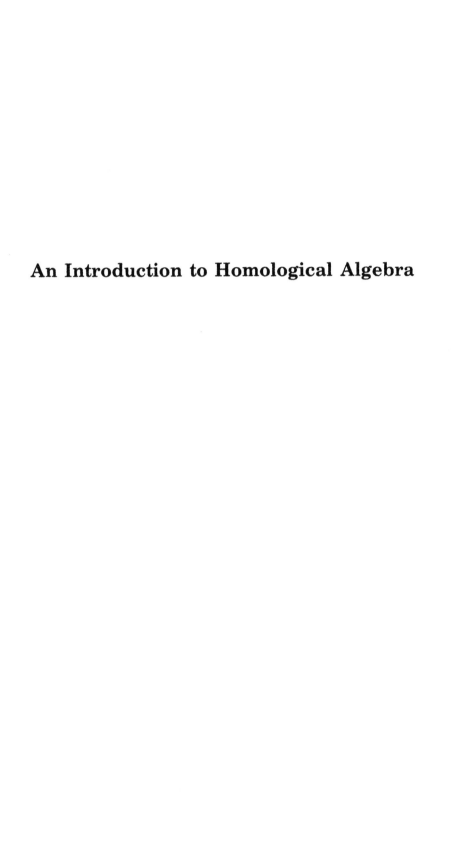

An Introduction to Homological Algebra

1 Introduction

LINE INTEGRALS AND INDEPENDENCE OF PATH

We begin with a brief discussion of line integrals in the plane. Assume X is an open disc in the plane with a finite number of points $z_1, \ldots, z_n$ deleted. Fix points a and b in X. Given a curve β in X from a to b, and given a pair of real-valued functions P and Q on X, one wants to evaluate $\int_\beta P \, dx + Q \, dy$. It is wise to regard β as a finite union of curves, for β may only be piecewise smooth, e.g., it may be a polygonal path. For the rest of this discussion, we ignore (needed) differentiability hypotheses. An important question is when the line integral above is independent of the path β; if β' is a second path in X from a to b, is $\int_{\beta'} P \, dx + Q \, dy = \int_\beta P \, dx + Q \, dy$?

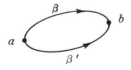

Plainly, if γ is the closed curve $\beta - \beta'$ (i.e., go from a to b via β and then go backward along β' to a), then the integral is independent of path if and only if $\int_\gamma P \, dx + Q \, dy = 0$. Thus, we are led to consideration of closed curves in X. Now this last line integral is affected by whether any of the "bad" points $z_1, \ldots, z_n$ lie inside γ (one of the functions P or Q may have a singularity

at some z_i). Consider

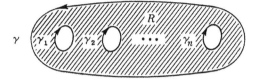

Each γ_i is a (simple) closed curve in X with the point z_i inside (and the other z's outside); γ is a (simple) closed curve in X having all the γ_i inside. If γ is oriented counterclockwise and each of the γ_i is oriented clockwise, then **Green's theorem** states that

$$\int_\gamma P\,dx + Q\,dy + \int_{\gamma_1} P\,dx + Q\,dy + \cdots + \int_{\gamma_n} P\,dx + Q\,dy$$
$$= \iint_R \left(\frac{\partial Q}{\partial x} - \frac{\partial P}{\partial y}\right) dx\,dy,$$

where R is the shaded 2-dimensional region in the picture. Of course, one is tempted to, and does, write the sum of line integrals more concisely as

$$\int_{\gamma + \gamma_1 + \cdots + \gamma_n} P\,dx + Q\,dy;$$

indeed, instead of mentioning the orientations explicitly, one often writes a formal linear combination of curves (negative coefficients being a convenient way to indicate how the orientations are aligned). Observe that for simple curves the notion of "inside" always makes sense. However, we may allow curves to wind about some z_i several times, assuming these curves are not so pathological as to fail to have an inside, and this introduces integer coefficients other than ± 1 in the formal linear combinations. Let us now restrict attention to those pairs of functions P and Q satisfying $\partial Q/\partial x = \partial P/\partial y$. For these function pairs, the double integral in Green's theorem vanishes, and we have $\int_{\gamma + \gamma_1 + \cdots + \gamma_n} P\,dx + Q\,dy = 0$. An equivalence relation on formal linear combinations (or, "oriented unions") of closed curves in X suggests itself: call two such combinations α and α' **equivalent** if the values of

$$\int_\alpha P\,dx + Q\,dy \qquad \text{and} \qquad \int_{\alpha'} P\,dx + Q\,dy$$

agree for all pairs P and Q as restricted above. In particular, note that combinations α whose curves comprise the boundary of a 2-dimensional region in X are trivial (the integrals have value 0). The equivalence classes of formal linear combinations of closed curves in X are called "homology classes", from the Greek "homologos" = "homo" + "logos" meaning "agreeing". In short, integration is always independent of paths lying in the same homology class. Visibly, these homology classes reflect the complexity of the ambient space X.

Our discussion mentioned three sorts of finite unions of oriented curves in X: arbitrary such (usually called "chains"); closed unions (usually called "cycles"); boundaries of 2-dimensional regions in X. We have just witnessed the origin of homology as an abelian group of classes of formal linear combinations of closed curves in X in which boundaries of 2-dimensional regions are trivial. The step from considering closed curves in a planar region X to generalized chains, cycles, and boundaries in any euclidean space, indeed, in any topological space, does not seem an awesome leap today; nevertheless, this leap was the birth of algebraic topology. We shall return to the picture of Green's theorem, for homological algebra was born when analogies between topological homology theory and algebra were recognized.

CATEGORIES AND FUNCTORS

Let us introduce some terms to facilitate our discussion. Recall a set-theoretical distinction: our primitive vocabulary contains the words "element" and "class"; the word "set" is reserved for those classes that are small enough to have a cardinal number. Thus, we may correctly say "the class of all rationals" or "the set of all rationals", but we may only say "the class of all groups" (this latter class cannot be a set, for there exist groups of any nonzero cardinal).

Definition: A **category** $\mathfrak{C}$ consists of a class of **objects**, obj $\mathfrak{C}$, pairwise disjoint sets of **morphisms**, $\text{Hom}_{\mathfrak{C}}(A, B)$, for every ordered pair of objects, and **compositions** $\text{Hom}_{\mathfrak{C}}(A, B) \times \text{Hom}_{\mathfrak{C}}(B, C) \to \text{Hom}_{\mathfrak{C}}(A, C)$, denoted $(f, g) \mapsto gf$, satisfying the following axioms:

(i) for each object A, there exists an **identity morphism** $1_A \in \text{Hom}_{\mathfrak{C}}(A, A)$ such that $f 1_A = f$ for all $f \in \text{Hom}_{\mathfrak{C}}(A, B)$ and $1_A g = g$ for all $g \in \text{Hom}_{\mathfrak{C}}(C, A)$;

(ii) associativity of composition holds whenever possible: if $f \in \text{Hom}_{\mathfrak{C}}(A, B)$, $g \in \text{Hom}_{\mathfrak{C}}(B, C)$ and $h \in \text{Hom}_{\mathfrak{C}}(C, D)$, then

$$h(gf) = (hg)f.$$

Several remarks are in order. First, the only requirement on $\text{Hom}(A, B)$ is that it be a set; it is allowed to be empty. Second, we usually write $f : A \to B$ instead of $f \in \text{Hom}(A, B)$; this explains our notation for composition. Finally, for each object A, the identity morphism 1_A is unique: if $2_A \in \text{Hom}(A, A)$ is a second morphism satisfying (i), then $2_A = 2_A 1_A = 1_A$. There is thus a one–one correspondence $A \mapsto 1_A$ between obj $\mathfrak{C}$ and the class of identity morphisms in $\mathfrak{C}$, so that one could describe $\mathfrak{C}$ solely in terms of morphisms and compositions. This is sometimes convenient for category-theorists, but we shall never forget objects.

Examples: 1. $\mathfrak{C} = \mathbf{Sets}$. Here objects are sets, morphisms are functions, and composition is the usual composition of functions.

2. $\mathfrak{C} = \mathbf{Top}$. Here objects are topological spaces, morphisms are continuous functions, and composition is the usual composition.

3. $\mathfrak{C} = {}_R\mathfrak{M}$. Here R is a ring (associative with 1), objects are left R-modules, morphisms are R-maps (i.e., R-homomorphisms), and usual composition.

Recall that a **left R-module** is an additive abelian group M equipped with an action of R making it a "vector space over R": there is a function $R \times M \to M$, denoted $(r, m) \mapsto rm$, satisfying:

(i) $r(m + m') = rm + rm'$;
(ii) $(r + r')m = rm + r'm$;
(iii) $(rr')m = r(r'm)$;
(iv) $1m = m$;

where $m, m' \in M$ and $1, r, r' \in R$.

A function $f: M \to N$ between left R-modules M and N is an R-**map** if

$$f(m + m') = f(m) + f(m') \qquad \text{and} \qquad f(rm) = rf(m).$$

We will often write ${}_R M$ to mean $M \in \mathrm{obj}\, {}_R\mathfrak{M}$.

4. If $R = \mathbf{Z}$, the ring of integers, then R-module = $\mathbf{Z}$-module = abelian group, while $\mathbf{Z}$-map = homomorphism. We usually denote the category of abelian groups by **Ab** rather than ${}_Z\mathfrak{M}$.

5. $\mathfrak{C} = \mathfrak{M}_R$. Here R is a ring, objects are right R-modules, morphisms are R-maps, and composition is the usual composition.

Recall that an abelian group M is a **right R-module** if there is a function $R \times M \to M$, denoted $(r, m) \mapsto rm$, satisfying all the axioms of a left R-module save (iii). The third axiom is replaced by

(iii)′ $(rr')m = r'(rm)$.

If the value of $R \times M \to M$ on (r, m) is denoted mr, then (iii)′ has the more appealing form

(iii)″ $m(rr') = (mr)r'$.

We shall consistently use the latter notation (having r on the right) when discussing right R-modules. A function $f: M \to N$ between right R-modules M and N is an R-**map** if $f(m + m') = f(m) + f(m')$ and $f(mr) = f(m)r$.

If R is commutative, we may regard every right R-module M as a left R-module by defining $rm = mr$ for all $r \in R$ and $m \in M$.

6.   Let S be a semigroup with (two-sided) unit, i.e., a **monoid**. We may construe S as a category $\mathfrak{C}$ by defining $\text{obj } \mathfrak{C} = \{*\}$ (i.e., $\mathfrak{C}$ has only one object, $*$), $\text{Hom}(*, *) = S$, and composition as the given multiplication in S.

There are two features of this last example worth noting. First, $\text{obj } \mathfrak{C}$ is a set, indeed, a rather small set. Second, morphisms are not functions between objects, for the object $*$ is not comprised of elements.

7.   Let X be a **quasi-ordered** set (this means X has a reflexive and transitive binary relation $\leq$). To say it another way, X would be partially ordered if, in addition, $\leq$ were antisymmetric. We may construe X as a category $\mathfrak{C}$ by defining $\text{obj } \mathfrak{C} = X$,

$$\text{Hom}(x, y) = \begin{cases} \{i_y^x\} & \text{if } x \leq y \quad \text{(we have invented a new symbol)} \\ \varnothing & \text{otherwise,} \end{cases}$$

and composition $i_z^y i_y^x = i_z^x$ whenever $x \leq y \leq z$.

This last example has the feature that some Hom sets may be empty.

The list of examples can be continued, and other categories will appear when necessary. There is a rough analogy between one's first meeting with point-set topology and one's first meeting with category theory. The definitions of topological space and of category are simple, abstract, and admit many examples, both natural and artificial. The *raison d'etre* of topological spaces is that continuous functions live on them, and it is continuity that is really interesting. Similarly, the *raison d'etre* of categories is that "functors" live on them; let us define these at once. If one views a category as a generalized monoid (forget the objects, so one is dealing with morphisms that can sometimes be multiplied), then a functor is just a homomorphism.

Definition: Let $\mathfrak{C}$ and $\mathfrak{D}$ be categories. A **functor** $F : \mathfrak{C} \to \mathfrak{D}$ is a function satisfying:

   (i)   If $A \in \text{obj } \mathfrak{C}$, then $FA \in \text{obj } \mathfrak{D}$;
   (ii)   If $f : A \to B$ is a morphism in $\mathfrak{C}$, then $Ff : FA \to FB$ is a morphism in $\mathfrak{D}$;
   (iii) if $A \xrightarrow{f} B \xrightarrow{g} C$ are morphisms in $\mathfrak{C}$, then

$$F(gf) = Fg\, Ff;$$

   (iv)   for every $A \in \text{obj } \mathfrak{C}$, we have $F(1_A) = 1_{FA}$.

Examples: 8.   The **identity functor** $F : \mathfrak{C} \to \mathfrak{C}$ defined by $FA = A$ and $Ff = f$.
   9.   The **Hom functors** $\mathfrak{C} \to$ **Sets**. Fix an object A in $\mathfrak{C}$ and define

$FC = \text{Hom}(A, C)$; if $f : C \to C'$ is a morphism in $\mathfrak{C}$, define $Ff : \text{Hom}(A, C) \to \text{Hom}(A, C')$ by $g \mapsto fg$. One usually denotes Ff by f_*.

10. If $\mathfrak{C} = {}_R\mathfrak{M}$, then the Hom functor $F = \text{Hom}_R(A, \ )$ actually takes values in **Ab**: $\text{Hom}_R(A, C) = \{\text{all } R\text{-maps } A \to C\}$ is an abelian group if we define $g + h$ by $a \mapsto g(a) + h(a)$; moreover, one easily checks that every f_* is a homomorphism of abelian groups.

11. A statement similar to that in Example 10 is true for $\mathfrak{C} = \mathfrak{M}_R$.

12. If D is a fixed object in $\mathfrak{D}$, define the **constant functor** $|\ | : \mathfrak{C} \to \mathfrak{D}$ by $|C| = D$ for every $C \in \text{obj } \mathfrak{C}$ and $|f| = 1_D$ for every morphism f in $\mathfrak{C}$.

Our description of the Hom functors is incomplete—what if we fix the second variable?

Definition: Let $\mathfrak{C}$ and $\mathfrak{D}$ be categories. A **contravariant functor** $F : \mathfrak{C} \to \mathfrak{D}$ is a function satisfying:

    (i)   If $A \in \text{obj } \mathfrak{C}$, then $FA \in \text{obj } \mathfrak{D}$;

    (ii)′   if $f : A \to B$ is a morphism in $\mathfrak{C}$, then $Ff : FB \to FA$ is a morphism in $\mathfrak{D}$ (thus F reverses arrows);

    (iii)′   if $A \xrightarrow{f} B \xrightarrow{g} C$ are morphisms in $\mathfrak{C}$, then

$$F(gf) = Ff \, Fg;$$

    (iv)   for every $A \in \text{obj } \mathfrak{C}$, we have $F(1_A) = 1_{FA}$.

Examples: 13. Fix an object B in $\mathfrak{C}$ and define $F : \mathfrak{C} \to \textbf{Sets}$ by $FA = \text{Hom}(A, B)$; if $f : A \to A'$ is a morphism in $\mathfrak{C}$, define $Ff : \text{Hom}(A', B) \to \text{Hom}(A, B)$ by $g \mapsto gf$. One usually denotes Ff by f^*.

14. If $\mathfrak{C} = {}_R\mathfrak{M}$, then $F = \text{Hom}_R(\ , B)$ actually takes its values in **Ab**. A similar statement is true for $\mathfrak{C} = \mathfrak{M}_R$.

A functor is often called a **covariant functor** to emphasize it preserves the direction of arrows.

Exercises: 1.1. A morphism $f : A \to B$ in a category $\mathfrak{C}$ is called an **equivalence** if there is a morphism $g : B \to A$ in $\mathfrak{C}$ such that $gf = 1_A$ and $fg = 1_B$. What are the equivalences in the categories described above?

1.2. If $F : \mathfrak{C} \to \mathfrak{D}$ is a functor (either variance) and f is an equivalence in $\mathfrak{C}$, then Ff is an equivalence in $\mathfrak{D}$.

Definition: A category $\mathfrak{C}$ is **pre-additive** if each $\text{Hom}_{\mathfrak{C}}(A, B)$ is an (additive) abelian group and the distributivity laws hold, when defined.

Distributivity relates the addition of morphisms to the given composition in $\mathfrak{C}$. As every nonempty set admits some abelian group structure, it would be foolish not to demand distributivity in the definition of pre-additive.

Of course, $_R\mathfrak{M}$ and $\mathfrak{M}_R$ are pre-additive. If $\mathfrak{C}$ and $\mathfrak{D}$ are pre-additive categories, then we say a functor (contra or co) $F:\mathfrak{C} \to \mathfrak{D}$ is **additive** if $F(f + g) = Ff + Fg$ for every pair of morphisms f and g lying in the same group. This says the function $\text{Hom}_{\mathfrak{C}}(C, C') \to \text{Hom}_{\mathfrak{D}}(FC, FC')$ given by $f \mapsto Ff$ is a homomorphism. Note that the Hom functors on module categories are additive.

Exercises: 1.3. In a pre-additive category, **zero morphisms** $0:A \to B$ are defined (namely, the zero element of $\text{Hom}(A, B)$). If $F:\mathfrak{C} \to \mathfrak{D}$ is an additive functor (either variance) between pre-additive categories, then F preserves zero morphisms. If $\mathfrak{C}$ and $\mathfrak{D}$ are module categories, then $F(0) = 0$, where 0 is the zero module.

1.4. If R is a commutative ring, then $\text{Hom}_R(M, N)$ is an R-module if we define rf by $m \mapsto r \cdot fm$. Moreover, either Hom functor takes values in $_R\mathfrak{M}$.

(A special case of Exercise 1.4 is R a field (so that R-modules are vector spaces and R-maps are linear transformations). The contravariant functor $\text{Hom}_R(, R)$ is the dual space functor that assigns to each space its space of linear functionals and to each linear transformation its transpose.)

1.5. If R is a ring, its **center** is the subring

$$Z(R) = \{x \in R : xr = rx \text{ for all } r \in R\}.$$

For every ring R, prove that $\text{Hom}_R(M, N)$ is a $Z(R)$-module and that the Hom functors take values in $_{Z(R)}\mathfrak{M}$.

1.6. Let A be a left R-module and let $r \in Z(R)$. Prove that the function $\mu_r : A \to A$ defined by $a \mapsto ra$ is an R-map. Give an example to show this may be false if $r \notin Z(R)$. (Hint: Take $A = {}_R R$.)

1.7. Let R and S be rings and $\varphi : R \to S$ a ring map. Every left S-module M may be construed as a left R-module if one defines $r \cdot m = \varphi(r)m$ (similarly for right modules).

TENSOR PRODUCTS

There is a second family of functors, tensor products, that are as fundamental in homological algebra as the Hom functors. Before describing them, we give a construction for abelian groups; it will be generalized to modules in Chapter 3.

Definition: Let G be an (additive) abelian group with subset X; we say G is a **free (abelian) group** with **basis** X if each $g \in G$ has a unique expression

of the form

$$g = \sum_{x \in X} m_x x,$$

where $m_x \in \mathbf{Z}$ and "almost all" $m_x = 0$ (i.e., all but a finite number of $m_x = 0$).

The analogy between free groups and vector spaces is a good one; the next result shows that the notions of basis in each are kindred.

Theorem 1.1: *Let G be free with basis X; let H be any abelian group and and let $f : X \to H$ be any function. Then there exists a unique homomorphism $\tilde{f} : G \to H$ with $\tilde{f}(x) = f(x)$ for all $x \in X$.*

Proof: If $g \in G$, then $g = \sum m_x x$. Define $\tilde{f} : G \to H$ by $\tilde{f} : g \mapsto \sum m_x f(x)$. Note that $\tilde{f}$ is well defined because the expression for g is unique and at most finitely many m_x are nonzero. That $\tilde{f}$ is unique follows from the fact that X generates G. ∎

The construction of $\tilde{f}$ from f is called **extending by linearity**.

Theorem 1.2: *Given a set X, there exists a free abelian group G having X as a basis.*

Proof: Let $\{\mathbf{Z}_x : x \in X\}$ be a family of copies of the integers $\mathbf{Z}$. Consider the cartesian product $\prod_{x \in X} \mathbf{Z}_x$; its elements are "vectors" (m_x)—this notation indicates that the xth coordinate is $m_x \in \mathbf{Z}$. The cartesian product is an abelian group under coordinatewise addition:

$$(m_x) + (n_x) = (m_x + n_x).$$

Define $X' = \{$all vectors having 1 as xth coordinate (fixed $x \in X$) and all other coordinates $0\}$. Clearly X' is in one–one correspondence with X. Define G' as the subgroup of $\prod \mathbf{Z}_x$ generated by X'. It is immediate that each $g \in G'$ is a linear combination of elements of X'; that such an expression for g is unique is the definition of equality in $\prod \mathbf{Z}_x$. Thus, G' is a free abelian group with basis X'.

If one actually wants X itself as a basis, define $G = (G' - X') \cup X$ (i.e., replace X' by X) and define the obvious addition on the new set G; the new group G is free abelian with X as basis. ∎

Here is a very formal but useful notion.

Definition: A diagram

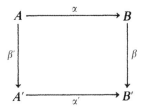

commutes if $\beta\alpha = \alpha'\beta'$; a diagram

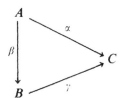

commutes if $\alpha = \gamma\beta$; a larger diagram comprised of squares and triangles **commutes** if each of its squares and triangles commutes.

(Of course, commutativity of a triangle is the special case of a commutative square one of whose sides is an identity map.)

Definition: Let R be an associative ring with 1. If A is a right R-module, B a left R-module, and G is an additive abelian group, then an R-**biadditive function**[1] is a function

$$f : A \times B \to G$$

such that for all a, $a' \in A$, b, $b' \in B$, and $r \in R$,

(i) $f(a + a', b) = f(a, b) + f(a', b)$;
(ii) $f(a, b + b') = f(a, b) + f(a, b')$;
(iii) $f(ar, b) = f(a, rb)$.

We define a tensor product of A and B as an abelian group $A \otimes_R B$ that converts R-biadditive functions $A \times B \to G$ to additive maps $A \otimes_R B \to G$.

Definition: A **tensor product** of $A \in \mathfrak{M}_R$ and $B \in {}_R\mathfrak{M}$ is an abelian group $A \otimes_R B$ and an R-biadditive function h which solves the following "universal

---

[1] There is a related but distinct notion, R-bilinearity, that we will consider later in this chapter.

mapping problem":

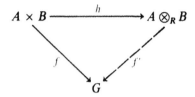

for every abelian group G and every R-biadditive f, there exists a unique homomorphism f' making the diagram commute.

Theorem 1.3:   *Any two tensor products of A and B are isomorphic.*

Proof: Suppose there is a second group X and an R-biadditive $k : A \times B \to X$ that also solves the universal mapping problem. There is a diagram

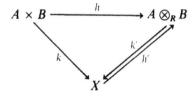

where k' and h' are homomorphisms such that

$$k'h = k \qquad \text{and} \qquad h'k = h.$$

There is a second diagram

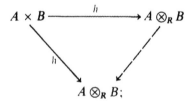

for the dashed arrow, we may choose either the identity or $h'k'$ (for $h'k'h = h'k = h$); uniqueness yields $1 = h'k'$. A similar argument shows that $k'h' = 1_X$, so that k' is an isomorphism $A \otimes_R B \xrightarrow{\sim} X$.   ∎

The proof of Theorem 1.3 should serve as a model for a more general fact: any two solutions of a universal mapping problem, if, indeed, there are any, are isomorphic. Having proved uniqueness of tensor product, we now prove its existence.

Theorem 1.4:   *The tensor product of a right R-module A and a left R-module B exists.*

Proof: Let F be the free abelian group with basis $A \times B$, i.e., F is the group of $\mathbf{Z}$-linear combinations of all ordered pairs (a, b). Define S as the subgroup of F generated by all elements of the following three forms:

$$(a + a', b) - (a, b) - (a', b), \qquad (a, b + b') - (a, b) - (a, b'),$$
$$(ar, b) - (a, rb).$$

Define $A \otimes_R B = F/S$. If we denote the coset $(a, b) + S$ by $a \otimes b$, then it is easy to check that $h: A \times B \to A \otimes_R B$ defined by $(a, b) \mapsto a \otimes b$ is R-biadditive.

Assume G is an abelian group, and we have the diagram

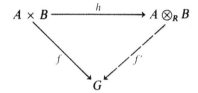

where f is R-biadditive. Since F is free on $A \times B$, there is a unique homomorphism $\varphi: F \to G$ with $\varphi((a, b)) = f(a, b)$. Moreover, f being R-biadditive implies $S \subset \ker \varphi$; it follows that φ induces a homomorphism $f': F/S \to G$ by $f': (a, b) + S \mapsto \varphi((a, b)) = f(a, b)$. This says that $f': a \otimes b \mapsto f(a, b)$, and this says $f'h = f$. We leave the proof of uniqueness of f' to the reader (use the fact that $A \otimes_R B$ is generated by all $a \otimes b$).   ∎

It is important to realize that a typical element of $A \otimes_R B$ has the form $\sum a_i \otimes b_i$ and may not have an expression of the form $a \otimes b$; furthermore, the expression $\sum a_i \otimes b_i$ is not unique. As a practical matter, one must be suspicious of a "homomorphism" g with domain $A \otimes_R B$ given by specifying its value on each $a \otimes b$; since these elements merely generate $A \otimes_R B$, the "map" g is not well defined unless it preserves all the relations. The safest scheme is to rely on the universal mapping problem, for it is easier to define an R-biadditive function on the cartesian product $A \times B$.

Theorem 1.5:   *Let $f: A \to A'$ be an R-map of right R-modules and let $g: B \to B'$ be an R-map of left R-modules. There is a unique homomorphism $A \otimes_R B \to A' \otimes_R B'$ with $a \otimes b \mapsto f(a) \otimes g(b)$.*

Proof: The function $A \times B \to A' \otimes_R B'$ defined by $(a, b) \mapsto f(a) \otimes g(b)$ is obviously R-biadditive; now use universality.   ∎

Definition: The map $A \otimes_R B \to A' \otimes_R B'$ sending $a \otimes b \mapsto fa \otimes gb$ is denoted

$$f \otimes g.$$

Theorem 1.6:   *Assume $A \xrightarrow{f} A' \xrightarrow{f'} A''$ are R-maps of right R-modules and $B \xrightarrow{g} B' \xrightarrow{g'} B''$ are R-maps of left R-modules. Then*

$$g' \otimes f' \circ g \otimes f = g'g \otimes f'f.$$

Proof: Use Theorem 1.5.   ∎

Corollary 1.7:   *If $A \in \mathfrak{M}_R$, then there is an additive functor $F: {}_R\mathfrak{M} \to \mathbf{Ab}$ defined by*

$$F(B) = A \otimes_R B;$$

if $f: B \to B'$ is an R-map between left R-modules B and B',

$$Ff = 1_A \otimes f.$$

Similarly, for fixed $B \in {}_R\mathfrak{M}$, there is an additive functor $G: \mathfrak{M}_R \to \mathbf{Ab}$ with $G(A) = A \otimes_R B$ and $Gg = g \otimes 1_B$ (where $g: A \to A'$ is an R-map).

Proof: It is clear that $F(1_B) = 1_A \otimes 1_B$ is the identity on $A \otimes_R B$, for it fixes the set of generators of the form $a \otimes b$. That F preserves composition follows immediately from Theorem 1.6, setting $g = g' = 1_A$. Finally, if $f, f': B \to B'$, then $1_A \otimes (f + f')$ sends $a \otimes b \mapsto a \otimes (f + f')b = a \otimes (fb + f'b) = a \otimes fb + a \otimes f'b = (1_A \otimes f + 1_A \otimes f')a \otimes b.$ ∎

We shall usually write $A \otimes_R$ instead of F and $\otimes_R B$ instead of G. When no confusion can arise, we suppress the subscript R.

In general, $A \otimes_R B$ is only an abelian group; however, there is a common circumstance in which it is a module.

Definition: Let R and S be rings. An abelian group B is an $(R–S)$ **bimodule**, denoted ${}_RB_S$, if B is a left R-module, a right S-module, and the two actions are related by an "associative law"

$$r(bs) = (rb)s \qquad \text{for all} \quad r \in R, \quad b \in B, \quad \text{and} \quad s \in S.$$

Remark: The last condition says that, for each $r \in R$, the map $B \to B$ given by $b \mapsto rb$ is an S-map, and, for each $s \in S$, the map $B \to B$ given by $b \mapsto bs$ is an R-map.

Examples: 15. R itself is an $(R–R)$ bimodule (which is merely a fancy restatement of the ordinary associative law).

16.   If R is commutative, every left R-module B may be construed as a right R-module by defining $br = rb$, in which case B is an $(R-R)$ bimodule.

17.   Every left R-module B is an $(R-\mathbf{Z})$ bimodule: $B = {}_R B_{\mathbf{Z}}$; similarly, every right R-module C is a $(\mathbf{Z}-R)$ bimodule: $C = {}_{\mathbf{Z}} C_R$.

18.   If R is commutative, then a ring S is an R-**algebra** if there is a ring map $\varphi: R \to Z(S)$, where $Z(S)$ is the center of S. Note that S becomes a left R-module (Exercise 1.7) if we set

$$r \cdot s = \varphi(r)s.$$

Since $\varphi(r) \in Z(S)$, this definition also makes S into a right R-module. Finally, S is a ring that is both an $(R-S)$ bimodule and an $(S-R)$ bimodule.

19.   Every ring is a $\mathbf{Z}$-algebra.

Example 16 is the special case of Example 18 in which $R = S$ and $\varphi = 1_R$. In older textbooks, one often assumes that S is an R-algebra only when R is a subring of $Z(S)$, i.e., $\varphi: R \to Z(S)$ is an inclusion. This tends to be unnecessarily restrictive; for example $\mathbf{Z}/n\mathbf{Z}$ is not a $\mathbf{Z}$-algebra with the older usage.

Theorem 1.8:   *If A is a right R-module and B is an $(R-S)$ bimodule, (briefly, A_R and ${}_R B_S$), then $A \otimes_R B$ is a right S-module, where*

$$(a \otimes b)s = a \otimes (bs).$$

Similarly, in the situation ${}_S A_R$ and ${}_R B$, then $A \otimes_R B$ is a left S-module, where

$$s(a \otimes b) = (sa) \otimes b.$$

Proof: For fixed $s \in S$, the function $\mu_s: B \to B$ defined by $b \mapsto bs$ is an R-map, for B is an $(R-S)$ bimodule. If F is the functor $A \otimes_R$, then $F(\mu_s): B \to B$ is a homomorphism (of groups). But $F(\mu_s) = 1_A \otimes \mu_s : a \otimes b \mapsto a \otimes (bs)$. Thus, the formula in the statement is well defined; it is mechanical that the module axioms are satisfied.   ∎

Corollary 1.9:   *Given A_R and ${}_R B_S$, the functor $\otimes_R B$ takes values in $\mathfrak{M}_S$; similarly, given ${}_S A_R$ and ${}_R B$, the functor $A \otimes_R$ takes values in ${}_S \mathfrak{M}$.*

Proof: All that remains to be shown is that if $f: A \to A'$ is an R-map, then $f \otimes 1_B$ is an S-map. This is a simple calculation on any generator $a \otimes b$, using the formulas we have established in Theorem 1.8.   ∎

Corollary 1.10:   *If R is commutative and S is an R-algebra, then $S \otimes_R B$ is a left S-module for every left R-module B.*

Proof: We know that S is an $(S-R)$ bimodule.   ∎

Tensoring an R-module by an R-algebra S is called **base change**, for it changes the ring of scalars from R to S. This remark should alert the reader that tensor product may behave in unexpected ways. For example, every abelian group G is a **Z**-module, and the rationals, **Q**, is a **Z**-algebra (as is any ring). Thus, $\mathbf{Q} \otimes_{\mathbf{Z}} G$ is a **Q**-module, i.e., a vector space over **Q**. Since every **Q**-space is torsion-free as an abelian group, the **Q**-space $\mathbf{Q} \otimes_{\mathbf{Z}} G$ cannot contain a copy of G if G has elements of finite order. It is thus possible that the map $G \to \mathbf{Q} \otimes_{\mathbf{Z}} G$ defined by $x \mapsto 1 \otimes x$ not be one–one! This curious phenomenon will be treated in some detail later.

The following special case of Theorem 1.8 is of sufficient importance to merit its own statement.

Corollary 1.11:   *If R is commutative and A and B are R-modules, then $A \otimes_R B$ is an R-module with $r(a \otimes b) = ra \otimes b = a \otimes rb$. Moreover, for fixed $r \in R$, if $\mu_r : S \to B$ is given by $b \mapsto rb$, then $1 \otimes \mu_r : A \otimes_R B \to A \otimes_R B$ is also multiplication by r (the same is true in the other variable).*

Theorem 1.12:   *If R is a ring and B a left R-module, then there is an R-isomorphism $R \otimes_R B \xrightarrow{\sim} B$ with $r \otimes b \mapsto rb$.*

Proof: First of all, $R \otimes_R B$ is a left R-module since R is an $(R$–$R)$ bimodule. The definition of left R-module says that the function $R \times B \to B$ with $(r, b) \mapsto rb$ is R-biadditive; the definition of tensor product provides a (group) homomorphism $\theta : R \otimes_R B \to B$ with $r \otimes b \mapsto rb$. The reader may show that θ is an R-isomorphism by observing that it has an inverse, namely, $b \mapsto 1 \otimes b$, which is an R-map.   ∎

Of course, there is a similar result on the other side: $A \otimes_R R \cong A$ as right R-modules.

Theorem 1.13:   *If R is commutative and A and B are R-modules, then there is an R-isomorphism $\tau : A \otimes_R B \to B \otimes_R A$ with $a \otimes b \mapsto b \otimes a$.*

Proof: The "twist" $t : A \times B \to B \otimes_R A$ defined by $(a, b) \mapsto b \otimes a$ is R-biadditive, hence induces a **Z**-homomorphism $\tau : A \otimes_R B \to B \otimes_R A$ that is easily seen to be an isomorphism (construct its inverse in a similar way). Theorem 1.8 shows that τ is, in fact, an R-map.   ∎

There is another consequence of Theorem 1.11 that should be mentioned.

Definition: If R is commutative and A, B, and G are R-modules, then a function $f : A \times B \to G$ is R-**bilinear** if f is R-biadditive and

$$rf(a, b) = f(ra, b) = f(a, rb) \qquad \text{all} \quad r \in R, \quad a \in A, \quad b \in B.$$

Exercise 1.8 gives an example of a biadditive function that is not bilinear. This new definition poses a new universal mapping problem:

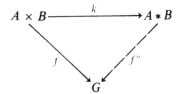

if A and B are R-modules (R commutative), is there an R-module $A * B$ and an R-bilinear function $k : A \times B \to A * B$ such that, for every R-module G and every R-bilinear $f : A \times B \to G$, there is a unique R-map f'' with $f''k = f$?

Corollary 1.14: *If R is commutative and A, B are R-modules, then $A \otimes_R B$ solves the universal mapping problem posed by R-bilinear functions as well as that posed by R-biadditive functions.*

Proof: Theorem 1.11 shows that $A \otimes_R B$ is an R-module and that $h : A \times B \to A \otimes_R B$ sending $(a, b) \mapsto a \otimes b$ is R-bilinear. Assume $f' : A \otimes_R B \to G$ is the unique $\mathbf{Z}$-map provided by R-biadditivity, so that $f'(a \otimes b) = f(a, b)$. It only remains to check that f' is an R-map. This is easy:

$$\begin{aligned}
f'(r(a \otimes b)) &= f'(ra \otimes b) &&\text{(by Theorem 1.11)} \\
&= f(ra, b) &&\text{(f is R-bilinear)} \\
&= rf(a, b) \\
&= rf'(a \otimes b). \quad \blacksquare
\end{aligned}$$

Exercise: 1.8. Let $R = \{m + n\sigma : m, n \in \mathbf{Z}, \sigma^2 = 1\}$. (A concrete realization of R is all 2×2 matrices over $\mathbf{Z}$ of the form $\left(\begin{smallmatrix} m & n \\ n & m \end{smallmatrix}\right)$). Let $A = B = \mathbf{Z}$ be made into an R-module by $(m + n\sigma)a = (m - n)a$, where $a \in \mathbf{Z}$. Let $G = \mathbf{Z}$ be made into an R-module by $(m + n\sigma)g = (m + n)g$, where $g \in G$. Finally, define $f : A \times B \to G$ to be ordinary multiplication of integers: $(a, b) \mapsto ab$. Show that f is R-biadditive but not R-bilinear.

The next exercises give associativity laws as well as a useful construction.

Definition: In the situation $(A_R, {}_R B_S, {}_S C)$, a **triadditive function**

$$f : A \times B \times C \to G$$

(where G is an abelian group) is a function that is additive in each variable and

$$f(ar, b, c) = f(a, rb, c) \qquad \text{and} \qquad f(a, bs, c) = f(a, b, sc).$$

Exercises: 1.9.   Prove that $(A \otimes_R B) \otimes_S C$ and $(a, b, c) \mapsto (a \otimes b) \otimes c$ is a solution to the universal mapping problem for triadditive functions from $A \times B \times C$. Prove that $A \otimes_R (B \otimes_S C)$ and $(a, b, c) \mapsto a \otimes (b \otimes c)$ is also a solution.

   1.10.   Using the technique of the proof of Theorem 1.3, prove the **associative law for tensor product**: there is an isomorphism $(A \otimes_R B) \otimes_S C \xrightarrow{\sim} A \otimes_R (B \otimes_S C)$ taking $(a \otimes b) \otimes c \mapsto a \otimes (b \otimes c)$. If either A or C is a bimodule, this isomorphism is a module isomorphism. (See Exercise 2.36.)

   1.11.   Prove the generalized associative law for tensor product by defining n-additive functions and showing that any association yields a solution to the corresponding universal mapping problem.

   1.12.   Let R be a commutative ring, I an ideal in R, and M and N R-modules. Prove that $M \otimes N / I(M \otimes N) \cong (M/IM) \otimes_{R/I} (N/IN)$. (Hint: Show that both modules solve the universal mapping problem for R/I-biadditive functions from $M/IM \times N/IN$.)

   1.13.   Let R and S be k-algebras. Prove that $R \otimes_k S$ is a k-algebra if one defines multiplication by $(r \otimes s)(r' \otimes s') = rr' \otimes ss'$. (Hint: This multiplication is well defined, being the composite

$$(R \otimes S) \otimes (R \otimes S) \xrightarrow{\sim} R \otimes (S \otimes R) \otimes S \xrightarrow{\;1 \otimes \tau \otimes 1\;}_{\sim} R \otimes (R \otimes S) \otimes S$$

$$\xrightarrow{\sim} (R \otimes R) \otimes (S \otimes S) \xrightarrow{\;\mu \otimes \nu\;} R \otimes S,$$

where τ is the twist map of Theorem 1.13 and μ, ν are the multiplications of R, S, respectively.)

   1.14.   Prove that $k[x] \otimes_k k[y] \cong k[x, y]$ as k-algebras.

Definition: If R is a k-algebra, define its **opposite** R^{op} as follows: as a k-module, there is a k-isomorphism $R \to R^{op}$ denoted $r \mapsto r^o$; multiplication $R^{op} \times R^{op} \to R^{op}$ is defined by $(r_1^o, r_2^o) \mapsto (r_2 r_1)^o$ (where $r_2 r_1$ is multiplication in R).

   1.15.   Every $(R–S)$ bimodule A is a left $R \otimes_k S^{op}$-module if one defines

$$(r \otimes s^o)a = (ra)s = r(as)$$

(we are assuming R and S are k-algebras).

   1.16.   Every right R-module M is a left R^{op}-module if one defines $r^o m = mr$.

   As anticipated in Exercise 1.4, the abelian group $\mathrm{Hom}_R(A, B)$ acquires a module structure if one of the variables is a bimodule. The proof of the next theorem is a dull exercise.

Theorem 1.15:   (i)   *Given $_R A_S$ and $_R B$, then $\mathrm{Hom}_R(A, B)$ is a left S-module if one defines $(sf)(a) = f(as)$;*

(ii)   *given* $_RA_S$ *and* B_S, *then* $\mathrm{Hom}_S(A, B)$ *is a right R-module if one defines* $(fr)(a) = f(ra)$;

(iii) *given* A_R *and* $_SB_R$, *then* $\mathrm{Hom}_R(A, B)$ *is a left S-module if one defines* $(sf)(a) = s(f(a))$;

(iv)   *given* $_SA$ *and* $_SB_R$, *then* $\mathrm{Hom}_S(A, B)$ *is a right R-module if one defines* $(fr)(a) = f(a)r$.

There is a mnemonic device for this theorem. Suppose one were to write maps on the side opposite to that on which the scalars are written. Thus, if A is a right module, write fa, but if A is a left module, write af. With this convention, each of the four parts of Theorem 1.15 becomes an associativity law.

There is an analogue of Theorem 1.12 holding in one variable of Hom.

Theorem 1.16:   *If B is a left R-module, then there is an R-isomorphism* $\mathrm{Hom}_R(R, B) \to B$ *with* $f \mapsto f(1)$. *Similarly, the same formula gives an R-isomorphism of right R-modules when B is a right R-module.*

Proof: Theorem 1.15 shows that the formula does define an R-map. If $b \in B$, define $f_b : R \to B$ by $r \mapsto rb$; then $b \mapsto f_b$ is the inverse. ∎

SINGULAR HOMOLOGY

Let us return to topology to construct the (singular) homology functors $H_n : \mathbf{Top} \to \mathbf{Ab}$, one functor for each $n \geq 0$; we provide more details for a generalization of our earlier discussion of curves in planar regions.

For each $n \geq 0$, consider euclidean n-space $\mathbf{R}^n$ imbedded in $\mathbf{R}^{n+1}$ as all vectors whose last coordinate is 0. Let v_0 denote the origin, and let $\{v_1, \ldots, v_n\}$ be the standard orthonormal basis of $\mathbf{R}^n$ (v_i has 1 in the ith coordinate and 0 elsewhere). For each $n \geq 0$, let $\Delta_n = \{(t_1, \ldots, t_n) : t_i \geq 0,$ all i, and $\sum t_i = 1\}$ be the convex set spanned by $\{v_0, \ldots, v_n\}$; Δ_n is called the **standard n-simplex** with **vertices** $\{v_0, \ldots, v_n\}$ and is also denoted $\Delta_n = [v_0, \ldots, v_n]$. Thus, $\Delta_0 = [v_0]$ is a point, $\Delta_1 = [v_0, v_1]$ is the unit interval $[0, 1]$; $\Delta_2 = [v_0, v_1, v_2]$ is the triangle (with interior) having vertices v_0, v_1, v_2; Δ_3 is a tetrahedron, and so forth. A **curve** in a topological space X is a continuous map $\sigma : \Delta_1 \to X$; a **closed curve** in X is a curve σ with $\sigma(0) = \sigma(1)$. The boundary of Δ_1 is $\{0, 1\}$; more generally, the boundary of Δ_n is $\bigcup_{i=0}^n [v_0, \ldots, \hat{v}_i, \ldots, v_n]$, where $\hat{}$ means "delete". However, we need an "oriented boundary" if we are to generalize the picture of Green's theorem.

Definition: An **orientation** of Δ_n is an ordering of its vertices.

It is clear that different orderings may give the "same" orientation. For example, consider Δ_2 with its vertices ordered $v_0 < v_1 < v_2$. A tour of the vertices shows that Δ_2 is oriented counterclockwise. Thus, the orderings

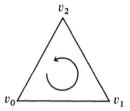

$v_1 < v_2 < v_0$ and $v_2 < v_0 < v_1$ give the same tour, while the other three permutations give a clockwise tour.

Definition: Two orientations of Δ_n are the **same** if, as permutations of $\{v_0, \ldots, v_n\}$, they have the same parity (both are even or both are odd); otherwise, the orientations are **opposite**.

After examining this definition for the tetrahedron Δ_3, the reader will be content with it.

Given an orientation of Δ_n, one may orient $[v_0, \ldots, \hat{v}_i, \ldots, v_n]$ in the sense $(-1)^i [v_0, \ldots, \hat{v}_i, \ldots, v_n]$, where $-[v_0, \ldots, \hat{v}_i, \ldots, v_n]$ means its orientation is opposite to that of $[v_0, \ldots, \hat{v}_i, \ldots, v_n]$ (vertices in displayed order). For example, consider Δ_2 oriented counterclockwise:

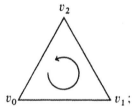

the natural way to orient the edges is:

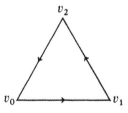

The edges are thus oriented $[v_0, v_1]$, $[v_1, v_2]$, and $[v_2, v_0]$. Since $[v_2, v_0] = -[v_0, v_2]$, the oriented boundary of Δ_2 is $[v_1, v_2] \cup -[v_0, v_2] \cup [v_0, v_1] =$

$[\hat{v}_0, v_1, v_2] \cup -[v_0, \hat{v}_1, v_2] \cup [v_0, v_1, \hat{v}_2]$. The oriented boundary of Δ_n should thus be $\bigcup_{i=0}^{n} (-1)^i [v_0, \ldots, \hat{v}_i, \ldots, v_n]$.

Definitions: If X is a topological space, an n-**simplex in** X is a continuous function $\sigma : \Delta_n \to X$. The n-**chains** in X comprise $S_n(X)$, the free abelian group with basis all n-simplexes in X. For convenience, set $S_{-1}(X) = 0$.

Observe that $S_1(X)$ is precisely the group of chains suggested by line integrals: all formal linear combinations of curves in X. The group $S_n(X)$ is the n-dimensional generalization of $S_1(X)$ we anticipated.

If $\sigma : \Delta_n \to X$, its boundary $\partial \sigma$ should be $\sum_{i=0}^{n} (-1)^i (\sigma | [v_0, \ldots, \hat{v}_i, \ldots, v_n])$. A technical problem arises. It would be nice if $\partial \sigma$ were an $(n-1)$-chain; it is not because the domain of $\sigma | [v_0, \ldots, \hat{v}_i, \ldots, v_n]$ is not the standard $(n-1)$-simplex Δ_{n-1}. To state the problem is to solve it. For each i, define $e_i : \Delta_{n-1} \to \Delta_n$ as the affine map sending the vertices $\{v_0, \ldots, v_{n-1}\}$ to the vertices $\{v_0, \ldots, \hat{v}_i, \ldots, v_n\}$ that preserves the displayed orderings:

$$e_i(t_1, \ldots, t_{n-1}) = (t_1, \ldots, t_{i-1}, 0, t_i, \ldots, t_{n-1}) \in \Delta_n.$$

Definition: If $\sigma : \Delta_n \to X$, then $\partial_n \sigma = \sum_{i=0}^{n} (-1)^i \sigma e_i \in S_{n-1}(X)$.

Theorem 1.17: *There is a unique homomorphism* $\partial_n : S_n(X) \to S_{n-1}(X)$ *with* $\partial_n \sigma = \sum_{i=0}^{n} (-1)^i \sigma e_i$ *for every n-simplex σ in X.*

Proof: Extend by linearity. ∎

The homomorphisms ∂_n are called **boundary operators**; usually one omits the subscript n.

We now have a sequence of homomorphisms

$$\cdots \to S_n(X) \overset{\partial_n}{\to} S_{n-1}(X) \to \cdots \to S_1(X) \overset{\partial_1}{\to} S_0(X) \to 0.$$

Let us denote $e_i : \Delta_{n-1} \to \Delta_n$ by $[v_0, \ldots, \hat{v}_i, \ldots, v_n]$.

Lemma 1.18: *The following formulas hold for* $e_i \circ e_j : \Delta_{n-2} \to \Delta_n$*: if* $i < j$*, then* $e_i \circ e_j = [v_0, \ldots, \hat{v}_i, \ldots, \hat{v}_j, \ldots, v_n]$*; if* $i \geq j$*, then* $e_i \circ e_j = [v_0, \ldots, \hat{v}_j, \ldots, \hat{v}_{i+1}, \ldots, v_n]$.

Proof: The maps e_i and e_j, hence their composite, are completely determined by their values on the vertices $\{v_0, \ldots, v_{n-2}\}$. The computation showing that the two displayed vertices are the deleted ones is routine. ∎

Theorem 1.19: *For each $n \geq 1$, we have $\partial_{n-1} \partial_n = 0$*

Proof: It suffices to show $\partial\,\partial\sigma = 0$ for every n-simplex $\sigma:\Delta_n \to X$.

$$\partial\,\partial\sigma = \partial(\sum(-1)^i\sigma e_i) = \sum(-1)^i\partial(\sigma e_i) = \sum_{i,j}(-1)^{i+j}\sigma e_i e_j$$

$$= \sum_{i<j}(-1)^{i+j}\sigma[v_0,\ldots,\hat{v}_i,\ldots,\hat{v}_j,\ldots,v_n]$$

$$+ \sum_{i\geq j}(-1)^{i+j}[v_0,\ldots,\hat{v}_j,\ldots,\hat{v}_{i+1},\ldots,v_n],$$

by Lemma 1.18. Now change variables in the second sum; set $l = j$ and $k = i + 1$. The second sum reads $\sum_{l<k}(-1)^{k+l-1}\sigma[v_0,\ldots,\hat{v}_l,\ldots,\hat{v}_k,\ldots,v_n]$. It is now clear that each term in the right sum occurs in the left sum with opposite sign. Therefore all cancels and $\partial\,\partial\sigma = 0$. ∎

Definition: An n-**cycle** is an element of $\ker\partial_n$; write $\ker\partial_n = Z_n(X)$. An n-**boundary** is an element of $\operatorname{im}\partial_{n+1}$; write $\operatorname{im}\partial_{n+1} = B_n(X)$.

Both $Z_n(X)$ and $B_n(X)$ are subgroups of $S_n(X)$. Our discussion of the oriented boundary of Δ_n should make the definition of $\partial_n\sigma$ appear reasonable. It is also reasonable that n-cycles are generalizations of closed curves; at the very least, this is so when $n = 1$. Assume $\sigma:\Delta_1 \to X$ is a closed curve, so that $\sigma(0) = \sigma(1)$. Now a 0-simplex in X can be identified with a point of X, so that $S_0(X)$ is the group of all formal linear combinations of the points of X. Furthermore, $\partial_1\sigma = \sigma(1) - \sigma(0)$, so a closed curve is a 1-cycle. As a second example, assume ρ, σ, and τ are curves forming a triangular path in X: say, $\rho(0) = x_0$, $\rho(1) = x_1 = \sigma(0)$, $\sigma(1) = x_2 = \tau(0)$, and $\tau(1) = x_0$. Then $\partial_1(\rho + \sigma + \tau) = (x_1 - x_0) + (x_2 - x_1) + (x_0 - x_2) = 0$, and $\rho + \sigma + \tau$ is a 1-cycle.

Corollary 1.20:   *For each $n \geq 0$, we have $B_n(X) \subset Z_n(X) \subset S_n(X)$.*

Proof: If $\beta \in B_n(X)$, then $\beta = \partial\gamma$ for some $\gamma \in S_{n+1}(X)$. Thus $\partial\beta = \partial\,\partial\gamma = 0$, by Theorem 1.19, whence $\beta \in \ker\partial_n = Z_n(X)$. ∎

Once we recall that Green's theorem tells us boundaries should be trivial, the next definition is forced on us.

Definition: The nth **homology group** of X is

$$H_n(X) = Z_n(X)/B_n(X).$$

The next few pages should be read without pausing to verify any particular assertion; more details will be provided when we study homology in a purely algebraic setting. At present, we merely wish to complete the topological tale.

For each fixed $n \geq 0$, we claim $H_n:\mathbf{Top} \to \mathbf{Ab}$ is a functor. It remains to define, for every continuous $f:X \to Y$ and every $n \geq 0$, homomorphisms

$H_n(f): H_n(X) \to H_n(Y)$. This is done as follows. As a preliminary step, define "chain" homomorphisms $f_\# : S_n(X) \to S_n(Y)$ by $\sigma \mapsto f \circ \sigma$ (where σ is an n-simplex in X) and extend by linearity. A simple calculation shows the following diagram commutes:

$$
\begin{array}{ccc}
S_n(X) & \xrightarrow{\;\partial\;} & S_{n-1}(X) \\
\downarrow{\scriptstyle f_\#} & & \downarrow{\scriptstyle f_\#} \\
S_n(Y) & \xrightarrow[\;\partial\;]{} & S_{n-1}(Y)
\end{array}
$$

i.e., $\partial f_\# = f_\# \partial$ (we have abused notation!). It follows easily that $f_\#(Z_n(X)) \subset Z_n(Y)$ and $f_\#(B_n(X)) \subset B_n(Y)$, so that $f_\#$ induces a well defined homomorphism between the quotients $H_n(X) \to H_n(Y)$ by

$$H_n(f): z_n + B_n(X) \mapsto f_\#(z_n) + B_n(Y),$$

where $z_n \in Z_n(X)$. Each H_n is, indeed, a functor.

Topological (and analytical) necessities require two modifications of this construction. If G is a fixed abelian group, replace the sequence

$$\cdots \to S_n(X) \xrightarrow{\partial_n} S_{n-1}(X) \to \cdots$$

by the sequence

$$\cdots \to S_n(X) \otimes_{\mathbf{Z}} G \xrightarrow{\;\partial_n \otimes 1_G\;} S_{n-1}(X) \otimes_{\mathbf{Z}} G \to \cdots .$$

By Theorem 1.6, the composite of adjacent maps is 0 so that we may, as above, define cycles, boundaries, and homology. The groups so obtained are denoted $H_n(X; G)$ and are called homology groups with **coefficients** G. In particular, Theorem 1.12 shows that our original construction yields the groups $H_n(X; \mathbf{Z})$.

The second modification constructs contravariant functors, called cohomology. If G is a fixed abelian group, replace the sequence

$$\cdots \to S_n(X) \xrightarrow{\partial_n} S_{n-1}(X) \to \cdots$$

by the sequence of "cochains"

$$\cdots \leftarrow \mathrm{Hom}_{\mathbf{Z}}(S_n(X), G) \xleftarrow{\;\partial_n^*\;} \mathrm{Hom}_{\mathbf{Z}}(S_{n-1}(X), G) \leftarrow \cdots .$$

The arrows have changed direction because $\mathrm{Hom}_{\mathbf{Z}}(\ , G)$ is contravariant. Again, additive functors preserve zero morphisms, so the composite of adjacent maps is still 0. Certain subgroups of $\mathrm{Hom}_{\mathbf{Z}}(S_n(X), G)$ are defined, "cocycles" and "coboundaries", and their quotient $H^n(X; G)$ is called the nth **cohomology group** of X with coefficients G. For each $n \geq 0$, $H^n(\ ; G): \mathbf{Top} \to \mathbf{Ab}$

is a contravariant functor. If one lets $G = \mathbf{R}$, the additive group of reals, this is the correct context in which to simultaneously view the Fundamental Theorem of Calculus, Green's theorem, Stokes' theorem, and higher-dimensional analogues (de Rham theorem).

We end this chapter by exhibiting an algebraic context in which one constructs a long sequence of modules and maps in which the composite of adjacent maps is 0 (details of this construction are in Chapter 3). Every module M can be described by generators and relations, i.e., there is a map $F_0 \to M$ of a "free" module F_0 onto M with kernel K_0, say. Now K_0, in turn, may also be so described: there is a map $F_1 \to K_0$ of a free module F_1 onto K_0 with kernel K_1, say. Link these together to get

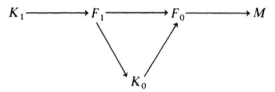

where $F_1 \to F_0$ is defined as the composite $F_1 \to K_0 \to F_0$. This procedure may be iterated indefinitely to give a sequence

$$\cdots \to F_n \to F_{n-1} \to \cdots \to F_0 \to M \to 0,$$

where each F_n is free and composites of adjacent maps are 0. Both of the topologists' modifications are available: for fixed module B, one may apply the functor $\otimes B$ to obtain a new sequence and construct homology functors; one may apply $\mathrm{Hom}(\ ,B)$ to obtain a new sequence and construct contravariant cohomology functors.

2 Hom and ⊗

Homological algebra studies a ring R by investigating its category of modules $_R\mathfrak{M}$; this category, in turn, is investigated by examining the behavior of certain functors on it, the most important of which are Hom, ⊗, and related functors derived from these.

There are at least two reasons why this approach should be successful. The fancier reason is a theorem of Morita: two commutative rings R and S are isomorphic if and only if the categories $_R\mathfrak{M}$ and $_S\mathfrak{M}$ are "equivalent"; actually, Morita's theorem gives a necessary and sufficient condition on any pair of (not necessarily commutative) rings R and S that their module categories be equivalent. This theorem thus shows that the category $_R\mathfrak{M}$ conveys much information about R. Of course, there is a much more elementary way to see this. Recall that a left R-module M is an abelian group with a scalar multiplication $\sigma: R \times M \to M$. The module axioms assert that σ is **Z**-biadditive. Thus, for every fixed $r \in R$, the function $\sigma_r: M \to M$ defined by $m \mapsto \sigma(r,m) = rm$ is a **Z**-homomorphism. Now $\operatorname{End}_{\mathbf{Z}}(M) = \operatorname{Hom}_{\mathbf{Z}}(M, M)$ is a ring if we define multiplication as composition, and it is easy to see that $\rho: R \to \operatorname{End}_{\mathbf{Z}}(M)$ defined by $r \mapsto \sigma_r$ is a ring map. Thus, every R-module M defines a **representation** of R in the endomorphism ring of an abelian group. Conversely, every such representation $\rho: R \to \operatorname{End}_{\mathbf{Z}}(M)$ makes the abelian group M into a left R-module by defining $\sigma: R \times M \to M$ by $(r,m) \mapsto \rho_r(m)$. Module theory is thus representation theory of rings.

MODULES

Let us now look at module categories. Our initial observations essentially say that the usual first properties of abelian groups and of vector spaces are also properties of more general modules.

Let R be a fixed ring (always associative with 1); we shall say "module" instead of "left R-module". Of course, all goes equally well for right modules, since Exercise 1.16 shows that every right R-module is a left R^{op}-module.

Definition: If M is a module, then a **submodule** M' of M is a subgroup that is closed under scalar multiplication:

$$m' \in M' \qquad \text{implies} \qquad rm' \in M', \qquad \text{all} \quad r \in R.$$

Examples: 1.   0 and M are submodules of M; any submodule $M' \neq M$ is called **proper**.

2.   If $M = R$, its submodules are precisely the left ideals.

3.   If I is a left ideal of R, then

$$IM = \{\textstyle\sum a_j m_j : a_j \in I, m_j \in M\}$$

is a submodule of M.

4.   If I is a two-sided ideal of R (so that R/I is a ring) and if M is a module with $IM = 0$, then M is an R/I-module (if $\bar{r} = r + I$, define $\bar{r}m = rm$).

5.   Let $f : M \to N$ be an R-map. Then

$$\ker f = \{m \in M : fm = 0\}$$

is a submodule of M, and

$$\text{im } f = f(M) = \{n \in N : n = f(m) \text{ for some } m \in M\}$$

is a submodule of N. Of course, we have abbreviated the words **kernel** and **image**.

6.   If M_1 and M_2 are submodules of M, then so is

$$M_1 + M_2 = \{m_1 + m_2 : m_1 \in M_1, m_2 \in M_2\}.$$

7.   If $\{M'_j : j \in J\}$ is a family of submodules of M, then $\bigcap_{j \in J} M'_j$ is also a submodule of M.

Definition: Let X be a subset of a module M. The submodule of M **generated** by X is $\bigcap_{j \in J} M'_j$, where $\{M'_j : j \in J\}$ is the family of all submodules of M that contain X. We denote this submodule by $\langle X \rangle$.

Theorem 2.1: *Let X be a subset of M. If $X = \varnothing$, then $\langle X \rangle = 0$; if $X \neq \varnothing$, then $\langle X \rangle = \{\sum r_i x_i : r_i \in R, x_i \in X\}$.*

Proof: If $X = \varnothing$, then 0 is a submodule of M containing X, from which it follows that $\langle \varnothing \rangle = 0$. If $X \neq \varnothing$, then the subset $S = \{\sum r_i x_i : r_i \in R, x_i \in X\}$ is defined (it is defined when $X = \varnothing$ if one enjoys summing over an empty index set). Since R contains 1, we have $X \subset S$. An easy check shows S is a submodule of M, so it follows at once that $\langle X \rangle \subset S$. For the reverse inclusion, it suffices to show that if M' is any submodule of M containing X, then $S \subset M'$ (for then S is contained in the intersection of all such M', which is $\langle X \rangle$). This is clear: $x_i \in M'$, all i, implies $\sum r_i x_i \in M'$ for all $r_i \in R$. ∎

Definition: A module M is **finitely generated** (**f.g.**) if there is a finite subset $\{x_1, \ldots, x_n\}$ of M with $\langle x_1, \ldots, x_n \rangle = M$; a module M is **cyclic** if there is a single element $x \in M$ with $\langle x \rangle = M$.

Definition: Let $f : M \to N$ be an R-map. We say f is **monic** (or is a **monomorphism**) if f is one–one; we say f is **epic** (or is an **epimorphism**) if f is onto.

Of course, f is an **isomorphism** if and only if f is both monic and epic.

Definition: If M' is a submodule of M, the **quotient module** M/M' is the quotient group M/M' made into an R-module by

$$r(m + M') = rm + M'.$$

One must assume M' is a submodule in order that the action of R on M/M' be well defined.

Examples: 8. If M' is a submodule of M, the inclusion $i : M' \to M$ is monic.

9. If M' is a submodule of M, the **natural map** $\pi : M \to M/M'$ defined by $m \mapsto m + M'$ is epic, and $\ker \pi = M'$.

10. If $f : M \to N$, then f is monic if and only if $\ker f = 0$.

11. If $f : M \to N$, then f is epic if and only if $\operatorname{coker} f = 0$ (**cokernel** f is defined as the quotient module $N/\operatorname{im} f$).

12. (**First Isomorphism Theorem**) If $f : M \to N$, then the map $m + \ker f \mapsto f(m)$ is an isomorphism $M/\ker f \xrightarrow{\sim} \operatorname{im} f$.

13. (**Second Isomorphism Theorem**) If M_1 and M_2 are submodules of M, then $m_1 + M_1 \cap M_2 \mapsto m_1 + M_2$ is an isomorphism

$$M_1/M_1 \cap M_2 \xrightarrow{\sim} (M_1 + M_2)/M_2.$$

The Second Isomorphism Theorem follows easily from the First: let $\pi : M \to M/M_1$ be the natural map, and let $f = \pi | M_1$. It is easy to see that $\ker f = M_1 \cap M_2$ and $\operatorname{im} f = (M_1 + M_2)/M_2$.

14. **(Third Isomorphism Theorem)**   If $M_2 \subset M_1$ are submodules of M, then $(M/M_2)/(M_1/M_2) \cong M/M_1$.

The Third Isomorphism Theorem also follows easily from the First: the map $f: M/M_2 \to M/M_1$ given by $m + M_2 \mapsto m + M_1$ is epic with kernel M_1/M_2.

15. **(Correspondence Theorem)**   If M' is a submodule of M, there is a one–one correspondence between the submodules S of M/M' and the "intermediate" submodules of M containing M' given by $S \mapsto \pi^{-1}(S)$ (where $\pi: M \to M/M'$ is the natural map).

Theorem 2.2:   *A module M is cyclic if and only if $M \cong R/I$ for some left ideal I. Moreover, if $M = \langle x \rangle$, then $I = \{r \in R : rx = 0\}$.*

Proof: First of all, R/I is cyclic with generator $1 + I$; if $f: R/I \to M$ is an isomorphism, then $M = \langle x \rangle$, where $x = f(1 + I)$. Conversely, assume $M = \langle x \rangle$. Define $f: R \to M$ by $f(r) = rx$. Since f is epic, $M \cong R/\ker f$. But $\ker f$ is a submodule of R, which is a left ideal; indeed, $\ker f = \{r \in R : rx = 0\}$.   ∎

Definition: Two maps

$$M' \xrightarrow{f} M \xrightarrow{g} M''$$

are **exact at** M if $\operatorname{im} f = \ker g$. A sequence of maps (perhaps infinitely long)

$$\cdots \to M_{n+1} \xrightarrow{f_{n+1}} M_n \xrightarrow{f_n} M_{n-1} \to \cdots$$

is **exact** if each adjacent pair of maps is exact.

Exercises: 2.1.   If $0 \to M' \xrightarrow{f} M$ is exact, then f is monic (there is no need to label the only possible map $0 \to M'$); if $M \xrightarrow{g} M'' \to 0$ is exact, then g is epic; if $0 \to M \xrightarrow{f} M' \to 0$ is exact, then f is an isomorphism.

2.2.   If $M' \xrightarrow{f} M \xrightarrow{g} M''$ is exact with f epic and g monic, then $M = 0$. Conclude that exactness of $0 \to M \to 0$ gives $M = 0$.

2.3.   Prove that a map φ is monic if and only if $\varphi f = \varphi g$ implies $f = g$ (the diagram is $A \underset{g}{\overset{f}{\rightrightarrows}} B \xrightarrow{\varphi} C$); prove that φ is epic if and only if $h\varphi = k\varphi$ implies $h = k$.

2.4.   If $M_1 \xrightarrow{f} M_2 \to M_3 \xrightarrow{g} M_4$ is exact, then f is epic if and only if g is monic.

2.5.   If $M_1 \xrightarrow{f} M_2 \to M_3 \to M_4 \xrightarrow{g} M_5$ is exact, then f epic and g monic imply $M_3 = 0$ (use Exercises 2.2 and 2.4).

2.6.   If $0 \to M' \xrightarrow{i} M \to M'' \to 0$ is exact, then $M' \cong iM'$ and $M/iM' \cong M''$. Such sequences are called **short exact sequences**.

2.7.   Consider the commutative diagram with exact rows

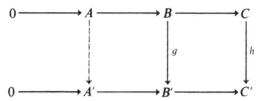

Prove that there exists a unique map $A \to A'$ making the diagram commute. Similarly, one can uniquely complete the commutative diagram with exact rows

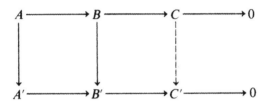

Remark: There is a categorical translation of Exercise 2.7. Let $\mathfrak{A}$ denote the category whose objects are all R-maps; define a morphism $\varphi: f \to g$ as a pair of maps $\varphi = (\varphi_1, \varphi_2)$ making the following diagram commute:

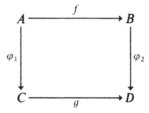

One may now see that ker and coker are functors $\mathfrak{A} \to {}_R\mathfrak{M}$.

Exercises: 2.8.   If $f: M \to N$ is a map, there is an exact sequence

$$0 \to \ker f \to M \xrightarrow{f} N \to \operatorname{coker} f \to 0.$$

2.9.   **(Restatement of Third Isomorphism Theorem)**   If $M_2 \subset M_1$ are submodules of M, there is a short exact sequence $0 \to M_1/M_2 \to M/M_2 \to M/M_1 \to 0$.

2.10.   **(Another Version of Third Isomorphism Theorem)**   Consider   the
commutative diagram

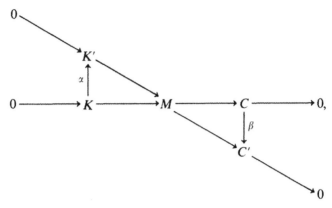

where α is monic and β is epic. Then ker $\beta \neq 0$ if and only if coker $\alpha \neq 0$,
i.e., αK is a proper submodule of K'. (Hint: Write $C = M/K$ and $C' = M/K'$,
so that ker $\beta = K'/K$.)

SUMS AND PRODUCTS

Definition: Let $\{A_j : j \in J\}$ be a family of modules. Their **product**, denoted
$\prod_{j \in J} A_j$, is the module whose underlying set is the cartesian product of the
A_j, i.e., all "vectors" $a = (a_j)$ where $a_j \in A_j$, and with module operations
defined by

$$(a_j) + (b_j) = (a_j + b_j),$$
$$r(a_j) = (ra_j).$$

Definition: The **sum** of the A_j, denoted $\coprod_{j \in J} A_j$, is the submodule of
$\prod_{j \in J} A_j$ consisting of all (a_j) "almost all" of whose coordinates $a_j = 0$ (i.e.,
all but a finite number of a_j are 0).

Other names and notations are common. Product is often called "direct
product", "complete direct sum", or "strong direct sum"; sum is often called
"direct sum" or "weak direct sum" and is usually denoted $\sum_{j \in J} A_j$. In
category theory, sums are called "coproducts".

If the index set J is finite with n elements, then $\prod_{j \in J} A_j = \coprod_{j \in J} A_j$; in
this case we also write $A_1 \oplus \cdots \oplus A_n$. If the index set J is infinite (and
infinitely many $A_j \neq 0$), then $\coprod A_j$ is a proper submodule of $\prod A_j$.

Exercises: 2.11.   If R is a field (or a division ring), every module is a sum
of copies of R (this is just the statement that every vector space has a basis).

2.12.   If M_1 and M_2 are submodules of M with $M_1 \cap M_2 = 0$ and $M_1 + M_2 = M$, then $M = M_1 \oplus M_2$. (In this case, we write $M = M_1 \oplus M_2$: "equal" instead of "isomorphic".)

Functors only recognize objects and morphisms, so our next task is to characterize product and sum in these terms. If $A = \prod A_j$, define **projections** $p_j : A \to A_j$ and **injections** $\lambda_j : A_j \to A$ by

$$p_j : (a_j) \mapsto a_j$$

and

$$\lambda_j(a_j) \quad \text{is that element having } a_j \text{ in the } j\text{th coordinate and 0 elsewhere.}$$

Note that $p_j\lambda_j = 1_{A_j}$ and $p_j\lambda_k = 0$ if $j \neq k$. These maps are also defined when A is a sum, in which case $\sum \lambda_j p_j = 1_A$; this means that for each $a \in A$, almost all $p_j a = 0$ and $\sum \lambda_j p_j a = a$.

Theorem 2.3:   *Let modules A and $\{A_j : j \in J\}$ be given. Then $A \cong \coprod_{j \in J} A_j$ if and only if there are maps $\lambda_j : A_j \to A$ such that, given any module X and any maps $f_j : A_j \to X$, there is a unique map $\varphi : A \to X$ with $\varphi\lambda_j = f_j$, all $j \in J$.*

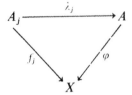

Proof: Suppose A is the sum of the A_j, and let $\{p_j : A \to A_j\}$ be the projections. Define $\varphi : A \to X$ by $\varphi a = \sum f_j p_j a$. Note that since A is a sum, almost all $p_j a$ are 0, so the formula for φa makes sense. The diagram above does commute:

$$\varphi\lambda_j a_j = \sum_k f_k p_k \lambda_j a_j = f_j a_j.$$

We claim φ is the unique such map. If $\psi : A \to X$ satisfies $\psi\lambda_j = f_j$, all j, then $\psi a = \sum \psi \lambda_j p_j a = \sum f_j p_j a = \varphi a$.

Suppose, conversely, that a unique $\varphi : A \to X$ always exists. Consider the diagrams

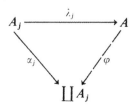   and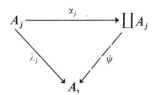

where $\alpha_j: A_j \to \coprod A_j$ is the jth injection. Now φ exists by hypothesis, while ψ exists by the first part of the proof. Assemble these diagrams into

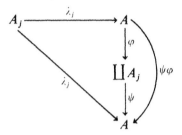

and note that we have commutativity: $\psi \varphi \lambda_j = \psi \alpha_j = \lambda_j$. The hypothesis applies to this last diagram if we set $X = A$ and $f_j = \lambda_j$, and gives uniqueness of $\psi \varphi$. But, visibly, $1_A: A \to A$ completes the same diagram, so that $\psi \varphi = 1_A$. If we assemble the original pair of diagrams a second way, then we obtain $\varphi \psi = 1_{\coprod A_j}$. Hence φ is an isomorphism. ∎

Remarks: 1. The maps λ_j must be monic.
  2. The proof above should remind the reader of the proof of Theorem 1.3. Indeed, we have shown that sum is the solution to a universal mapping problem.

Theorem 2.4: *If λ_j is the jth injection $A_j \to \coprod A_j$ and if B is a module, then the map*

$$\theta: \mathrm{Hom}(\coprod A_j, B) \to \prod \mathrm{Hom}(A_j, B)$$

given by

$$\varphi \mapsto (\varphi \lambda_j)$$

is an isomorphism.

Proof: It is clear that θ is a homomorphism (of abelian groups). θ is epic: if $(f_j) \in \prod \mathrm{Hom}(A_j, B)$, then $f_j: A_j \to B$ for each j; by Theorem 2.3, there is a map $\varphi: \coprod A_j \to B$ with $\varphi \lambda_j = f_j$, all j, i.e., $\theta(\varphi) = (f_j)$. θ is monic: suppose $\theta(\varphi) = 0 = (\varphi \lambda_j)$, i.e., each $\varphi \lambda_j = 0$. Thus, φ completes the diagrams

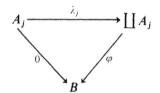

Since any completion is unique and since the zero map also completes, we have $\varphi = 0$. ∎

In terms of functors, we have just seen that $\text{Hom}(\ , B)$ converts sums to products; in particular, it preserves *finite* sums.

Exercises: 2.13.   If all the A_j are bimodules or if B is a bimodule, then the isomorphism θ of Theorem 2.4 is a map of modules.

   2.14   Give an example in which $\text{Hom}(\coprod A_j, B) \not\cong \coprod \text{Hom}(A_j, B)$.

   2.15   Give an example in which $\text{Hom}(\coprod A_j, B) \not\cong \prod \text{Hom}(A_j, B)$.

   2.16   If k is a field and V is a vector space over k, let $V^* = \text{Hom}(V, k)$, its dual space. Assume $\dim V = m$. If m is finite, then $\dim V^* = \dim V$ (whence $V^* \cong V$); if m is infinite, then $\dim V^* = |k|^m$ (whence $V^* \not\cong V$).

We are also able to describe product as the solution to a universal mapping problem.

Theorem 2.5:   *Let modules A and $\{A_j : j \in J\}$ be given. Then $A \cong \prod_{j \in J} A_j$ if and only if there are maps $p_j : A \to A_j$ such that, given any module X and any maps $f_j : X \to A_j$, there is a unique map $\varphi : X \to A$ with $p_j \varphi = f_j$, all $j \in J$.*

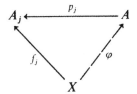

Proof: If A is a product, then one easily shows that the projections p_j have the property described. The converse is proved exactly as Theorem 2.3 if one merely reverses direction of all arrows. ∎

One may prove that the maps p_j must be epic.

Theorem 2.6:   *If p_j is the jth projection $\prod A_j \to A_j$ and if B is a module, then the map*

$$\theta : \text{Hom}(B, \prod A_j) \to \prod \text{Hom}(B, A_j)$$

given by

$$\varphi \mapsto (p_j \varphi)$$

is an isomorphism.

Proof: Repeat the proof of Theorem 2.4 *mutatis mutandis*. ∎

We have just seen that Hom(B,) preserves products; in particular, it preserves *finite* sums.

Exercises: 2.17.   If all the A_j are bimodules or if B is a bimodule, then the isomorphism θ of Theorem 2.6 is a module isomorphism.
  2.18.   Give an example in which $\mathrm{Hom}(B, \coprod A_j) \not\cong \coprod \mathrm{Hom}(B, A_j)$.
  2.19.   Give an example in which $\mathrm{Hom}(B, \coprod A_j) \not\cong \prod \mathrm{Hom}(B, A_j)$.

We have seen that the universal mapping problems describing sum and product differ only in that all arrows have reversed direction. This phenomenon is common enough to deserve a definition.

Recall that a statement S (in the first order propositional calculus) is a well formed formula involving the usual logical connectives ("and", "or", "not", "implies", "equality") and the quantifiers $\exists$ and $\forall$. If $\mathfrak{A}$ is a category, then we say S is a statement about $\mathfrak{A}$ if the variables occurring in S are interpreted as objects and morphisms of $\mathfrak{A}$.

Definition: If S is a statement about a category $\mathfrak{A}$, then its **dual** S^* is the statement about $\mathfrak{A}$ obtained by reversing the direction of each morphism (i.e., interchange "domain" and "target") and replacing each composite $\alpha\beta$ of morphisms by $\beta\alpha$.

The notion of "dual" extends in an obvious way to diagrams; reverse every arrow and retain commutativity of any square or triangle.

Exercises: 2.20.   Show that the zero module is a solution to the universal mapping problem: for every module X, there is a unique map $X \to 0$.
  2.21.   Show that the zero module is also a solution to the dual universal mapping problem: for each module X, there is a unique map $0 \to X$. Conclude that 0 is "self-dual".

It is easy to see "identity morphism" and "isomorphism" are self-dual. In Exercises 2.41 and 2.49, it is shown that "cokernel" and "kernel" are dual. From this, it follows that "monic" and "epic" are dual. Finally, "exact" is self-dual, for we may describe exactness of $A \xrightarrow{f} B \xrightarrow{g} C$ as $\ker(B \to \mathrm{coker} f) = \ker g$ or as $\mathrm{coker}(\ker f \to B) = \mathrm{coker} g$.

It is not true that the dual of a theorem must be a theorem. However, should both be true, it is often (but not always) the case that the proofs of each are dual. Two such examples are the dual pairs: Theorems 2.3 and 2.5; Theorems 2.4 and 2.6. In future instances, we will supply only one proof (as we did for Theorems 2.3 and 2.4) and merely say the other proof is dual.

The special case of the sum of two modules merits more discussion.

Theorem 2.7: *Given A and B with $i: A \to B$ monic, then A is a summand of B (i.e., $B = iA \oplus C$ for some submodule C of B) if and only if there is a map $p: B \to A$ with $pi = 1_A$.*

Proof: If A is a summand of B, define $p: B \to A$ as the projection on A having $C = \ker p$. Conversely, given $p: B \to A$ with $pi = 1_A$, define $C = \ker p$. We claim $B = iA \oplus C$. If $b \in B$, then

$$b = ipb + (b - ipb).$$

The first term is in iA, and the second is in C, for $p(b - ipb) = 0$. Finally, it is clear that $iA \cap C = 0$, so Exercise 2.12 completes the argument. ∎

Definition: A short exact sequence $0 \to A \overset{i}{\to} B \overset{p}{\to} C \to 0$ is **split** if there is a map $j: C \to B$ with $pj = 1_C$.

Exercises: 2.22. The short exact sequence above is split if and only if there is a map $q: B \to A$ with $qi = 1_A$. Conclude that in a split short exact sequence, A is a summand of B and C is a summand of B.

    2.23. Assume $A = B \oplus C$ and $B \subset M \subset A$. Prove that $M = B \oplus (M \cap C)$.

    2.24. If $F: {}_R\mathfrak{M} \to \mathbf{Ab}$ is an additive functor (either variance) and if $0 \to A \to B \to C \to 0$ is a split exact sequence, then FA is a summand of FB. Conclude that F preserves finite sums. (Hint: $1_B = jp + iq$, where the notation is as in Exercise 2.22.)

Having seen how the Hom functors treat sums, let us now see how tensor product behaves.

Theorem 2.8: *Let A be a right R-module and let $\{B_j : j \in J\}$ be left R-modules. The map*

$$\theta : A \otimes_R \coprod B_j \to \coprod (A \otimes_R B_j)$$

given by

$$a \otimes (b_j) \mapsto (a \otimes b_j)$$

is an isomorphism. There is a similar isomorphism if the sum is in the first variable.

Proof: Observe first that θ is well defined, for the function $A \times \coprod B_j \to \coprod (A \otimes B_j)$ given by $(a, (b_j)) \mapsto (a \otimes b_j)$ is R-biadditive. Let $\lambda_j : B_j \to \coprod B_j$ be the jth injection. In order to define a map $\varphi : \coprod (A \otimes B_j) \to A \otimes \coprod B_j$, it suffices to give maps $f_j : A \otimes B_j \to A \otimes \coprod B_j$. Define f_j so that $a \otimes b_j \mapsto a \otimes \lambda_j b_j$ (i.e., $f_j = 1_A \otimes \lambda_j$). If μ_j is the jth injection $A \otimes B_j \to \coprod (A \otimes B_j)$,

Theorem 2.3 gives a map φ with $\varphi\mu_j = f_j$, all j. It is a straightforward check that $\theta\varphi$ and $\varphi\theta$ are identities. ∎

We have just seen that the functor $A \otimes_R$ preserves sums; the functor $\otimes_R B$ also preserves sums.

Exercises: 2.25. If either A or each B_j is a bimodule, then the isomorphism θ in Theorem 2.8 is a map of modules.

2.26. Give an example of modules A and $\{B_j : j \in J\}$ for which $A \otimes_R (\prod B_j) \not\cong \prod (A \otimes_R B_j)$. (Hint: Take $R = \mathbf{Z}$ and $p \in \mathbf{Z}$ a prime; define $A = \mathbf{Q}$ and $B_n = \mathbf{Z}/p^n\mathbf{Z}$, $n = 1, 2, \ldots$.)

EXACTNESS

The next question is how functors affect exactness. All functors are additive functors between categories of modules.

Definition: A functor F is **left exact** if exactness of

$$0 \to A \xrightarrow{\alpha} B \xrightarrow{\beta} C$$

implies exactness of

$$0 \to FA \xrightarrow{F\alpha} FB \xrightarrow{F\beta} FC;$$

a functor F is **right exact** if exactness of

$$A \xrightarrow{\alpha} B \xrightarrow{\beta} C \to 0$$

implies exactness of

$$FA \xrightarrow{F\alpha} FB \xrightarrow{F\beta} FC \to 0.$$

If F is left exact, then $F\alpha : FA \to FB$ is monic and $\operatorname{im} F\alpha = \ker F\beta$; thus F preserves monomorphisms and F "preserves kernels" in the sense that $F\alpha : F(\ker \beta) \xrightarrow{\sim} \ker F\beta$. Similarly, if F is right exact, then F preserves epimorphisms and F "preserves cokernels". It follows that $FC \cong FB/\operatorname{im} F\alpha$.

There are analogous definitions for contravariant functors.

Definition: A contravariant functor F is **left exact** if exactness of

$$A \xrightarrow{\alpha} B \xrightarrow{\beta} C \to 0$$

implies exactness of

$$0 \to FC \xrightarrow{F\beta} FB \xrightarrow{F\alpha} FA;$$

a contravariant functor F is **right exact** if exactness of

$$0 \to A \xrightarrow{\alpha} B \xrightarrow{\beta} C$$

implies exactness of $FC \xrightarrow{F\beta} FB \xrightarrow{F\alpha} FA \to 0$.

Thus, a contravariant functor F is left exact if and only if $F(\operatorname{coker} \alpha) \cong$ ker $F\alpha$ via $F\beta$, and is right exact if and only if $F(\ker \beta) \cong \operatorname{coker} F\beta$ via $F\alpha$.

Definition: A functor is **exact** if it is both left exact and right exact.

Clearly a left exact functor that preserves epimorphisms is exact, as is a right exact functor that preserves monomorphisms. Now that we have the words, we may state the theorems.

Theorem 2.9: $\operatorname{Hom}(M, \ )$ *is a left exact functor and* $\operatorname{Hom}(\ , M)$ *is a left exact contravariant functor, for every module* M.

Proof: We only show $F = \operatorname{Hom}(M, \ )$ is left exact; the proof for $\operatorname{Hom}(\ , M)$ is similar. If $0 \to A \xrightarrow{\alpha} B \xrightarrow{\beta} C$ is exact, we must show exactness of

$$0 \to \operatorname{Hom}(M, A) \xrightarrow{F\alpha} \operatorname{Hom}(M, B) \xrightarrow{F\beta} \operatorname{Hom}(M, C),$$

where $F\alpha: f \mapsto \alpha f$ and $F\beta: g \mapsto \beta g$.

  (i) $F\alpha$ is monic. If $(F\alpha)f = 0$, then $\alpha f = 0$ and $\alpha f(m) = 0$ for all $m \in M$. Since α is monic, $f(m) = 0$ for all $m \in M$, whence $f = 0$.

  (ii) im $F\alpha \subset \ker F\beta$. Suppose $g \in \operatorname{im} F\alpha$, so that $g = \alpha f$ for some $f \in \operatorname{Hom}(M, A)$. Then $(F\beta)g = \beta g = \beta \alpha f = 0$, since $\beta \alpha = 0$.

  (iii) ker $F\beta \subset \operatorname{im} F\alpha$. Suppose $g \in \operatorname{Hom}(M, B)$ and $\beta g = 0$; we must show $g = \alpha f$ for some $f : M \to A$. If $m \in M$, then $\beta g m = 0$ and $gm \in \ker \beta = \operatorname{im} \alpha$; hence there is a unique $a \in A$ with $\alpha a = gm$ (since α is monic). Define $f : M \to A$ by $f(m) = a = \alpha^{-1} g(m)$; clearly $\alpha f = g$. ∎

Examples: Let $R = \mathbf{Z}$, so that we deal with abelian groups, and consider the exact sequence $0 \to \mathbf{Z} \xrightarrow{\alpha} \mathbf{Q} \xrightarrow{\beta} \mathbf{Q}/\mathbf{Z} \to 0$.

  16. $F = \operatorname{Hom}(M, \ )$ need not be right exact. Choose $M = \mathbf{Z}/2\mathbf{Z}$. Note that $\operatorname{Hom}(M, \mathbf{Q}) = 0$ and $\operatorname{Hom}(M, \mathbf{Q}/\mathbf{Z}) \neq 0$, so that

$$F\beta : \operatorname{Hom}(M, \mathbf{Q}) \to \operatorname{Hom}(M, \mathbf{Q}/\mathbf{Z})$$

cannot be epic.

  17. $F = \operatorname{Hom}(\ , N)$ need not be right exact. Choose $N = \mathbf{Z}$. Note that $\operatorname{Hom}(\mathbf{Q}, \mathbf{Z}) = 0$ and $\operatorname{Hom}(\mathbf{Z}, \mathbf{Z}) \neq 0$, so that $F\alpha : \operatorname{Hom}(\mathbf{Q}, \mathbf{Z}) \to \operatorname{Hom}(\mathbf{Z}, \mathbf{Z})$ cannot be epic.

Theorem 2.10: *The functors* $M \otimes_R$ *and* $_R \otimes N$ *are right exact functors.*

Proof: We only give the proof for $M \otimes$; the other proof is similar. If $A \xrightarrow{\alpha} B \xrightarrow{\beta} C \to 0$ is exact, then we must show

$$M \otimes A \xrightarrow{1 \otimes \alpha} M \otimes B \xrightarrow{1 \otimes \beta} M \otimes C \to 0$$

is exact.

(i) im $1 \otimes \alpha \subset \ker 1 \otimes \beta$. It suffices to prove $(1 \otimes \beta)(1 \otimes \alpha) = 0$. But $(1 \otimes \beta)(1 \otimes \alpha) = 1 \otimes \beta\alpha = 1 \otimes 0 = 0$.

(ii) $\ker 1 \otimes \beta \subset \operatorname{im} 1 \otimes \alpha$. If $E = \operatorname{im}(1 \otimes \alpha)$, then $1 \otimes \beta$ induces a map $\bar{\beta}: (M \otimes B)/E \to M \otimes C$ given by $m \otimes b + E \mapsto m \otimes \beta b$ (for $E \subset \ker 1 \otimes \beta$, by (i)). It is easily seen that $1 \otimes \beta = \bar{\beta}\pi$, where $\pi: M \otimes B \to (M \otimes B)/E$ is the natural map.

Suppose we show $\bar{\beta}$ is an isomorphism. Then

$$\ker 1 \otimes \beta = \ker \bar{\beta}\pi = \ker \pi = E = \operatorname{im} 1 \otimes \alpha$$

and we are done. We construct a map $M \otimes C \to (M \otimes B)/E$ inverse to $\bar{\beta}$. The function $f: M \times C \to (M \otimes B)/E$ given by

$$f(m, c) = m \otimes b + E, \qquad \text{where} \quad \beta b = c$$

is well defined: such an element b exists because β is epic; if $\beta b' = c = \beta b$, then $\beta(b' - b) = 0$ and there is an element $a \in A$ with $\alpha a = b' - b$; it follows that $m \otimes b' - m \otimes b = m \otimes (b' - b) = (1 \otimes \alpha)(m \otimes a) \in \operatorname{im} 1 \otimes \alpha = E$. Since f is biadditive, the universal property of tensor product gives a map $\bar{f}: M \otimes C \to (M \otimes B)/E$ with $\bar{f}(m \otimes c) = m \otimes b + E$. Visibly, $\bar{f}$ and $\bar{\beta}$ are inverse functions.

(iii) $1 \otimes \beta$ is epic. Let $\sum m_i \otimes c_i \in M \otimes C$. Since β is epic, there are $b_i \in B$ with $\beta b_i = c_i$, all i. Hence $1 \otimes \beta(\sum m_i \otimes b_i) = \sum m_i \otimes c_i$. ∎

Example: 18. The functor $M \otimes_R$ need not be left exact.

Let $R = \mathbf{Z}$ and let $M = \langle x \rangle = \mathbf{Z}/2\mathbf{Z}$. Exactness of $0 \to \mathbf{Z} \to \mathbf{Q} \to \mathbf{Q}/\mathbf{Z} \to 0$ does not give exactness of $0 \to M \otimes \mathbf{Z} \to M \otimes \mathbf{Q}$, for $M \otimes \mathbf{Z} \cong M \neq 0$, by Theorem 1.12, while $M \otimes \mathbf{Q} = 0$ $(x \otimes q = x \otimes 2q/2 = 2x \otimes q/2 = 0 \otimes q/2 = 0)$.

Exercises: 2.27. If M is a **torsion** group (i.e., every element has finite order), prove that $M \otimes \mathbf{Q} = 0$.

2.28. Prove that $\mathbf{Q}/\mathbf{Z} \otimes \mathbf{Q}/\mathbf{Z} = 0$.

ADJOINTS

There is a remarkable relationship between Hom and $\otimes$, which has a very simple explanation: every function of two variables $f: A \times B \to C$ can be regarded as a one-parameter family of functions of one variable $f_a: B \to C$, where $a \in A$, merely by fixing the first variable. For technical reasons, see below, fixing the second variable leads to a slightly different description.

Theorem 2.11 **(Adjoint Isomorphism):** *For rings* R *and* S, *consider the situation* $(_R A, {}_S B_R, {}_S C)$. *Then there is an isomorphism*

$$\tau: \mathrm{Hom}_S(B \otimes_R A, C) \xrightarrow{\sim} \mathrm{Hom}_R(A, \mathrm{Hom}_S(B, C)).$$

In the situation $(A_R, {}_R B_S, C_S)$, *there is an isomorphism*

$$\tau': \mathrm{Hom}_S(A \otimes_R B, C) \xrightarrow{\sim} \mathrm{Hom}_R(A, \mathrm{Hom}_S(B, C)).$$

Proof: We only discuss the first isomorphism; the second is similar. Since B is a bimodule, Theorem 1.8 shows $B \otimes_R A$ is a left S-module $[s(b \otimes a) = (sb) \otimes a]$ and Theorem 1.15 shows $\mathrm{Hom}_S(B, C)$ is a left R-module $[(rf)(b) = f(br)]$. Thus, both sides make sense.

If $f: B \otimes_R A \to C$ is an S-map, for each $a \in A$ define $f_a: B \to C$ by $f_a(b) = b \otimes a$; one checks easily that $a \mapsto f_a$ is an R-map $A \to \mathrm{Hom}_S(B, C)$, which we denote $\bar{f}$. One further checks that $\tau: f \mapsto \bar{f}$ is a homomorphism. To see that τ is an isomorphism, we exhibit its inverse. Assume $g: A \to \mathrm{Hom}_S(B, C)$ is an R-map. Define $g': B \otimes_R A \to C$ by $b \otimes a \mapsto g_a(b)$. Again computations are left to the reader: g' is a well defined S-map; $g \mapsto g'$ is the inverse of τ. ∎

Theorem 2.11 is the proper context in which to view the relation between R-modules and representations of R. Every abelian group M is a $(\mathbf{Z} - \mathbf{Z})$ bimodule, so the second isomorphism above gives

$$\tau': \mathrm{Hom}_{\mathbf{Z}}(R \otimes_{\mathbf{Z}} M, M) \xrightarrow{\sim} \mathrm{Hom}_{\mathbf{Z}}(R, \mathrm{Hom}_{\mathbf{Z}}(M, M)).$$

If $\sigma: R \times M \to M$ is the scalar multiplication defining the module structure on M, then this $\mathbf{Z}$-biadditive function determines a map $\tilde{\sigma}: R \otimes_{\mathbf{Z}} M \to M$, and $\tau': \tilde{\sigma} \mapsto \rho$, where $\rho: R \to \mathrm{Hom}_{\mathbf{Z}}(M, M) = \mathrm{End}_{\mathbf{Z}}(M)$ is a representation.

The reason the isomorphism of Theorem 2.11 is called "adjoint" is quite formal. Let $F = B \otimes_R$ and $G = \mathrm{Hom}_S(B,)$. With this notation, Theorem 2.11 becomes $\mathrm{Hom}_S(FA, C) \cong \mathrm{Hom}_R(A, GC)$. If one pretends that $\mathrm{Hom}(,)$ is an inner product, then, in the parlance of linear algebra, F and G are adjoint.

We remind the reader of a convenient notation introduced in Chapter 1. If $\alpha: A \to B$ is a map and $F = \mathrm{Hom}(X,)$, then let us write $F\alpha = \alpha_*$; if $G = \mathrm{Hom}(, X)$, let us write $G\alpha = \alpha^*$. Thus, lower star is the map induced by the covariant Hom functor, upper star is the map induced by the contravariant Hom functor,

$$\alpha_*(f) = \alpha f, \qquad \text{and} \qquad \alpha^*(g) = g\alpha.$$

Definition: Let $F: \mathfrak{A} \to \mathfrak{C}$ and $G: \mathfrak{C} \to \mathfrak{A}$ be functors. The ordered pair (F, G) is an **adjoint pair** if, for each $A \in \mathrm{obj}\, \mathfrak{A}$ and $C \in \mathrm{obj}\, \mathfrak{C}$ there is a bijection

$$\tau = \tau_{A,C}: \mathrm{Hom}_{\mathfrak{C}}(FA, C) \to \mathrm{Hom}_{\mathfrak{A}}(A, GC)$$

which is "natural"[1] in each variable, i.e., the following diagrams commute:

$$
\begin{array}{ccc}
\mathrm{Hom}_{\mathfrak{C}}(FA,C) & \xrightarrow{\ (Ff)^*\ } & \mathrm{Hom}_{\mathfrak{C}}(FA',C) \\
\downarrow{\scriptstyle\tau} & & \downarrow{\scriptstyle\tau} \\
\mathrm{Hom}_{\mathfrak{A}}(A,GC) & \xrightarrow{\ f^*\ } & \mathrm{Hom}_{\mathfrak{A}}(A',GC)
\end{array}
\qquad \text{all}\quad f:A'\to A \quad \text{in } \mathfrak{A};
$$

$$
\begin{array}{ccc}
\mathrm{Hom}_{\mathfrak{C}}(FA,C) & \xrightarrow{\ g_*\ } & \mathrm{Hom}_{\mathfrak{C}}(FA,C') \\
\downarrow{\scriptstyle\tau} & & \downarrow{\scriptstyle\tau} \\
\mathrm{Hom}_{\mathfrak{A}}(A,GC) & \xrightarrow{\ (Gg)_*\ } & \mathrm{Hom}_{\mathfrak{A}}(A,GC')
\end{array}
\qquad \text{all}\quad g:C\to C' \quad \text{in } \mathfrak{C}.
$$

Theorem 2.12: *If $B = {}_S B_R$ is a bimodule, then $(B \otimes_R \ , \ \mathrm{Hom}_S(B, \ ))$ is an adjoint pair.*

Proof: For each A and C, we have constructed the isomorphism $\tau = \tau_{A,C}$ in Theorem 2.11. It only remains to check naturality, and this is straightforward because there is an explicit formula for every map. ∎

Before we give an application of adjointness, we give a converse of Theorem 2.9.

Lemma 2.13: *Let $B' \xrightarrow{\alpha} B \xrightarrow{\beta} B'' \to 0$ be a sequence (not necessarily exact); suppose that for every module M, we have exactness of*

$$0 \to \mathrm{Hom}_R(B'', M) \xrightarrow{\beta^*} \mathrm{Hom}_R(B, M) \xrightarrow{\alpha^*} \mathrm{Hom}_R(B', M).$$

Then the original sequence is exact.

Remarks: 1. A dual statement is true if we apply covariant Hom functors to $0 \to B' \to B \to B''$.
   2. We shall have a better version of this in Lemma 3.51.

Proof: (i) β is epic. Let $M = \mathrm{coker}\,\beta = B''/\mathrm{im}\,\beta$ and let $f:B'' \to M$ be the natural map. Then $\beta^* f = f\beta = 0$, so that $f = 0$ (since β^* is monic). It follows that $\mathrm{coker}\,\beta = 0$, whence β is epic.
   (ii) $\mathrm{im}\,\alpha \subset \ker\beta$. We know $\alpha^*\beta^* = 0$. In particular, if $M = B''$ and $f = 1_{B''}$, then $\alpha^*\beta^*(f) = 0$. But $\alpha^*\beta^* f = f\beta\alpha = \beta\alpha$, since $f = 1_{B''}$, and so $\beta\alpha = 0$.
   (iii) $\ker\beta \subset \mathrm{im}\,\alpha$. Choose $M = \mathrm{coker}\,\alpha = B/\mathrm{im}\,\alpha$, and let $g:B \to M$ be the natural map. As before, $\alpha^* g = g\alpha = 0$, so there is $f \in \mathrm{Hom}(B'', M)$ with

---

[1] A precise definition of naturality is given later in this chapter.

$\beta^*f = g$, i.e., $f\beta = g$. If im $\alpha \subsetneqq \ker \beta$, there is an element $b \in B$ with $\beta b = 0$ and $b \notin \text{im } \alpha$; this second fact gives $gb \neq 0$. But $gb = f\beta b = 0$, a contradiction. ∎

Theorem 2.14: *Let* $F:_R\mathfrak{M} \to {}_S\mathfrak{M}$ *and* $G:_S\mathfrak{M} \to {}_R\mathfrak{M}$ *be functors. If* (F, G) *is an adjoint pair, then* F *is right exact and* G *is left exact.*

Proof: Assume $A' \to A \to A'' \to 0$ is exact; we must show $FA' \to FA \to FA'' \to 0$ is exact. Consider the commutative diagram (the naturality condition)

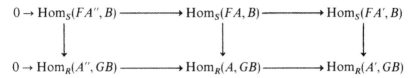

where B is any S-module. Since $\text{Hom}_R(\ , GB)$ is contravariant left exact, the bottom row is exact; since all the downward arrows are isomorphisms, commutativity gives exactness of the top row. As B is arbitrary, Lemma 2.13 applies to give exactness of $FA' \to FA \to FA'' \to 0$.

The proof of the left exactness of G is similar, using the other naturality condition in the definition of adjoint pair and the dual of Lemma 2.13. ∎

Corollary 2.15: $B \otimes_R$ *is right exact for any right R-module B.*

Proof: We may consider the right R-module B as a bimodule $_ZB_R$, so that $(B \otimes_R \ , \text{Hom}_Z(B, \ ))$ is an adjoint pair. ∎

There are two remarks necessary here. First, there are many other examples of adjoint pairs. Second, if (F, G) is an adjoint pair, much more is true than Theorem 2.14. Actually, F preserves all direct limits and G preserves all inverse limits. In order to prove this, and because the constructions are useful, we now investigate limits.

DIRECT LIMITS

Definition: Let I be a quasi-ordered set and $\mathfrak{C}$ a category. A **direct system** in $\mathfrak{C}$ with **index set** I is a functor $F:I \to \mathfrak{C}$ (of course, we construe I as a category as in Example 7 of Chapter 1).

Let us elaborate this definition. For each $i \in I$, there is an object F_i and, whenever $i, j \in I$ satisfy $i \leq j$, there is a morphism $\varphi^i_j:F_i \to F_j$ such that:

    (i) $\varphi^i_i:F_i \to F_i$ is the identity for every $i \in I$;

(ii) if $i \le j \le k$, there is a commutative diagram

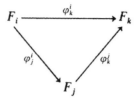

Remarks: 1. There is no reason to restrict oneself to quasi-ordered sets; one could speak of direct systems with an index category.

2. After our discussion of direct systems and direct limits, we shall consider the dual notions of inverse systems and inverse limits.

In the following examples, the category $\mathfrak{C}$ will be a category of modules (and so will not be mentioned).

Examples: 19. For any I, fix a module A and set $A_i = A$, all $i \in I$, and $\varphi_j^i = 1_A$ for all $i \le j$. This is the **constant** direct system with index set I, denoted $|A|$.

20. Let I have the **trivial** quasi-order: $i \le j$ if and only if $i = j$. A direct system with index set I is an indexed family of modules $\{F_i : i \in I\}$.

21. Let I have exactly three elements, $I = \{1, 2, 3\}$, with quasi-order $1 < 2$ and $1 < 3$. A direct system with index set I is a diagram

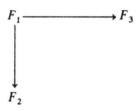

22. Let I be the positive integers with the usual (quasi-) order. A direct system is a sequence $A_1 \to A_2 \to A_3 \to \cdots$.

23. If A is a module, then the family of all f.g. submodules of A is quasi-ordered by inclusion. This family together with all possible inclusion maps is a direct system (over itself).

24. If $A = \coprod_{j \in J} A_j$, then the family of all finite "partial sums" $A_{j_1} \oplus \cdots \oplus A_{j_n}$ is quasi-ordered by inclusion; this family together with all inclusion maps is a direct system (over itself).

25. Let R be a domain with quotient field Q. The family of all cyclic R-submodules of Q of the form $\langle 1/r \rangle$ is quasi-ordered by $\langle 1/r \rangle \le \langle 1/s \rangle$ if and only if $r \mid s$, i.e., $rr' = s$ for some $r' \in R$. This family with inclusions is a direct system (over itself).

Definition: Let $F = \{F_i, \varphi_j^i\}$ be a direct system in $\mathfrak{C}$. The **direct limit** of this system, denoted $\varinjlim F_i$, is an object and a family of morphisms $\alpha_i : F_i \to \varinjlim F_i$ with $\alpha_i = \alpha_j \varphi_j^i$ whenever $i \leq j$ satisfying the following universal mapping problem:

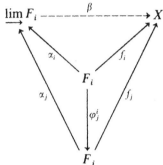

for every object X and every family of morphisms $f_i : F_i \to X$ with $f_i = f_j \varphi_j^i$ whenever $i \leq j$, there is a unique morphism $\beta : \varinjlim F_i \to X$ making the above diagram commute.

As usual, $\varinjlim F_i$ is unique to isomorphism if it exists. Although $\varinjlim F_i$ is the standard notation, we admit it is deficient in that it omits the morphisms φ_j^i.

Theorem 2.16: *The direct limit of a direct system of modules* $\{F_i, \varphi_j^i\}$ *exists.*

Proof: For each $i \in I$, let $\lambda_i : F_i \to \coprod F_i$ be the ith injection into the sum. Define

$$\varinjlim F_i = (\coprod F_i)/S,$$

where S is the submodule generated by all elements $\lambda_j \varphi_j^i a_i - \lambda_i a_i$, where $a_i \in F_i$ and $i \leq j$. If one further defines $\alpha_i : F_i \to \varinjlim F_i$ by $a_i \mapsto \lambda_i a_i + S$, then he may routinely verify that we have solved the universal problem. ∎

Let us now determine the direct limits of the particular direct systems above.

19′. The direct limit of the constant direct system $|A|$ is A (if the index set satisfies a mild hypothesis [see Exercise 2.45]).

20′. If I has the trivial quasi-order, then $\varinjlim F_i = \coprod F_i$. There are two ways to see this. Since there are no φ_j^i with $i \neq j$, the universal mapping problem is exactly that of Theorem 2.3. Alternatively, the submodule S in the construction in Theorem 2.16 is just 0, for φ_i^i is the identity.

21′. If I is the three point quasi-ordered set above, then the direct limit is usually called the **pushout**. Let us restate the universal mapping

problem. Given a pair of maps

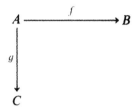

the pushout is a module L and maps α, β such that

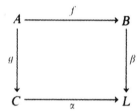

commutes; moreover, given any other pair of maps α', β' making the diagram commute, there exists a unique θ as below making all commute:

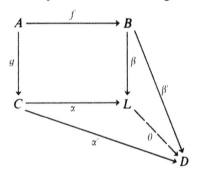

The construction of L as in Theorem 2.16 yields a certain quotient of $A \oplus B \oplus C$.

Exercises: 2.29.   Given

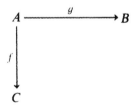

prove that the following diagram is a pushout:

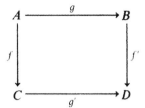

where $D = (C \oplus B)/W$, $W = \{(fa, -ga): a \in A\}$, $f': b \mapsto (0, b) + W$, and $g': c \mapsto (c, 0) + W$. (This is not the construction of Theorem 2.16.)

2.30. In the pushout diagram in Exercise 2.29, if g is monic (or epic), then g' is monic (or epic). Moreover, parallel arrows have isomorphic cokernels.

2.31. Assume I is the positive integers and $A_1 \subset A_2 \subset \cdots$ is an ascending sequence of subsets of a set X. Prove that $\varinjlim A_i \cong \bigcup_{i=1}^{\infty} A_i$.

2.32. Prove every module $A \cong \varinjlim A'$, where A' ranges over all f.g. submodules of A.

2.33. Prove $\coprod A_i \cong \varinjlim (A_{i_1} \oplus \cdots \oplus A_{i_k})$, the direct limit of all finite "partial sums."

2.34. If R is a domain with quotient field Q, then $Q \cong \varinjlim \langle 1/r \rangle$, $r \neq 0$, so that Q is a direct limit of cyclic submodules each isomorphic to R.

Definition: Let E and F be functors: $E: \mathfrak{A} \to \mathfrak{B}$ and $F: \mathfrak{A} \to \mathfrak{B}$. A **natural transformation** $t: E \to F$ is a class of morphisms $t_A: EA \to FA$, one for each $A \in \operatorname{obj} \mathfrak{A}$, giving commutativity of

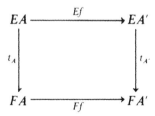

for every $f: A \to A'$ in $\mathfrak{A}$. There is a similar definition if both E and F are contravariant.

Exercises: 2.35. (i) The maps $t_A: A \to R \otimes_R A$ defined by $a \mapsto 1 \otimes a$ constitute a natural transformation $t: 1 \to R \otimes_R$, where 1 is the identity functor. (ii) The maps $s_A: A \to \operatorname{Hom}_R(R, A)$ defined by $a \mapsto f_a$ (where $f_a(r) = ra$) constitute a natural transformation $s: 1 \to \operatorname{Hom}_R(R,)$.

2.36. Fixing two variables in the expression $A \otimes B \otimes C$ gives a functor of the remaining variable. Prove that the associativity isomorphism of Exercise 1.10 is natural in each variable.

2.37. If $\mathfrak{A}$ is the category of all vector spaces over a field, then $t_V: V \to V^{**}$ (second dual) defined by $x \mapsto \langle \, , x \rangle$ (= evaluation at x) constitute a natural transformation $1 \to {}^{**}$.

2.38. Let (F, G) be an adjoint pair, where $F: \mathfrak{A} \to \mathfrak{C}$ and $G: \mathfrak{C} \to \mathfrak{A}$, and let $\tau: \mathrm{Hom}(FA, C) \to \mathrm{Hom}(A, GC)$ be the natural bijection. Now set $C = FA$, so that $\tau: \mathrm{Hom}(FA, FA) \to \mathrm{Hom}(A, GFA)$. If one defines $t_A = \tau(1_{FA}): A \to GFA$, prove that the t_A constitute a natural transformation $1 \to GF$. A similar construction yields a natural transformation $FG \to 1$.

2.39. Let 1 denote the identity functor on ${}_R\mathfrak{M}$. Show that the class of all natural transformations $1 \to 1$ is a ring isomorphic to $Z(R)$, the center of R.

2.40. Show that all functors $\mathfrak{A} \to \mathfrak{B}$ form a category if we define $\mathrm{Hom}(F, G) = $ all natural transformations $F \to G$. (This is not quite correct, for $\mathrm{Hom}(F, G)$ may be a class and not a set. To eliminate this difficulty, one usually assumes that $\mathfrak{A}$ is a **small category**, i.e., the class of all morphisms in $\mathfrak{A}$ is a set.)

We continue our discussion of direct systems with index set I. Since I is a small category, Exercise 2.40 shows that all such direct systems form a category, **Dir**(I). We repeat the definition of a morphism (natural transformation) $t: \{F_i, \varphi_j^i\} \to \{G_i, \psi_j^i\}$; t is a family of maps $t_i: F_i \to G_i$ making all the following diagrams commute (when $i \le j$):

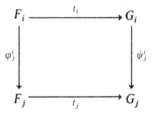

We claim that $\varinjlim: \mathbf{Dir}(I) \to {}_R\mathfrak{M}$ is a functor. We already know $\varinjlim(\{F_i, \varphi_j^i\}) = (\coprod F_i)/S$. If $t: \{F_i, \varphi_j^i\} \to \{G_i, \psi_j^i\}$ and if $\varinjlim(\{G_i, \psi_j^i\}) = (\coprod G_i)/S'$, then define $\bar{t}: \varinjlim F_i \to \varinjlim G_i$ by $\sum \lambda_i a_i + S \mapsto \sum \lambda_i' t_i a_i + S'$ (where $a_i \in F_i$, and λ_i, λ_i' are injections into $\coprod F_i$, $\coprod G_i$ respectively).

Definition: A quasi-ordered set I is **directed** if, for each $i, j \in I$, there exists $k \in I$ with $i \le k$ and $j \le k$.

Our first two examples of quasi-ordered sets (trivial and three point) are not directed; the other examples are directed. In particular, a sum $\coprod A_i$ is a direct limit of the A_i; if one wants the index set directed, however, then $\coprod A_i$ must be taken as the direct limit of its finite partial sums.

The reason we have introduced directed index sets is twofold: we shall have a simpler description of $\varinjlim F_i$; $\varinjlim : \mathbf{Dir}(I) \to {}_R\mathfrak{M}$ is an exact functor.

Theorem 2.17: *Let $\{A_i, \varphi^i_j\}$ be a direct system with directed index set I, let λ_i be the ith injection $\lambda_i : A_i \to \coprod A_i$, and let $\varinjlim A_i = (\coprod A_i)/S$ (as in Theorem 2.16).*

(i) $\varinjlim A_i$ *consists of all* $\lambda_i a_i + S$;

(ii) $\lambda_i a_i + S$ *is 0 if and only if* $\varphi^i_t a_i = 0$ *for some* $t \geq i$.

Proof: Let us observe at once that Theorem 2.16 tells us that $\varinjlim A_i$ consists of all elements of the form $x = \sum \lambda_i a_i + S$. Since I is directed, there is an index $j \geq i$, all i occurring in the sum for x. For each such i, define $b^i = \varphi^i_j a_i \in A_j$, so that $b = \sum b^i \in A_j$. It follows that

$$\sum \lambda_i a_i - \lambda_j b = \sum (\lambda_i a_i - \lambda_j b^i) = \sum (\lambda_i a_i - \lambda_j \varphi^i_j a_i) \in S,$$

so that $x = \sum \lambda_i a_i + S = \lambda_j b + S$; this proves (i).

If $\varphi^i_t a_i = 0$, then $\lambda_i a_i = \lambda_i a_i - \lambda_t \varphi^i_t a_i \in S$, so that $\lambda_i a_i + S = 0$. For the converse, assume $\lambda_i a_i + S = 0$ (by (i), $\lambda_i a_i + S$ is a typical element of $\varinjlim A_i$), i.e., $\lambda_i a_i \in S$. Since any scalar multiple of a relator $\lambda_k \varphi^j_k a_j - \lambda_j a_j$ (where $j \leq k$) has the same form, there is an expression

$$\lambda_i a_i = \sum_j (\lambda_k \varphi^j_k a_j - \lambda_j a_j) \in S.$$

Choose an index $t \in I$ larger than any of the indices in this expression. Clearly,

$$\lambda_t \varphi^i_t a_i = (\lambda_t \varphi^i_t a_i - \lambda_i a_i) + \lambda_i a_i$$
$$= (\lambda_t \varphi^i_t a_i - \lambda_i a_i) + \sum_j (\lambda_k \varphi^j_k a_j - \lambda_j a_j).$$

We may rewrite each of the terms on the right as a sum of relators in which the lower index is t:

$$\lambda_k \varphi^j_k a_j - \lambda_j a_j = (\lambda_t \varphi^j_t a_j - \lambda_j a_j) + [\lambda_t \varphi^k_t (-\varphi^j_k a_j) - \lambda_k (-\varphi^j_k a_j)],$$

for $\varphi^k_t \varphi^j_k = \varphi^j_t$, by definition of direct system. Therefore, we may write

$$\lambda_t \varphi^i_t a_i = \sum (\lambda_t \varphi^j_t a_j - \lambda_j a_j)$$

after a harmless change of notation; moreover, we may assume that we have combined all terms having the same top index j (for the sum of such terms has the same form). This last equation relates elements in $\coprod A_i$, where each element has a unique expression of the form $\sum \lambda_p a_p$. We conclude that if

$j \neq t$, then $\lambda_j a_j = 0$, hence $a_j = 0$, for the element $\lambda_t \varphi_t^i a_i$ has jth coordinate 0; if $j = t$, then $\lambda_t \varphi_t^i a_t - \lambda_t a_t = 0$ because φ_t^i is the identity. Thus, each term on the right is 0, so that $\lambda_t \varphi_t^i a_i = 0$, whence $\varphi_t^i a_i = 0$. ∎

It is now a simple matter to verify that the following gives an alternative description of direct limit when the index set is directed. First of all, recall the standard set-theoretic construction of **disjoint union** $\bigcup A_i$: given a family of sets $\{A_i : i \in I\}$, one may pretend no two of them overlap, $A_i \cap A_j = \varnothing$ for $i \neq j$, and then take their union.

Definition: Let I be a directed index set and let $\{A_i, \varphi_j^i\}$ be a direct system over I. If X is the disjoint union $\bigcup A_i$, define an equivalence relation on X by

$$a_i \sim a_j, \qquad a_i \in A_i, \quad a_j \in A_j,$$

if there exists an index $k \geq i, j$ with $\varphi_k^i a_i = \varphi_k^j a_j$. The equivalence class of a_i is denoted $[a_i]$.

If $\{A_i, \varphi_j^i\}$ is a direct system over a directed set I, define an R-module L as follows. The elements of L are the equivalence classes $[a_i]$ defined above,

$$r[a_i] = [r a_i] \qquad \text{if} \quad r \in R,$$

and

$$[a_i] + [a_j'] = [a_k + a_k'],$$

where $k \geq i, j$, $a_k = \varphi_k^i a_i$, and $a_k' = \varphi_k^j a_j'$.

Using Theorem 2.17, the reader may check that when the index set I is directed, the map $\varinjlim A_i \to L$ defined by $\lambda_i a_i + S \mapsto [a_i]$ is an isomorphism.

Theorem 2.18: *Let I be a directed quasi-ordered set. Suppose there are morphisms of direct systems over I*

$$\{A_i, \varphi_j^i\} \xrightarrow{t} \{B_i, \psi_j^i\} \xrightarrow{s} \{C_i, \theta_j^i\}$$

such that

$$0 \to A_i \xrightarrow{t_i} B_i \xrightarrow{s_i} C_i \to 0$$

is exact for each $i \in I$. Then there is an exact sequence of modules

$$0 \to \varinjlim A_i \xrightarrow{\tilde{t}} \varinjlim B_i \xrightarrow{\tilde{s}} \varinjlim C_i \to 0.$$

Remark: The hypothesis is the appropriate notion of exactness in **Dir**(I).

Proof: The only possible difficulty (and the only place where one must assume I is directed) is showing $\tilde{t}$ is monic; therefore, we prove this and leave the remainder to the reader (but see the first paragraph following this proof).

Assume $x \in \varinjlim A_i$ and $\bar{t}x = 0$ in $\varinjlim B_i$. Let us set $\varinjlim A_i = (\coprod A_i)/S$ and set $\lambda_i : A_i \to \coprod A_i$ the ith injection; let us set $\varinjlim B_i = (\coprod B_i)/T$ and set $\mu_i : B_i \to \coprod B_i$ the ith injection. Thus, $x = \lambda_i a_i + S$, by Theorem 2.17(i), and $\bar{t}x = \mu_i t_i a_i + T$. Since $\bar{t}x = 0$, Theorem 2.17(ii) says there is some $j \geq i$ with $\psi_j^i t_i a_i = 0$. Since t is a morphism between direct systems, we have $t_j \varphi_j^i a_i = 0$. But t_j is monic, whence $\varphi_j^i a_i = 0$, and this gives $x = \lambda_i a_i + S = 0$. ∎

Here is a sketch of an argument proving that $\varinjlim$ is right exact (the tedious part of Theorem 2.18 that we left to the reader). "Constant" is a functor $| \; | : {}_R\mathfrak{M} \to \mathbf{Dir}(I)$: to each module A, it assigns the constant direct system with index set I having $A_i = A$, all $i \in I$; if $f : A \to A'$, then $|f| : |A| \to |A'|$ is the natural transformation $|f|_i = f$ for all $i \in I$. Another way of stating that $\varinjlim$ is a solution to a universal mapping problem is to say there is a natural bijection

$$\mathrm{Hom}_R(\varinjlim A_i, C) \to \mathrm{Hom}_{\mathbf{Dir}(I)}(\{A_i, \varphi_j^i\}, |C|)$$

(it is instructive to give all necessary details). But this just says that $(\varinjlim, | \; |)$ is an adjoint pair. Therefore $\varinjlim$ is right exact, by a simple generalization of Theorem 2.14 replacing module categories by more general categories (e.g., $\mathbf{Dir}(I)$). Indeed, the next theorem is such a generalization.

Theorem 2.19: *Let $\mathfrak{A}$ and $\mathfrak{C}$ be categories, and let $F : \mathfrak{A} \to \mathfrak{C}$ and $G : \mathfrak{C} \to \mathfrak{A}$ be functors. If (F, G) is an adjoint pair, then F preserves all direct limits (with any, not necessarily directed, index set).*

Remark: Once we introduce inverse limits, we shall see that G preserves them.

Proof: If I is a quasi-ordered set and $\{A_i, \varphi_j^i\}$ is a direct system in $\mathfrak{A}$ with index set I, then it is easy to see $\{FA_i, F\varphi_j^i\}$ is a direct system in $\mathfrak{C}$ with index set I. Consider the commutative diagram in $\mathfrak{C}$

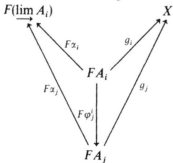

where $\alpha_i : A_i \to \varinjlim A_i$ are provided by the definition of $\varinjlim$. We require a unique morphism $\gamma : F(\varinjlim A_i) \to X$ making all commute. Since (F, G) is an adjoint pair, there is a natural bijection $\tau : \operatorname{Hom}_{\mathfrak{C}}(F \varinjlim A_i, X) \to \operatorname{Hom}_{\mathfrak{A}}(\varinjlim A_i, GX)$. By Exercise 2.38, the following diagram in $\mathfrak{A}$: commutes:

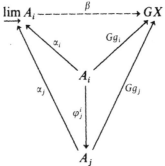

By definition of $\varinjlim$, there is a unique $\beta \in \operatorname{Hom}_{\mathfrak{A}}(\varinjlim A_i, GX)$ making the diagram commute. Define $\gamma \in \operatorname{Hom}_{\mathfrak{C}}(F \varinjlim A_i, X)$ by $\gamma = \tau^{-1}(\beta)$. Since τ (hence τ^{-1}) is natural, γ makes the original diagram commute. Finally, γ must be unique: if γ' were a second such morphism, then $\tau(\gamma')$ would be a second morphism $\varinjlim A_i \to GX$, contradicting the uniqueness of β. Thus $F(\varinjlim A_i)$ and $\varinjlim FA_i$ both solve the same universal mapping problem, and so they are isomorphic. ∎

Corollary 2.20: *For any right R-module B, the functor $B \otimes_R$ preserves direct limits.*

Proof: As in Corollary 2.15, $(B \otimes_R , \operatorname{Hom}_{\mathbf{Z}}(B,))$ is an adjoint pair. ∎

We now have a second proof of Theorem 2.8 that $B \otimes_R$ preserves sums, for sum is a direct limit. We can also reprove Theorem 2.15 that $B \otimes_R$ is right exact, i.e., $B \otimes_R$ preserves cokernels. It only remains to show that cokernel is a pushout, for pushout is a direct limit.

Exercises: 2.41. Prove that if $f : A \to B$, then coker f is the pushout of the diagram

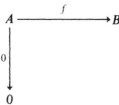

2.42. If F is a right exact functor that preserves sums and if F is additive (so that $F(0) = 0$), prove that F preserves direct limits. (Hint: Regard a direct limit as a cokernel of a suitable map between sums.)

2.43. A subset K of a directed set I is **cofinal** if, for each $i \in I$, there exists $k \in K$ with $i \le k$. If $\{A_i, i \in I, \varphi^i_j\}$ is a direct system with index set I, then $\{A_i, i \in K, \varphi^i_j\}$ is a direct system with index set K; moreover, the direct limits of these two systems are isomorphic. Show this may be false if the index set is not directed. (Hint: Pushout.)

2.44. If $\{A_i, \varphi^i_j\}$ is a direct system with index set I, and if I has a **top element**, ∞ (i.e., $\infty \in I$, $i \le \infty$ for all $i \in I$, and ∞ is the unique such element), then $\varinjlim A_i \cong A_\infty$. (Hint: I is automatically directed.)

2.45. If $|A|$ is the constant direct system with index set I, then $\varinjlim |A| \cong A$ if I is directed.

The reader may wonder why we were not content to prove Theorem 2.19 for module categories. Aside from the fact that the proof of the more general version given is the same as its particular case, there is a nice application to modules.

Theorem 2.21: *Any two direct limits* (*perhaps with distinct index sets*) *commute.*

Proof: We have already noted that $(\varinjlim, | \; |)$ is an adjoint pair of functors (of course, the domains and ranges of the functors are not module categories). Theorem 2.19 applies to show $\varinjlim$ preserves all direct limits. ∎

We have refrained from giving precise notation for Theorem 2.21, for it would make a simple result appear complicated. We only make two remarks to aid the reader. First, if I and K are quasi-ordered sets, then the modules in the direct system should be doubly indexed F_{ik}. Second, here is a modest example of Theorem 2.21. If $A_1 \subset A_2 \subset A_3 \subset \cdots$ is a sequence of submodules of A, then the cokernel of $\bigcup A_i$ (namely, $A/\bigcup A_i$) is isomorphic to the "union" of the cokernels $\varinjlim A/A_i$.

INVERSE LIMITS

Let us now consider inverse limits, which will appear as the dual of direct limits. Since all the proofs of the next theorems are dual to the ones just given, we leave the details to the reader's mirror.

Definition: Let I be a quasi-ordered set and $\mathfrak{C}$ a category. An **inverse system** in $\mathfrak{C}$ with **index set** I is a contravariant functor $F:I \to \mathfrak{C}$ (as usual, regard I as a category).

In more detail, for each $i \in I$ there is an object F_i and whenever $i \leq j$ there is a morphism $\psi_i^j : F_j \to F_i$ such that

(i) $\psi_i^i : F_i \to F_i$ is the identity for every $i \in I$;

(ii) if $i \leq j \leq k$, there is a commutative diagram

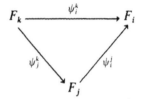

In each of the following examples, the category $\mathfrak{C}$ is a category of modules (and so will not be mentioned).

26.   For any I, the constant direct system $|A|$ with index set I (where A is a module) is also an inverse system with index set I.

27.   If I has the trivial quasi-order, then an inverse system with index set I is a family of modules $\{A_i : i \in I\}$ (so again this is the same as the corresponding direct system over I).

28.   If I is the three point quasi-ordered set we have been considering, an inverse system over I is a diagram

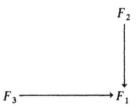

29.   If I is the positive integers with the usual (quasi-) order, then an inverse system over I is a sequence

$$A_1 \leftarrow A_2 \leftarrow A_3 \leftarrow \cdots.$$

Definition: Let $F = \{F_i, \psi_i^j\}$ be an inverse system in $\mathfrak{C}$. The **inverse limit** of this system, denoted $\varprojlim F_i$, is an object and a family of morphisms $\alpha_i : \varprojlim F_i \to F_i$ with $\alpha_i = \psi_i^j \alpha_j$ whenever $i \leq j$ satisfying the following universal mapping problem:

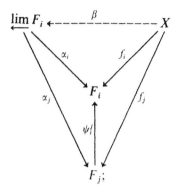

for every object X and morphisms $f_i : X \to F_i$ making the diagram commute whenever $i \leq j$, there is a unique morphism $\beta : X \to \varprojlim F_i$ making the diagram commute.

As usual $\varprojlim F_i$ is unique to isomorphism if it exists. Although $\varprojlim F_i$ is the standard notation, it is deficient in that it omits the morphisms ψ_i^j.

Theorem 2.22: *The inverse limit of an inverse system of modules $\{F_i, \psi_i^j\}$ exists.*

Proof: Since direct limit is a quotient of a sum, the dual notion of inverse limit should be a submodule of a product. For each $i \in I$, let p_i be the ith projection $p_i : \prod F_i \to F_i$. Define

$$\varprojlim F_i = \{(a_i) \in \prod F_i : a_i = \psi_i^j a_j, \text{ whenever } i \leq j\}.$$

If one further defines $\alpha_i : \varprojlim F_i \to F_i$ as the restriction $p_i | \varprojlim F_i$, then he may routinely verify that he has solved the universal problem. ∎

Let us now determine the inverse limits of some particular inverse systems.

26′. The inverse limit of the constant system $|A|$ is A if the index set is directed.

27′. If I has the trivial quasi-order, then $\varprojlim F_i = \prod F_i$. There are two ways to see this. Since there are no ψ_i^j with $j \neq i$, the universal mapping problem is exactly that of Theorem 2.5. Alternatively, the submodule of $\prod F_i$ is everything, for there are no constraints on the coordinates.

28′. If I is the three point quasi-ordered set above, then the inverse limit is usually called the **pullback** (or **fiber product**). Let us restate the universal mapping problem. Given a pair of maps

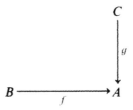

the pullback is a module L and maps α, β such that

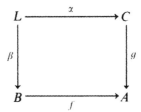

commutes; moreover, given any other pair of maps α', β' making the diagram commute, there exists a unique θ as below making all commute:

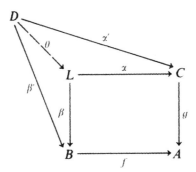

The construction of L as in Theorem 2.22 yields a certain submodule of $A \oplus B \oplus C$.

Exercises: 2.46.   Given

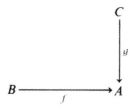

prove that the following diagram is a pullback:

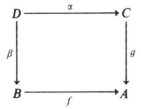

where $D = \{(b, c) \in B \oplus C : fb = gc\}$, $\alpha : (b, c) \mapsto c$ and $\beta : (b, c) \mapsto b$. (This is not the construction of Theorem 2.22.)

2.47. In the pullback diagram in Exercise 2.46, if g is monic (or epic), then β is monic (or epic). Moreover, parallel arrows have isomorphic kernels.

2.48. In the pullback diagram of Exercise 2.46, if both f and g are monic, then we may identify D with $B \cap C$. Moreover, in this case, there is a pushout diagram

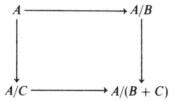

2.49. If $f : B \to A$, then

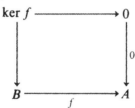

is a pullback diagram. (Thus, cokernel is a pushout, hence a direct limit, while kernel is a pullback, hence an inverse limit.)

2.50. Let $\{A_i : i \in I\}$ be a family of submodules of a module M. If one quasi-orders $\{A_i : i \in I\}$ by reverse inclusion, then this family and the various inclusion maps form an inverse system, and $\varprojlim A_i \cong \bigcap_{i \in I} A_i$ when the system is directed.

2.51. If K is a cofinal subset of a directed quasi-ordered set, and if $\{A_i, \psi_i^j\}$ is an inverse system with index set I, then $\varprojlim A_i$ is isomorphic to the inverse limit of the corresponding system with index set K (see Exercise 2.43).

2.52. If $\{A_i, \psi_i^j\}$ is an inverse system with index set I, and if I has a top element ∞, then $\varprojlim A_i \cong A_\infty$.

2.53.   Show that Exercise 2.51 may be false if the index set is not directed. (Hint: Pullback.)

It would be a sin not to mention completions. If I is an ideal in a commutative ring R, then

$$I^n = \{\textstyle\sum a_1 a_2 \cdots a_n : a_i \in I\};$$

it is easy to see I^n is an ideal and that $I = I^1 \supset I^2 \supset \cdots$. If M is an R-module, then $M \supset IM \supset I^2M \supset \cdots$. The family of quotient modules M/I^iM, $i = 1, 2, 3, \ldots$, and maps $\psi_i^j : M/I^jM \to M/I^iM$, for $j \geq i$, given by $x + I^jM \mapsto x + I^iM$, is an inverse system indexed by the positive integers.

Definition: The I-adic **completion** of M, denoted $\hat{M}$, is $\varprojlim M/I^iM$.

Consider the diagram

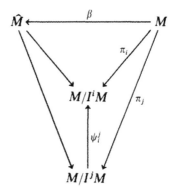

where π_i is the natural map, and β is provided by the definition of $\varprojlim$. It is a fact that β is monic if $\bigcap I^iM = 0$; it is another fact that if $M = R$, then $\hat{R}$ is a commutative ring and every I-adic completion is an $\hat{R}$-module.

There is a more down-to-earth description of $\hat{M}$ (we leave details to the interested reader). Assume $\bigcap I^iM = 0$. If $x \in M$, and $x \neq 0$, there exists i with $x \in I^iM$ but $x \notin I^{i+1}M$; define $\|x\| = 2^{-i}$; if $x = 0$, define $\|x\| = 0$. It is easy to see that $\|x - y\|$ is a metric on M. It turns out that $\hat{M}$ is the completion of M in the sense of metric spaces (M is a dense subspace of $\hat{M}$ and every Cauchy sequence in M converges in $\hat{M}$).

Given two contravariant functors $\mathfrak{A} \to \mathfrak{C}$, one may define a natural transformation between them (no surprises). Let us only give the special instance of a morphism between inverse systems.

Definition: If I is a quasi-ordered set, a **morphism** $t : \{F_i, \psi_i^j\} \to \{G_i, \varphi_i^j\}$ between inverse systems over I is a family of maps $t_i : F_i \to G_i$ making the following diagrams commute when $i \leq j$:

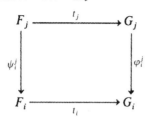

It is easy to see that all inverse systems with index set I and their morphisms form a category, **Inv**(I). We claim that $\varprojlim : \textbf{Inv}(I) \to {}_R\mathfrak{M}$ is a functor. It remains only to define $\bar{t} : \varprojlim F_i \to \varprojlim G_i$ for a morphism $t : \{F_i, \psi_i^j\} \to \{G_i, \varphi_i^j\}$, and this is done by

$$\bar{t} : (a_i) \mapsto (t_i a_i).$$

We have already noted that we may regard $|\ | : {}_R\mathfrak{M} \to \textbf{Inv}(I)$, for the constant system is, at our pleasure, either a direct system or an inverse system.

Theorem 2.23: $(|\ |, \varprojlim)$ *is an adjoint pair of functors.*

Proof: Dual to the argument that $(\varinjlim, |\ |)$ is an adjoint pair. $\blacksquare$

Theorem 2.24: *Let $\mathfrak{A}$ and $\mathfrak{C}$ be categories, and let $F : \mathfrak{A} \to \mathfrak{C}$ and $G : \mathfrak{C} \to \mathfrak{A}$ be functors. If (F, G) is an adjoint pair, then G preserves all inverse limits.*

Proof: Dual to the proof of Theorem 2.19. $\blacksquare$

Corollary 2.25: *If B is a left R-module, then $\text{Hom}_R(B, \ )$ preserves inverse limits.*

Proof: We may regard B as a bimodule ${}_R B_{\mathbb{Z}}$, so Theorem 2.12 gives $(B \otimes_{\mathbb{Z}} \ , \text{Hom}_R(B, \ ))$ an adjoint pair. $\blacksquare$

There is another way to prove Corollary 2.25, dual to Exercise 2.42.

Exercises: 2.54. If F is a left exact functor that preserves products, then F preserves inverse limits. (Hint: Regard an inverse limit as the kernel of a suitable map between products.)

2.55. If K and L are submodules of M, then $\text{Hom}(B, K)$ and $\text{Hom}(B, L)$ may be regarded as subgroups of $\text{Hom}(B, M)$ and

$$\text{Hom}(B, K \cap L) \cong \text{Hom}(B, K) \cap \text{Hom}(B, L).$$

(Hint: Use pullback.)

2.56. $\varprojlim$ is left exact (in the sense of Theorem 2.18). One need not assume the index set is directed.

Theorem 2.26: *Any two inverse limits (perhaps with distinct index sets) commute.*

Proof: Dual to the proof of Theorem 2.21. ∎

Theorem 2.27: *For any module B, $\mathrm{Hom}(\varinjlim A_j, B) \cong \varprojlim \mathrm{Hom}(A_j, B)$.*

Proof: Use the analogue of Exercise 2.54: if F is a left exact contravariant functor converting sums to products, then F converts direct limits to inverse limits. ∎

One last remark about limits: usually, direct limit and inverse limit do not commute. We give an easy example which the reader may formalize. Construct an example of three subgroups A, B, C of an abelian group for which

$$A \cap (B \oplus C) \neq (A \cap B) \oplus (A \cap C),$$

and note that intersection is an inverse limit (even a pullback in this case) and sum is a direct limit. A more interesting example is that the completion of a product is the product of the completions, but this is not so if one replaces "product" by "sum".

3 Projectives, Injectives, and Flats

FREE MODULES

We begin by generalizing the notion of free abelian group to modules.

Definition: A left R-module A is **free** if it is a sum of copies of R. If $Ra_i \cong R$ and $A = \coprod_{i \in I} Ra_i$, then the set $\{a_i : i \in I\}$ is called a **basis** of A.

From the definition of sum as "vectors" almost all of whose coordinates are 0, it follows at once that given a basis $\{a_i : i \in I\}$, each $a \in A$ has a unique expression

$$a = \sum r_i a_i,$$

where $r_i \in R$ and almost all $r_i = 0$.

Theorem 3.1: Let $X = \{a_i : i \in I\}$ be a basis of a free module A. Given any module B and any function $f : X \to B$, there is a unique map $\tilde{f} : A \to B$ extending f.

Proof: For fixed $i \in I$, define $f_i : Ra_i \to B$ by $ra_i \mapsto rf(a_i)$. Since $A = \coprod Ra_i$, Theorem 2.3 gives a unique map $\tilde{f} : A \to B$ with $\tilde{f}a_i = f_i a_i = fa_i$, all i. ∎

Note that it is easy to describe $\tilde{f}$ explicitly: $\tilde{f} : \sum r_i a_i \mapsto \sum r_i f(a_i)$.

Theorem 3.2:   *If X is a set, there exists a free module A having X as a basis.*

Proof: Exactly as Theorem 1.2 (with the advantage that all necessary constructions are already available).   ∎

Theorem 3.3:   *Every module M is a quotient of a free module.*

Proof: Let UM be the underlying set of M (forget addition and scalar multiplication), and let A be the free module with basis UM. Define $f: UM \to M$ by $m \mapsto m$. By Theorem 3.1, there is a map $\tilde{f}: A \to M$ extending f, and $\tilde{f}$ is epic because f is onto.   ∎

Of course, the free module with basis UM may be extravagantly large; there may be many, smaller, free modules mapping onto M. In down-to-earth language, Theorem 3.3 says that M may be described by **generators and relations**: if A is free with basis X and $f: A \to M$ is epic, then X is called a set of **generators** of M and ker f is called its submodule of **relations**. This definition conflicts with our previous notion of generators of M; the earlier usage (which we shall maintain) says M is generated by $f(X)$.

Exercises: 3.1.   Let $U: {}_R\mathfrak{M} \to$ **Sets** be the "forgetful" functor: regard every module as a mere set and every map as a mere function. Let $F:$ **Sets** $\to {}_R\mathfrak{M}$ be the "free functor": FX is the free module with basis X; if $f: X \to Y$ is a function, $Ff = \tilde{f}$ defined as in Theorem 3.1. Prove that U and F are functors.

3.2.   Prove that (F, U) is an adjoint pair. (This really says that module-valued functions on a set X may be identified with homomorphisms from the free module with basis X.)

Before pursuing generators and relations further, let us consider an obvious question. Do every two bases of a free module have the same cardinal? We do know this is so for left R-modules when R is a field or R is a division ring.

Definition: A ring R has **IBN (invariant basis number)** if, for every free left R-module A, every two bases of A have the same cardinal. In this case, **rank** A is defined as the cardinal of some basis of A.

Theorem 3.4:   *Every commutative ring R has* IBN.

Proof: By Zorn's lemma, R has a maximal ideal M, whence R/M is a field. Let A be free with basis $\{a_i: i \in I\}$. By Example 4 of Chapter 2, the quotient module A/MA is an R/M-module (being annihilated by M), i.e., A/MA is a vector space over R/M. Since $MA = \coprod Ma_i$, we see that $A/MA \cong \coprod (Ra_i/Ma_i)$. Since $Ra_i/Ma_i \cong R/M$, it follows that A/MA has dimension card(I). If $\{b_j : j \in J\}$ is a second basis of A, the same argument gives dim $A/MA = \operatorname{card}(J)$. Since fields have IBN, we conclude that card(I) = card(J). ∎

Exercises: 3.3. A ring R has IBN if and only if, for every f.g. free left R-module A, every two bases of A have the same cardinal.

3.4. If R has IBN and $\varphi : S \to R$ is a ring map (by definition, $\varphi(1) = 1$), then S has IBN. (Hint: If A is a free left S-module, consider $R \otimes_S A$.)

3.5. Let k be a field and V an infinite-dimensional vector space over k. Prove that $R = \operatorname{End}_k(V)$ does not have IBN. (Hint: Show that, as left R-modules, $R \cong R \oplus R$ by applying $\operatorname{Hom}_k(V, \ )$ to an isomorphism $V \cong V \oplus V$.)

3.6. Assume R has IBN, and there is an exact sequence

$$0 \to F_n \to F_{n-1} \to \cdots \to F_0 \to 0,$$

where each F_i is a f.g. free left R-module. Prove that $\sum_{i=0}^{n} (-1)^i \operatorname{rank} F_i = 0$.

In the next chapter, we shall introduce a large class of (not necessarily commutative) rings having IBN: left noetherian rings. It is fair to say that most rings one encounters do have IBN.

Here is a standard use of determinants.

Lemma 3.5: *Let S be commutative and let M be a f.g. S-module. If $M = JM$ for some ideal J of S, then $(1 - a)M = 0$ for some $a \in J$.*

Proof: If $\{x_1, \ldots, x_n\}$ generates M, then each $x_i = \sum a_{ij} x_j$, where $a_{ij} \in J$. In terms of matrices, these equations become $BX = 0$, where X is the column vector of the x_i's and B is the $n \times n$ matrix $B = I - (a_{ij})$ (of course, I is the $n \times n$ identity matrix). Let B^* be the classical adjoint of B, i.e., the matrix of cofactors, so that $B^*B = (\det B)I$. Then $B^*BX = (\det B)X$ and $B^*BX = 0$ (since $BX = 0$). We conclude that $(\det B)x_i = 0$ for all i, hence $(\det B)M = 0$. But obviously $\det B = 1 - a$ for $a \in J$, because each $a_{ij} \in J$. ∎

Theorem 3.6: *Let R be commutative and let M be a f.g. R-module. If $\beta : M \to M$ is epic, then β is an isomorphism.*

Proof: Let $S = R[x]$, the polynomial ring in one variable x, and make M into an S-module by defining

$$(r_n x^n + \cdots + r_1 x + r_0)m = r_n \beta^n m + \cdots + r_1 \beta m + r_0 m,$$

where β^i is the composite of β with itself i times. If $J = Sx$, then β epic implies $JM = M$. By Lemma 3.5, $(1 - a)M = 0$ for some $a \in J$. Since $a = \sum_{i=1}^{t} r_i \beta^i$, where $r_i \in R$, it follows that $\gamma = \sum_{i=1}^{t} r_i \beta^{i-1}$ is β^{-1} (for β and γ commute, and $1 - \gamma\beta = 0$). ∎

Corollary 3.7: *Let R be commutative and let A be a free R-module of finite rank n. If $\{a_1, \ldots, a_n\}$ generates A, then it is a basis of A.*

Proof: Let F be free with basis $\{x_1, \ldots, x_n\}$ and define $\varphi : F \to A$ by $x_i \mapsto a_i$; it suffices to prove φ is an isomorphism. Now φ is epic because A is generated by $\{a_1, \ldots, a_n\}$. There is thus an exact sequence

$$0 \to K \to F \xrightarrow{\varphi} A \to 0,$$

where $K = \ker \varphi$. Since A is free, hence projective, this sequence splits. Therefore, $F = K \oplus A'$, where $A' \cong A \cong F$ (both A and F are free of rank n). Define $\beta : F \to F$ as the composite $F \xrightarrow{p} A' \to F$, where p is the projection having kernel K. Clearly β is epic with kernel K. By Theorem 3.6, $K = 0$ and $\varphi : F \to A$ is an isomorphism. ∎

In the next chapter, we shall see that Corollary 3.7 holds when R is left noetherian. Indeed, it is not difficult to see that a ring R satisfying Corollary 3.7 must have IBN (where one replaces "A of rank n" by "A a sum of n copies of R"). The converse is false [P. M. Cohn, 1966].

Let us return to generators and relations. In the next chapter we shall prove that every subgroup S of a free abelian group F is itself free; moreover, rank $S \leq$ rank F. If a f.g. abelian group G is given as F/S, where F is f.g. free (and Theorem 3.3 says G can always be so given), then one may use the freeness of S to obtain information about G. The "Simultaneous Basis Theorem" [Fuchs, 1970, p. 78] states that if $\{x_1, \ldots, x_n\}$ is a basis of F, then there is a basis $\{y_1, \ldots, y_m\}$ of S, $m \leq n$, and integers $\{k_1, \ldots, k_m\}$ such that $k_i x_i = y_i$, all i, and each k_i divides k_{i+1}. Of course, the Simultaneous Basis Theorem implies that G is a sum of cyclic groups (in which the order of any summand divides the order of the next summand). If one considers R-modules instead of $\mathbf{Z}$-modules (abelian groups), then submodules of free modules need not be free; also, f.g. R-modules need not be sums of cyclic modules. Rather than surrender, we iterate the process of generators and relations.

Definition: A **free resolution** of a module M is an exact sequence

$$\cdots \to F_n \xrightarrow{d_n} F_{n-1} \to \cdots \to F_1 \xrightarrow{d_1} F_0 \xrightarrow{\varepsilon} M \to 0$$

in which each F_n is a free module.

Theorem 3.8: *Every module M has a free resolution.*

Proof: By Theorem 3.3, there is a free module F_0 and an exact sequence

$$0 \to S_0 \to F_0 \xrightarrow{\varepsilon} M \to 0.$$

Similarly, there is a free module F_1 and an exact sequence

$$0 \to S_1 \to F_1 \to S_0 \to 0,$$

and, by induction, a free module F_n and an exact sequence

$$0 \to S_n \to F_n \to S_{n-1} \to 0.$$

Assemble all these sequences into the diagram

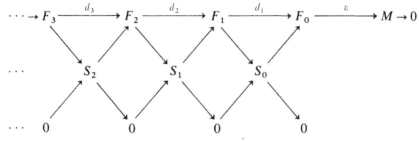

where the maps d_n are just the indicated composites. For every n, $\ker d_n = S_n$ and $\operatorname{im} d_n = S_{n-1}$. Hence $\operatorname{im} d_{n+1} = \ker d_n$ and the top row is exact. ∎

The diagram above makes it plain how to decompose a long exact sequence into short exact sequences.

We emphasize that a module has many free resolutions. Let us return to free modules.

Theorem 3.9: *Consider the diagram with β epic:*

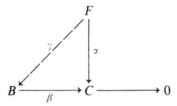

If F is free and $\alpha : F \to C$ is any map, then there exists $\gamma : F \to B$ with $\alpha = \beta\gamma$.

Remark: We do not assert γ is unique.

Proof: Let $X = \{x_i : i \in I\}$ be a basis of F. Since β is epic, each αx_i can be lifted, i.e., there is an element $b_i \in B$ with $\beta b_i = \alpha x_i$. The axiom of choice provides a function $\varphi : X \to B$ with $\varphi x_i = b_i$, all $i \in I$. By Theorem 3.1, there is a map $\gamma : F \to B$ with $\gamma x_i = \varphi x_i$, all i. To see that $\alpha = \beta\gamma$, it suffices to check each on X, and $\beta\gamma(x_i) = \beta\varphi(x_i) = \beta b_i = \alpha x_i$. ∎

Corollary 3.10: *If F is free, then the functor $\operatorname{Hom}(F, \ )$ is exact.*

Proof: Since Hom(F,) is left exact (Theorem 2.9), we need only show that it preserves epimorphisms. If $B \xrightarrow{\beta} C \to 0$ is exact, then we must show Hom(F, B) $\xrightarrow{\beta_*}$ Hom(F, C) $\to 0$ is exact, i.e., if $f \in$ Hom(F, C), then $f = \beta_*(g) = \beta g$ for some $g \in$ Hom(F, B). But this is precisely the situation of Theorem 3.9:

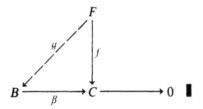

PROJECTIVE MODULES

Definition: A module P is **projective** if it behaves as F does in Theorem 3.9:

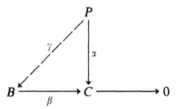

Theorem 3.11:   *A module P is projective if and only if* Hom(P,) *is exact.*

Proof: The proof of Corollary 3.10 shows that Hom(P,) is exact if P is projective. Assume, conversely, that Hom(P,) is exact. Consider the diagram

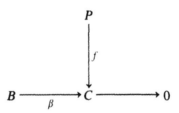

Since β_*: Hom(P, B) $\to$ Hom(P, C) is epic, there is $g \in$ Hom(P, B) with $f = \beta_*(g) = \beta g$; this says that P is projective.   ∎

Is every projective module free? Let us first gather a bit more information.

Theorem 3.12:   *If P is projective and $\beta: B \to P$ is epic, then $B = \ker \beta \oplus P'$, where $P' \cong P$.*

Proof: Consider the diagram

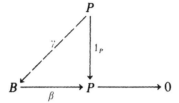

Since P is projective, there is a map $\gamma:P \to B$ with $\beta\gamma = 1_P$. By set theory, γ is monic, and so Theorem 2.7 gives the theorem. ∎

It is useful to restate Theorem 3.12 in two ways.

Corollary 3.13:   *If A is a submodule of B with B/A projective, then A is a summand of B. Every short exact sequence $0 \to A \to B \to P \to 0$, with P projective, is split.*

Theorem 3.14:   *A module P is projective if and only if it is a summand of a free module. Moreover, every summand of a projective is projective.*

Proof: By Theorem 3.3, there is a free module F and an epimorphism $F \to P$. By Theorem 3.12, P is (isomorphic to) a summand of F.

For the converse, we show that any summand of a projective module F is itself projective. There are maps $i:P \to F$ and $p:F \to P$ with $pi = 1_P$. Consider the diagram

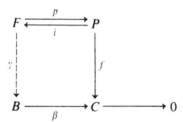

Since F is projective, there is a map $\gamma:F \to B$ with $\beta\gamma = fp$. Define $g:P \to B$ by $g = \gamma i$. Since $\beta g = \beta\gamma i = fpi = f$, it follows that P is projective. ∎

We can now give an example of a ring R for which projectives may not be free. If $R = \mathbf{Z}/6\mathbf{Z}$, then $R = \mathbf{Z}/2\mathbf{Z} \oplus \mathbf{Z}/3\mathbf{Z}$, so that $\mathbf{Z}/2\mathbf{Z}$ is projective, being a summand of the free module R. But $\mathbf{Z}/2\mathbf{Z}$ is not free, for every nonzero free R-module has at least 6 elements. We shall consider the question when projectives are free in the next chapter.

The next theorem is often useful; it characterizes projectives by means of "linear functionals" that act like coordinate functions.

Theorem 3.15   (Projective Basis):   *A module A is projective if and only if there exist elements* $\{a_k : k \in K\} \subset A$ *and R-maps* $\{\varphi_k : A \to R : k \in K\}$ *such that*

(i)   *if* $x \in A$, *then almost all* $\varphi_k x = 0$;
(ii)   *if* $x \in A$, *then* $x = \sum_{k \in K} (\varphi_k x) a_k$.

Moreover, A is then generated by $\{a_k : k \in K\}$.

Proof: Assume A is projective, and let $\psi : F \to A$ be an epimorphism from some free module F. By Theorem 3.12, there is a map $\varphi : A \to F$ with $\psi\varphi = 1_A$. Let $\{e_k : k \in K\}$ be a basis of F. If $x \in A$, then φx has a unique expression

$$\varphi x = \sum r_k e_k,$$

where $r_k \in R$ and almost all $r_k = 0$. Define $\varphi_k : A \to R$ by $\varphi_k x = r_k$. Visibly $\varphi_k x = 0$ for almost all k. If we define $a_k = \psi e_k$, then ψ epic implies $\{a_k : k \in K\}$ generates A. Moreover, if $x \in A$, then

$$x = \psi\varphi x = \psi(\sum r_k e_k) = \sum r_k \psi e_k = \sum (\varphi_k x)\psi e_k = \sum (\varphi_k x) a_k.$$

Conversely, assume the existence of $\{a_k : k \in K\}$ and $\{\varphi_k : A \to R : k \in K\}$. Let F be a free module with basis $\{e_k : k \in K\}$, and define a map $\psi : F \to A$ by $e_k \mapsto a_k$. It suffices to exhibit a map $\varphi : A \to F$ with $\psi\varphi = 1_A$. Define $\varphi : A \to F$ by $x \mapsto \sum (\varphi_k x) e_k$; this sum is finite, by condition (i), so φ is well defined. By condition (ii),

$$\psi\varphi x = \psi\sum (\varphi_k x) e_k = \sum (\varphi_k x)\psi e_k = \sum (\varphi_k x) a_k = x.$$

Therefore, $\psi\varphi = 1_A$.   ∎

Exercises: 3.7.   A module P is projective if and only if every short exact sequence $0 \to A \to B \to P \to 0$ splits.

3.8.   If $\{P_j : j \in J\}$ is a family of projective modules, then $\coprod P_j$ is projective. (We shall see in the next chapter that a product of projectives need not be projective.)

3.9.   If R is commutative and P and Q are projective R-modules, then $P \otimes_R Q$ is a projective R-module.

3.10.   Assume R and S are rings, with S an $(S–R)$ bimodule. If P is a projective R-module, then $S \otimes_R P$ is a projective S-module.

3.11.   Assume P is a projective left R-module and I is a two-sided ideal in R. Prove that P/IP is a projective left R/I-module.

3.12.   A **complement** of a projective module P is a module Q (necessarily projective) with $P \oplus Q$ free. Prove that every f.g. projective has a f.g. complement.

3.13   **(Eilenberg)**   Every projective module P has a free complement. (Hint: First, prove this when $R = \mathbf{Z}/6\mathbf{Z}$ and $P = \mathbf{Z}/2\mathbf{Z}$.)

INJECTIVE MODULES

We now dualize the notion of projective.

Definition: A module E is **injective** if, for every module B and every submodule A of B, every $f: A \to E$ can be extended to a map $g: B \to E$. The diagram is

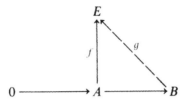

There are no obvious examples of injective modules, but we shall see that they exist in abundance. Let us derive their first properties; one should expect duals of the first properties of projectives.

Theorem 3.16: *A module E is injective if and only if the functor* $\mathrm{Hom}(\ , E)$ *is exact.*

Proof. Assume E is injective. Since $\mathrm{Hom}(\ , E)$ is contravariant left exact, it suffices to show it converts monomorphisms to epimorphisms: if $\alpha: A \to B$ is monic, then $\alpha^*: \mathrm{Hom}(B, E) \to \mathrm{Hom}(A, E)$ is epic. Therefore, we must show that if $f \in \mathrm{Hom}(A, E)$, there exists $g \in \mathrm{Hom}(B, E)$ with $\alpha^*(g) = f$. Since $\alpha^*(g) = g\alpha$, this is precisely the definition of injectivity of E.

The converse is just as easy and is left to the reader. ∎

Theorem 3.17: *If* $\{E_j : j \in J\}$ *is a family of injective modules, then* $\prod E_j$ *is injective.*

Proof: Let λ_j and p_j be the injections and projections of the product $\prod E_j$. Consider the diagram

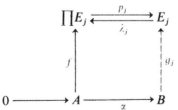

Since E_j is injective, there is a map $g_j: B \to E_j$ with $g_j\alpha = p_j f$. Define $h: B \to \prod E_j$ by $b \mapsto (g_j b)$. Then

$$h\alpha a = (g_j \alpha a) = (p_j fa) = fa,$$

so that $h\alpha = f$ and $\prod E_j$ is injective. ∎

We shall see in the next chapter that a sum of injectives need not be injective.

Theorem 3.18: *Every summand D of an injective module E is itself injective.*

Proof: Consider the diagram

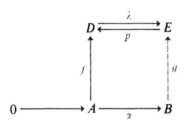

where λ and p are injection and projection, respectively. Since E is injective, there is a map $g: B \to E$ with $g\alpha = \lambda f$. Define $h: B \to D$ by $h = pg$. Then

$$h\alpha = pg\alpha = p\lambda f = f$$

since $p\lambda = 1_D$. Thus, D is injective. ∎

Theorem 3.19: *A module E is injective if and only if every short exact sequence $0 \to E \xrightarrow{i} B \to C \to 0$ splits. In particular, E is a summand of B.*

Proof: Assume E is injective; consider the diagram

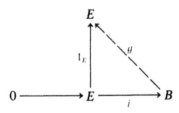

There exists a map $g: B \to E$ with $gi = 1_E$, so the sequence splits.
For the converse, consider the diagram

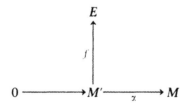

Construct the pushout diagram

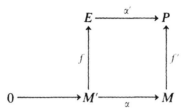

By Exercise 2.30, the map $\alpha':E \to P$ is monic. By hypothesis, there is a map $\beta:P \to E$ with $\beta\alpha' = 1_E$. Define $g:M \to E$ by $g = \beta f'$. Then

$$g\alpha = \beta f'\alpha = \beta\alpha' f = f,$$

and so E is injective. ∎

Now Theorem 3.19 is the dual of Theorem 3.14. If one compares the proofs, one sees that we could give a second proof of Theorem 3.19 if we knew the dual of the statement "Every module is a quotient of a projective" were true. The dual, "Every module can be imbedded in an injective," is, indeed, true; it is an important result (being the dual of "generators and relations") and we now begin working toward it.

Theorem 3.20 (**Baer Criterion**): *An R-module E is injective if and only if every map $f:I \to E$, where I is a left ideal of R, can be extended to R.*

Proof: Suppose E is injective. As a left ideal is a submodule of R, the hypothesis is just a special case of the definition of injective.

Suppose we have the diagram

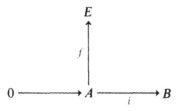

For notational convenience, let us assume i is an inclusion map. We approximate a map $g:B \to E$ by looking at all modules between A and B that do possess an extension of f. More precisely, let $\mathfrak{E}$ consist of all pairs (A',g'), where $A \subset A' \subset B$ and $g':A' \to E$ extends f. Note that $\mathfrak{E} \neq \emptyset$, for $(A,f) \in \mathfrak{E}$. Partially order $\mathfrak{E}$ by saying $(A',g') \leq (A'',g'')$ if $A' \subset A''$ and g'' extends g'. By Zorn's lemma, there is a maximal pair (A_0,g_0) in $\mathfrak{E}$. If $A_0 = B$, we are done.

Suppose $A_0 \neq B$ and let $x \in B - A_0$. If $I = \{r \in R : rx \in A_0\}$, then I is a left ideal in R. Define $h : I \to E$ by $h(r) = g_0(rx)$. By hypothesis, there is a map $h' : R \to E$ extending h. Define $A_1 = A_0 + Rx$, and define $g_1 : A_1 \to E$ by $a_0 + rx \mapsto g_0 a_0 + rh'(1)$, where $r \in R$. First, g_1 is well defined: if $a_0 + rx = a'_0 + r'x$, then $(r - r')x = a'_0 - a_0 \in A_0$ and $r - r' \in I$. Therefore, $g_0((r - r')x)$ and $h(r - r')$ are defined, and we have

$$g_0(a'_0 - a_0) = g_0((r - r')x) = h(r - r') = h'(r - r') = (r - r')h'(1).$$

Thus, $g_0(a'_0) - g_0(a_0) = rh'(1) - r'h'(1)$ and $g_0(a'_0) + r'h'(1) = g_0(a_0) + rh'(1)$. Second, g_1 does extend g_0, for $g_1 a_0 = g_0 a_0$ for all $a_0 \in A_0$. The pair (A_1, g_1) lies in $\mathfrak{E}$ and is larger than the maximal pair (A_0, g_0), a contradiction. Therefore, $A_0 = B$ and E is injective. ∎

The following diagram lemma will be useful.

Lemma 3.21:   *The diagram with exact row*

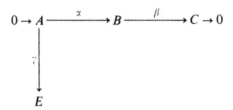

can be completed to a commutative diagram with exact rows

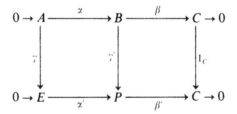

in which the first square is a pushout.

Proof :   Augment the given diagram by forming the pushout of the arrows emanating from A; note that α' is monic, by Exercise 2.30. Recall (Exercise 2.29) that we may assume $P = (E \oplus B)/W$, where $W = \{(\gamma a, -\alpha a) : a \in A\}$, $\gamma' : b \mapsto (0, b) + W$, and $\alpha' : e \mapsto (e, 0) + W$. Define $\beta' : P \to C$ by $(e, b) + W \mapsto \beta b$. We let the reader check that β' is well defined, the diagram commutes, and the bottom row is exact. ∎

Theorem 3.22:   *A module E is injective if and only if every short exact sequence $0 \to E \to B \to C \to 0$, with C cyclic, splits.*

Proof: If E is injective, Theorem 3.19 says the sequence splits for every (not necessarily cyclic) module C.

For the converse, let I be a left ideal of R and let $f : I \to E$ be an R-map. By Lemma 3.21, there is a commutative diagram with exact rows

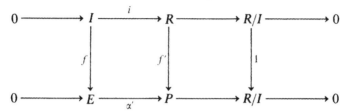

Since R/I is cyclic, our hypothesis is that the bottom row splits. There is thus a map $s : P \to E$ with $s\alpha' = 1_E$. If we define $g : R \to E$ by $g = sf'$, then it is easy to see $gi = f$. By Theorem 3.20, E is injective.   ∎

Definition: Let M be an R-module, $m \in M$, and $r \in R$. We say m is **divisible** by r if $rm' = m$ for some $m' \in M$; we say the module M is **divisible** if each $m \in M$ is divisible by every non-zero-divisor[1] $r \in R$ (i.e., there is no nonzero $s \in R$ with $sr = 0$).

Example: 1.   The additive group of rationals, **Q**, is a divisible abelian group ($=$ **Z**-module).

Theorem 3.23:   *Every injective module E is divisible.*

Proof: Let $m \in E$ and let $r_0 \in R$ be a non-zero-divisor. Define $f : Rr_0 \to E$ by $f(rr_0) = rm$; note that f is well defined because r_0 is not a zero divisor. Since E is injective, there is a map $g : R \to E$ extending f. In particular,

$$m = f(r_0) = g(r_0) = r_0 g(1),$$

so that m is divisible by r_0.   ∎

Exercises: 3.14.   Every quotient of a divisible module is divisible.

3.15.   Every summand of a divisible module is divisible.

3.16.   Every product of divisible modules is divisible.

3.17.   The sum of divisible modules is divisible.

---

[1] It is clear that r must be restricted: for example, we must exclude $r = 0$. The proof of Theorem 3.23 shows why zero divisors are excluded.

3.18.   Let R be a domain with quotient field Q. Prove that Q is divisible.

3.19.   A module M over a domain R is **torsion-free** if $rm = 0$ implies $r = 0$ or $m = 0$ (where $r \in R$ and $m \in M$). Prove that a torsion-free module M is injective if and only if M is divisible.

3.20.   Let Q be the quotient field of a domain R. Prove that a torsion-free module is divisible if and only if it is a vector space over Q.

It is obvious that divisible modules share many properties of injectives; however, the converse of Theorem 3.23 only holds for a restricted class of rings.

Theorem 3.24: *If R is a principal ideal domain, then an R-module D is divisible if and only if it is injective.*

Proof: By Theorem 3.20, it suffices to extend every map $f : I \to D$ to R, where I is an ideal. Since R is a PID, we know $I = Rr_0$; clearly we may assume $r_0 \neq 0$, and thus r_0 is not a zero divisor. Since D is divisible, there is an element $d \in D$ with $r_0 d = f(r_0)$. Define $g : R \to D$ by $r \mapsto rd$, and note that g extends f.

The converse is Theorem 3.23.   ∎

We finally have some examples of injective modules, at least for principal ideal domains. Here is a special case of the imbedding theorem we seek.

Theorem 3.25: *Every abelian group G can be imbedded in an injective abelian group.*

Proof: Write $G = F/S$, where F is a free abelian group; thus, $F = \coprod \mathbf{Z}$. If we imbed each copy of $\mathbf{Z}$ in a copy of the rationals, $\mathbf{Q}$, then we have

$$G = F/S = (\coprod \mathbf{Z})/S \subset (\coprod \mathbf{Q})/S.$$

Since $\mathbf{Q}$ is divisible, so are $\coprod \mathbf{Q}$ and $(\coprod \mathbf{Q})/S$, by Exercises 3.17 and 3.14. By Theorem 3.24, $(\coprod \mathbf{Q})/S$ is injective as a $\mathbf{Z}$-module.   ∎

Theorem 3.26: *If D is a divisible abelian group, then $\operatorname{Hom}_{\mathbf{Z}}(R, D)$ is an injective left R-module.*

Proof: First of all, R is a bimodule: $R = {}_{\mathbf{Z}}R_R$; by Theorem 1.15(i), $\operatorname{Hom}_{\mathbf{Z}}(R, D)$ is a left R-module via $rf : r' \mapsto f(r'r)$.

We shall show that the contravariant functor $\operatorname{Hom}_R(\ , \operatorname{Hom}_{\mathbf{Z}}(R, D))$ is exact; it is only necessary to show it converts monomorphisms to epimorphisms. The adjoint isomorphism (Theorems 2.11 and 2.12) gives a commutative diagram whenever $0 \to A \to B$ is an R-monomorphism:

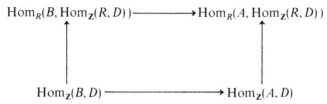

(we have identified $R \otimes_R B$ with B and $R \otimes_R A$ with A). Since D is divisible, it is $\mathbf{Z}$-injective, and so the bottom arrow is epic, by Theorem 3.16. It follows that the top arrow is also epic (by commutativity and the fact that the vertical arrows are isomorphisms). ∎

Theorem 3.27: *Every left R-module M can be imbedded in an injective module.*

Proof: If we regard M only as an abelian group, there is a $\mathbf{Z}$-monomorphism $0 \to M \xrightarrow{i} D$ for some divisible group D (Theorem 3.25). If $m \in M$, define $f_m : R \to M$ by $r \mapsto rm$. It is easy to see that $\varphi : M \to \mathrm{Hom}_{\mathbf{Z}}(R, D)$ given by $m \mapsto if_m$ is an R-map that is monic. ∎

Exercises: 3.21. Fix a module E. Use Theorem 3.27 to prove that if $0 \to E \to B \to C \to 0$ always splits, then E is injective.

3.22. If S is commutative, D an injective S-module, and R is an S-algebra, then $\mathrm{Hom}_S(R, D)$ is an injective left R-module.

3.23 Assume R is a left principal ideal domain (i.e., R has no zero divisors and every left ideal is principal). Prove that every divisible left R-module is injective.

We now dualize the notion of generators and relations.

Definition: An **injective resolution** of a module M is an exact sequence

$$0 \to M \to E^0 \to E^1 \to \cdots \to E^n \to E^{n+1} \to \cdots$$

in which each E^n is injective.

Theorem 3.28: *Every module M has an injective resolution.*

Proof: Dual to the proof of Theorem 3.8, using Theorem 3.27. ∎

We may also define a **projective resolution** of a module M in the obvious way; since free modules are projective, free resolutions are a special kind of projective resolution. Therefore, every module has projective resolutions.

A natural question is whether there is a "smallest" injective containing a given module. The affirmative answer was first discovered by Baer (who was the first to discover injectives), and the characterization below was first given by Eckmann and Schöpf. Our exposition follows that of Lambek [1966].

Definition: An **essential extension** of a module M is a module E containing M such that every nonzero submodule of E meets M (i.e., if $S \subset M$ and $S \neq 0$, then $S \cap M \neq 0$); if, in addition, $M \subsetneqq E$, we say E is a **proper essential extension** of M.

The additive group of rationals, $\mathbf{Q}$, is an essential extension of the integers, $\mathbf{Z}$; indeed, every intermediate subgroup G with $\mathbf{Z} \subset G \subset \mathbf{Q}$ is an essential extension of $\mathbf{Z}$. Here are some easy exercises we shall soon use.

Exercises: 3.24.   Let $M \subset E \subset E_1$; if E is essential over M and E_1 essential over E, then E_1 is essential over M.

3.25.   Let $M \subset E$. Show that E is an essential extension of M if and only if, for each $e \in E$, either $e = 0$ or there is an $r \in R$ with $re \in M$ and $re \neq 0$.

3.26.   Let $M \subset E$, and let $\{E_i : i \in I\}$ be a simply ordered (under inclusion) family of submodules of E, each E_i being essential over M. Prove that $\bigcup E_i$ is essential over M. (Hint: Use Exercise 3.25.)

3.27.   If $M \subset E' \subset E$ and both E' and E are essential over M, then E is essential over E'.

3.28.   If E is essential over M and $\varphi : E \to D$ is a map with $\varphi | M$ monic, then φ is monic. (Hint: $M \cap \ker \varphi = 0$.)

Theorem 3.29:   *A module M is injective if and only if M has no proper essential extensions.*

Proof: Assume M is injective and $M \subsetneqq E$, where E is an essential extension. By Theorem 3.19, M is a summand of E; there is thus a nonzero submodule N of E with $E = M \oplus N$. As $N \cap M = 0$, this contradicts E being an essential extension.

Assume M has no proper essential extensions, and let E be an injective module containing M. By Zorn's lemma, there is a submodule N of E maximal with $M \cap N = 0$. The composite $M \hookrightarrow E \to E/N$ is monic (since $N \cap M = 0$). Indeed, E/N is essential over M: if S/N is a nonzero submodule of E/N, then $S \supsetneqq N$ and maximality of N yields $S \cap M \neq 0$; it follows easily that S/N meets M. By hypothesis, the composite $M \to E \to E/N$ is an isomorphism, whence $E = M + N$. As $M \cap N = 0$, we have M a summand of E, and so M is injective.   ∎

Theorem 3.30:   *The following conditions on a module E containing a module M are equivalent:*

(i)   *E is a maximal essential extension of M (i.e., no proper extension of E is an essential extension of M)*;

(ii)   *E is an essential extension of M and E is injective*;

(iii)   *E is injective and there is no injective E' with $M \subset E' \subsetneqq E$.*

Moreover, such a module E exists.

Proof: (i) ⇒ (ii)   Since "an essential extension of" is a transitive relation (Exercise 3.24), the hypothesis is that E has no proper essential extensions. By Theorem 3.29, E is injective.

(ii) ⇒ (iii)   If such an injective E' exists, then E' would be a summand of E, say, $E = E' \oplus E''$. Since $M \subset E'$, we have $E' \cap E'' = 0$ implying $M \cap E'' = 0$, contradicting E being an essential extension of M.

(iii) ⇒ (i)   Assume only that E is an injective containing M and consider the family of all essential extensions of M that lie in E. By Exercise 3.26, Zorn's lemma yields a maximal such, say, E'. We claim E' is a maximal essential extension of M. Assume N is an essential extension of M that contains E'. There is a diagram

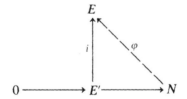

where i is the inclusion. Since E is injective, there is a map $\varphi: N \to E$ with $\varphi | E' = i$; thus, φ fixes E', hence M, pointwise. By Exercise 3.28, φ is monic (for N must be an essential extension of E', by Exercise 3.27). But now $\varphi(N)$ is an essential extension of M contained in E. Maximality of E' forces $\varphi(N) = E'$, whence $N = E'$. We conclude that E' is a maximal essential extension of M (as defined in statement (i)) and that E' has no proper essential extensions. By Theorem 3.29, E' is injective. If we now invoke the full hypothesis of (iii), then $E' = E$, as desired.

To prove the existence of E, imbed M in an injective module and take the submodule E' just constructed.   ∎

Definition: A module E satisfying any of the equivalent conditions of Theorem 3.30 is called an **injective envelope** (or **injective hull**) of M.

We have already proven existence of injective envelopes; we now prove uniqueness.

Theorem 3.31:   *Let E be an injective envelope of a module M.*

(i)   *If D is an injective containing M, then there is a monic $\varphi: E \to D$ fixing M pointwise;*

(ii)   *any two injective envelopes of M are isomorphic (by an isomorphism fixing M pointwise).*

Proof: (i)   Injectivity of *D* allows us to complete the diagram

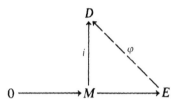

where *i* is the inclusion. Exercise 3.28 gives φ monic, since *E* is an essential extension of *M*.

(ii)   Assume *D* is an injective envelope of *M*. Using the notation of part (i), we claim φ is an isomorphism, i.e., φ is also epic. Were φ not epic, then $\varphi(E)$ would be a summand of *D* containing *M*, and this would contradict *D* being an essential extension of *M*.   ∎

One may thus speak of *the* injective envelope of *M*.

Exercises: 3.29.   If *R* is a domain, considered as a module over itself, then its injective envelope is its quotient field.

3.30.   Theorem 3.31(ii) characterizes injective envelopes: a monomorphism $i: M \to E$, where *E* is injective, is an injective envelope if and only if a monic dashed arrow always exists below

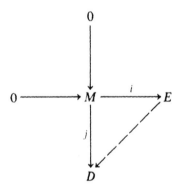

whenever *j* is an imbedding of *M* into an injective *D*.

Exercise 3.30 allows us to dualize the notion of injective envelope; also see Exercise 3.31 for the usual (equivalent) description.

Definition: A **projective cover** of M is an epimorphism $\varepsilon: P \to M$, where P is projective, so that an epic dashed arrow always exists below

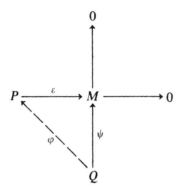

whenever ψ is an epimorphism from a projective Q.

Exercises: 3.31.   **(Usual Definition of Projective Cover)**   A submodule N of M is **superfluous** if, whenever L is a submodule of M with $L + N = M$, then $L = M$. Prove that an epimorphism $\varepsilon: P \to M$, where P is projective, is a projective cover if and only if ker ε is superfluous. (Hint: $P = \text{im } \varphi + \text{ker } \varepsilon$.) Prove that any two projective covers of a module M are isomorphic.

3.32.   Prove that the $\mathbf{Z}$-module $\mathbf{Z}/2\mathbf{Z}$ has no projective cover. Conclude that projective covers may not exist.

3.33.   If $\varepsilon_i: P_i \to M_i$ are projective covers, $i = 1, \ldots, n$, then

$$\coprod \varepsilon_i : \coprod P_i \to \coprod M_i$$

is a projective cover, where $\coprod \varepsilon_i : (p_1, \ldots, p_n) \mapsto (\varepsilon_1 p_1, \ldots, \varepsilon_n p_n)$.

WATTS' THEOREMS

There is a lovely application of frees and injectives in the coming proofs, due to Watts, of characterizations of the Hom and tensor functors. First, we need the appropriate notion of isomorphism of functors.

Definition: If $F, G: \mathfrak{A} \to \mathfrak{B}$ are functors of the same variance, then F and G are **naturally equivalent**, denoted $F \cong G$, if there is a natural transformation $\tau: F \to G$ with each $\tau_A: FA \to GA$ an equivalence.

Of course, if $\mathfrak{B}$ is a category of modules, each τ_A is an isomorphism.

Examples: 2.   The natural transformation $1 \to R \otimes_R$ of Exercise 2.35 is a natural equivalence.

3.   The natural transformation $1 \to \mathrm{Hom}_R(R, \ )$ of Exercise 2.36 is a natural equivalence.

4.   If $\mathfrak{A}$ is the category of finite-dimensional vector spaces over a field k, and if $* = \mathrm{Hom}_k(\ , k)$, then the usual evaluation $1 \to {}^{**}$ (see Exercise 2.37) is a natural equivalence.

5.   If $\mathfrak{A}$ is the category of finite abelian groups and $* = \mathrm{Hom}_{\mathbf{Z}}(\ , \mathbf{Q}/\mathbf{Z})$, then there is a natural equivalence $1 \to {}^{**}$ (this is a special case of Pontrjagin duality of locally compact abelian groups).

6.   Let R be a ring and $r \in Z(R)$ (so that multiplication by r is an R-map). Define $F : {}_R\mathfrak{M} \to {}_R\mathfrak{M}$ by $F(M) = M/rM$ and, if $f : M \to N$, then $Ff : M/rM \to N/rN$ is given by $m + rM \mapsto fm + rN$. If we apply $\otimes_R M$ to the exact sequence

$$R \xrightarrow{r} R \to R/Rr \to 0,$$

we obtain an exact sequence

$$M \xrightarrow{r} M \to (R/Rr) \otimes_R M \to 0.$$

There is thus an isomorphism $\sigma_M : M/rM \to (R/Rr) \otimes_R M$ (first isomorphism theorem), and $\sigma : F \to (R/Rr) \otimes_R$ is a natural equivalence.

This seems an appropriate opportunity to introduce an elementary and common technique, called **diagram-chasing**. At the moment, we will only need a special case of the coming lemma, but we give a general form for later use as well as to illustrate the technique. The important thing to observe is that each step of the proof is automatic in the sense that there is only one, obvious thing to do.

Lemma 3.32   (Five Lemma):   *Consider the commutative diagram with exact rows*

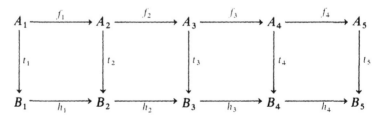

(i)   *If t_2 and t_4 are epic and t_5 is monic, then t_3 is epic;*

(ii)   *If t_2 and t_4 are monic and t_1 is epic, then t_3 is monic.*

In particular, if t_1, t_2, t_4, t_5 are isomorphisms, then t_3 is an isomorphism.

Proof: We only prove (i); the proof of the dual (ii) is similar.

Let $b_3 \in B_3$. Since t_4 is epic, $\exists a_4$ with $t_4 a_4 = h_3 b_3$. Commutativity of the last square gives $t_5 f_4 a_4 = h_4 t_4 a_4 = h_4 h_3 b_3 = 0$ (exactness at B_4). Since t_5

is monic, $f_4 a_4 = 0$, so $a_4 \in \ker f_4 = \operatorname{im} f_3$. Hence $\exists a_3$ with $a_4 = f_3 a_3$, and so $h_3 t_3 a_3 = t_4 f_3 a_3 = t_4 a_4 = h_3 b_3$ (definition of a_4). Therefore $b_3 - t_3 a_3 \in \ker h_3 = \operatorname{im} h_2$, whence $\exists b_2$ with $h_2 b_2 = b_3 - t_3 a_3$. Since t_2 is epic, $\exists a_2$ with $t_2 a_2 = b_2$. Commutativity gives $t_3 f_2 a_2 = h_2 t_2 a_2 = h_2 b_2 = b_3 - t_3 a_3$. Therefore

$$t_3(f_2 a_2 + a_3) = b_3$$

and t_3 is epic. ∎

Theorem 3.33 (Watts): *Let $F: \mathfrak{M}_R \to \mathbf{Ab}$ be a right exact functor that preserves sums. Then $F \cong \otimes_R B$ for some left R-module B. Moreover, we may choose $B = F(R)$.*

Proof: First of all, let us make the abelian group FR into a left R-module. For $M = M_R$ and $m \in M$, define $\varphi_m: R \to M$ by $r \mapsto mr$. Since φ_m is an R-map, we have $F\varphi_m: FR \to FM$. Define $\tau_M: M \times FR \to FM$ by $(m, x) \mapsto (F\varphi_m)x$, where $x \in FR$. If we set $M = R$, then $\tau_R: R \times FR \to FR$ may easily be seen to define a left R-module structure on FR. The same calculation now shows that $\tau_M: M \times FR \to FM$ is R-biadditive, and τ_M induces a map $\sigma_M: M \otimes_R FR \to FM$. If we denote FR by B, then one checks that $\sigma: \otimes_R B \to F$ is a natural transformation.

Now $\sigma_R: R \otimes_R B \to FR$ is an isomorphism (for $B = FR$); moreover, since both $\otimes_R B$ and F preserve sums, $\sigma_A: A \otimes_R B \to FA$ is an isomorphism for every free R-module A. Let M be any right R-module. By Theorem 3.8, there is an exact sequence $C \to A \to M \to 0$, where both C and A are free. Since both $\otimes_R B$ and F are right exact, there is a commutative diagram with exact rows:

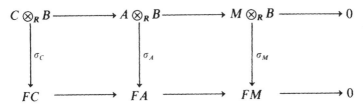

Since σ_C and σ_A are isomorphisms, the Five Lemma gives σ_M an isomorphism. Therefore, $\sigma: \otimes_R B \to F$ is a natural equivalence. ∎

Remark: If $F: \mathfrak{M}_R \to \mathfrak{M}_S$, then one may modify the first paragraph of the proof to see that the right S-module FR can be construed as a bimodule; thus the theorem remains true if we replace **Ab** by $\mathfrak{M}_S$.

Example 6 illustrates this theorem: the functor F there (F = "dividing by r") is easily seen to be right exact and sum preserving, and $F \cong FR \otimes_R = (R/Rr) \otimes_R$.

Corollary 3.34: *Assume R is a ring for which every submodule of a f.g. module is f.g. If $\mathfrak{F}_R$ is the category of all f.g. right R-modules and if $F:\mathfrak{F}_R \to \mathfrak{M}_S$ is right exact additive, then there is a bimodule $B = {}_R B_S$ and a natural equivalence $\otimes_R B \cong F$. Moreover, we may choose $B = F(R)$.*

Proof: The proof is almost the same as that of Theorem 3.33 (together with our remark at its end). As we are looking at $\mathfrak{F}_R$, we need only consider f.g. free R-modules A; moreover, the hypothesis on the ring R guarantees that there is an exact sequence $C \to A \to M \to 0$ in which both C and A are free and f.g. (were C not f.g., then FC would not be defined). Finally, we need not assume F preserves sums, for every additive functor preserves finite sums (Exercise 2.24). ∎

In the next chapter, we shall identify the rings R satisfying the hypothesis of the Corollary (left noetherian again!). The following theorem ties several ideas together.

Theorem 3.35: *The following are equivalent for a functor $F:\mathfrak{M}_R \to \mathbf{Ab}$:*

   (i)   *F preserves direct limits;*
  (ii)   *F is right exact and preserves sums;*
 (iii)   *$F \cong \otimes_R B$ for some left R-module B;*
 (iv)   *there is a functor $G:\mathbf{Ab} \to \mathfrak{M}_R$ so that (F, G) is an adjoint pair.*

Proof: (i) $\Rightarrow$ (ii)   Cokernel and sum are direct limits.
          (ii) $\Rightarrow$ (iii)   Theorem 3.33.
         (iii) $\Rightarrow$ (iv)   Theorem 2.12; set $G = \mathrm{Hom}_Z(B, \ )$.
        (iv) $\Rightarrow$ (i)   Theorem 2.19. ∎

Once we give a characterization of covariant Hom functors, we shall have the dual version of Theorem 3.35. These theorems are special cases of "Adjoint Functor Theorems" [MacLane, 1971, pp. 116–127] which give necessary and sufficient conditions that a functor between more general categories be half of an adjoint pair. Before we do this, however, let us characterize contravariant Hom functors, for the proof is so similar to that of Theorem 3.33.

Theorem 3.36 (Watts): *Let $F:{}_R\mathfrak{M} \to \mathbf{Ab}$ be a contravariant left exact functor converting sums to products. Then $F \cong \mathrm{Hom}_R(\ , B)$ for some left R-module B. Moreover, we may choose $B = F(R)$.*

Proof: First of all, we make the abelian group FR into a left R-module. If $M = {}_R M$ and $m \in M$, define $\varphi_m:R \to M$ by $\varphi_m(r) = rm$. Since φ_m is an R-map, $F\varphi_m:FM \to FR$ is defined. Define $\tau_M:M \times FM \to FR$ by $(m, x) \mapsto (F\varphi_m)x$, where $x \in FM$. In particular, $\tau_R:R \times FR \to FR$ equips FR with a left R-

module structure: note that if $r, s \in R$, then $\varphi_{rs} = \varphi_s \varphi_r : R \to R$ and contravariance of F gives $F\varphi_{rs} = F\varphi_r F\varphi_s$.

Define $\sigma_M : FM \to \operatorname{Hom}_R(M, FR)$ by $\sigma_M(x) : m \mapsto (F\varphi_m)x$. It is easy to check that $\sigma : F \to \operatorname{Hom}_R(\ , FR)$ is a natural transformation, and that σ_R is an isomorphism. The remainder of the proof proceeds, *mutatis mutandis*, as that of Theorem 3.33. ∎

Remarks: 1. If one assumes $F:{}_R\mathfrak{M} \to {}_S\mathfrak{M}$, then one may modify the proof above by showing FR is a bimodule.

 2. The analogue of Corollary 3.34 is also true, if one assumes the same condition on the ring R that appears there.

Definition: A module C is a **cogenerator** for ${}_R\mathfrak{M}$ if, for every module M and every nonzero $m \in M$, there exists $f : M \to C$ with $fm \neq 0$.

Lemma 3.37: *There exists an injective cogenerator for ${}_R\mathfrak{M}$.*

Proof: Define $B = \coprod (R/I)$, where the sum is over all left ideals I in R, and define C to be some injective module containing B. Let M be a module with $m \in M$, $m \neq 0$. Now $\langle m \rangle$ is cyclic, so that $\langle m \rangle \cong R/I$ for some left ideal I (Theorem 2.2). Since R/I is contained in C, there is a monomorphism

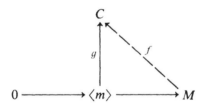

$g : \langle m \rangle \to C$; since C is injective, there is a map $f : M \to C$ with $fm = gm \neq 0$. ∎

The construction above of an injective cogenerator may not be the most efficient. For example, we will see later that $\mathbf{Q}/\mathbf{Z}$ is an injective cogenerator for **Ab** (Lemma 3.50).

Theorem 3.38 (Watts): *Let $G:{}_R\mathfrak{M} \to \mathbf{Ab}$ be a functor preserving inverse limits. Then $G \cong \operatorname{Hom}_R(B, \ )$ for some left R-module B.*

Remark: The module B has no easy description as in Theorems 3.33 and 3.36.

Proof: For a module A and a set X, let A^X denote the product of copies of A indexed by X. We may regard A^X as all functions $X \to A$; in particular, we may regard $e = 1_A \in A^A$. If $x \in A$, then the xth coordinate of e is x, so that

$$p_x e = x,$$

where $p_x: A^A \to A$ is the xth projection. Choose an injective cogenerator C for $_R\mathfrak{M}$, and let $\Pi = C^{GC}$. Since products are inverse limits, G preserves products and

$$e = 1_{GC} \in G\Pi = (GC)^{GC}.$$

Define $\tau: \operatorname{Hom}(\Pi, C) \to GC$ by $f \mapsto (Gf)e$. The map τ is epic, for if $x \in GC$, then the xth projection $p_x: \Pi \to C$ exists. Since G preserves products, Gp_x is the xth projection $G\Pi = GC^{GC} \to GC$, so that $(Gp_x)e = x$. Therefore $\tau(p_x) = x$ and τ is epic.

We now describe $\ker \tau$. If S is a submodule of Π with inclusion $i: S \to \Pi$, we may identify GS with the submodule $\operatorname{im}(Gi)$ of $G\Pi$, for G is left exact (kernel is an inverse limit). Define B as the intersection of all submodules S of Π for which $e \in GS$. By Exercise 2.50, $B = \varprojlim S$; it follows that $GB = \varprojlim GS = \bigcap GS$, using our identification and Exercise 2.50 again; we observe that $e \in GB$. The following are equivalent for $f: \Pi \to C: f \in \ker \tau$; $0 = \tau(f) = (Gf)e$; $e \in \ker Gf$; $e \in G(\ker f)$ (G is left exact); $B \subset \ker f$ (definition of B). If $j: B \to \Pi$ is the inclusion, then $j^*: \operatorname{Hom}(\Pi, C) \to \operatorname{Hom}(B, C)$, and $f \in \ker j^*$ if and only if $j^*(f) = fj = 0$. Thus $f \in \ker j^*$ if and only if $B = \operatorname{im} j \subset \ker f$, so that $\ker j^* = \ker \tau$.

Exactness of $0 \to B \xrightarrow{j} \Pi \to \Pi/B \to 0$ and injectivity of C give exactness of the top row of the diagram

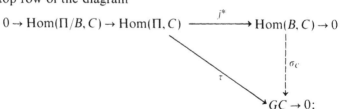

our previous calculation shows the lower sequence is exact. Since $\ker \tau = \ker j^*$, the two cokernels are isomorphic via

$$\sigma_C: \operatorname{Hom}(B, C) \to GC$$

given by $f \mapsto (Gf)e$ (this is a general fact: if $\alpha: A \to X$ and $\beta: A \to Y$ are epimorphisms with the same kernel, then $X \cong Y$ via $x \mapsto \beta(\alpha^{-1}x)$). It is easily checked that $\sigma: \operatorname{Hom}(B, \ ) \to G$ is a natural transformation, where $\sigma_M: \operatorname{Hom}(B, M) \to GM$ is given by $f \mapsto (Gf)e$.

For any module M, there is an imbedding $0 \to M \to C^{\operatorname{Hom}(M,C)}$ given by $m \mapsto (fm)$, that "vector" whose fth coordinate is fm (here we use the fact that

C is a cogenerator). There is thus an exact sequence

$$0 \to M \to C^X \to C^Y$$

for suitable sets X and Y. Since G and $\operatorname{Hom}(B, \ )$ are left exact, there is a commutative diagram with exact rows

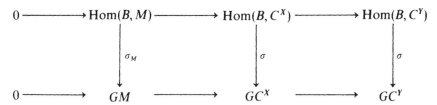

The vertical maps σ are isomorphisms, since σ_C is an isomorphism and both functors preserve products; the Five Lemma shows σ_M is an isomorphism. We conclude that $\operatorname{Hom}(B, \ ) \cong G$. ∎

Corollary 3.39: *The following are equivalent for a functor $G: {}_R\mathfrak{M} \to \mathbf{Ab}$.*

   (i)   *G preserves inverse limits;*
  (ii) *$G \cong \operatorname{Hom}_R(B, \ )$ for some left R-module B;*
 (iii) *there is a functor $F: \mathbf{Ab} \to {}_R\mathfrak{M}$ so that (F, G) is an adjoint pair.*

Proof: (i) $\Rightarrow$ (ii)   Theorem 3.38.
          (ii) $\Rightarrow$ (iii)   Define $F = B \otimes_{\mathbf{Z}}$, noting that B is an $(R - \mathbf{Z})$ bimodule.
          (iii) $\Rightarrow$ (i)   Theorem 2.24. ∎

Remark: Exercise 2.54 gives a fourth condition equivalent to (i):

 (iv) *G is left exact and preserves products.*

There are several comments, whose proofs are more appropriate in a book emphasizing category theory, that should be made. First, the hypothesis in Corollary 3.34 that F be additive is not necessary in the other Watts' theorems, for any functor between module categories that preserves finite sums must be additive. Second, the more general adjoint functor theorems stress characterization of Hom functors, for these always exist, by the very definition of category. Third, one may give a precise definition of a solution to a universal mapping problem, "universal construction", and this is intimately related to adjoint functors. There are also two uniqueness results. If (F, G) and (F, G') are adjoint pairs, then $G \cong G'$; the same result holds in the other variable. Second, if $\operatorname{Hom}(B, \ )$ and $\operatorname{Hom}(B', \ )$ are naturally equivalent, then $B \cong B'$. For details, the reader may consult [MacLane, 1971].

There is one more result in this circle of ideas which is within our reach and which is rather interesting.

Definition: Two categories $\mathfrak{C}$ and $\mathfrak{D}$ are **equivalent**, denoted $\mathfrak{C} \cong \mathfrak{D}$, if there are functors $F : \mathfrak{C} \to \mathfrak{D}$ and $G : \mathfrak{D} \to \mathfrak{C}$ such that $GF \cong 1_{\mathfrak{C}}$ and $FG \cong 1_{\mathfrak{D}}$.

We consider whether it is possible that $_R\mathfrak{M} \cong {}_S\mathfrak{M}$. Should this obtain, then homological algebra cannot distinguish between the rings R and S.

Exercise: 3.34.   If $_R\mathfrak{M} \cong {}_S\mathfrak{M}$, then $Z(R) \cong Z(S)$. (Hint: Use Exercise 2.39.) Conclude that if R and S are commutative rings and $_R\mathfrak{M} \cong {}_S\mathfrak{M}$, then $R \cong S$.

Definition: A module P is a **generator** for $_R\mathfrak{M}$ if every module M is a quotient of a sum of copies of P.

Clearly R itself is a generator for $_R\mathfrak{M}$, for every module is a quotient of a free module. Also, the sum of n copies of R, denoted R^n, is a generator for $_R\mathfrak{M}$.

Exercise: 3.35.   Prove that a module Q is a cogenerator for $_R\mathfrak{M}$ if and only if every module can be imbedded in a product of copies of Q. Conclude that generator and cogenerator are dual.

Definition: A module P is **small** if $\mathrm{Hom}(P, \ )$ preserves sums.

Note that R^n is a small projective generator for $_R\mathfrak{M}$.

Theorem 3.40:   *Let P be a small projective generator for $_R\mathfrak{M}$ and let $S = \mathrm{End}_R(P)$. Then $_R\mathfrak{M} \cong {}_S\mathfrak{M}$.*

Proof: Clearly P is a right S-module (if we write maps on the right); moreover, P is an $(R\text{-}S)$ bimodule, for if $x \in P$, then $(rx)f = r(xf)$ is just the condition that f is an R-map. By Theorem 1.15, the functor $G = \mathrm{Hom}_R(P, \ ) : {}_R\mathfrak{M} \to \mathbf{Ab}$ actually takes values in $_S\mathfrak{M}$. Define $F : {}_S\mathfrak{M} \to {}_R\mathfrak{M}$ by $F = P \otimes_S \ $. Since (F, G) is an adjoint pair, Exercise 2.38 provides natural transformations $1_S \to GF$ and $FG \to 1_R$, where 1_R is the identity functor on $_R\mathfrak{M}$ and 1_S is the identity functor on $_S\mathfrak{M}$. It suffices to prove these natural transformations are equivalences.

Since P is projective, $G = \mathrm{Hom}(P, \ )$ is exact; since P is small, G preserves sums. The functor F is right exact and preserves sums. It follows that the composites GF and FG preserve sums and are right exact.

Let us evaluate:

$$FG(P) = F(\mathrm{Hom}(P, P)) = F(S) = P \otimes_S S \cong P.$$

If M is any R-module, there is an exact sequence of the form

$$\coprod P \to \coprod P \to M \to 0$$

(where the index sets of the sums may be distinct) because P is a generator for $_R\mathfrak{M}$. There is thus a commutative diagram with exact rows

$$
\begin{array}{ccccccc}
\coprod P & \to & \coprod P & \to & M & \to 0 \\
\uparrow & & \uparrow & & \uparrow & \\
\coprod FG(P) & \to & \coprod FG(P) & \to & FG(M) & \to 0
\end{array}
$$

We know the first two vertical maps are isomorphisms, so the Five Lemma gives $M \cong FG(M)$. This gives $1_R \cong FG$.

For the other composite, note that

$$GF(S) = G(P \otimes_S S) \cong G(P) = \operatorname{Hom}_R(P, P) = S.$$

If N is any S-module, there is an exact sequence of the form

$$\coprod S \to \coprod S \to N \to 0,$$

for every module is a quotient of a free module. The argument concludes just as for the first composite. ∎

Corollary 3.41: *If R is a ring and $S = R_n$, all $n \times n$ matrices with entries in R, then $_R\mathfrak{M} \cong {}_S\mathfrak{M}$.*

Proof: We know that R^n is a small projective generator for $_R\mathfrak{M}$, and $\operatorname{End}_R(R^n) \cong R_n$. ∎

Corollary 3.42: *Let P be a small projective generator for $_R\mathfrak{M}$ and let $S = \operatorname{End}_R(P)$. If $F = \operatorname{Hom}_R(P,): {}_R\mathfrak{M} \to {}_S\mathfrak{M}$, then F is an exact functor preserving projectives and injectives.*

Proof: Assume Q is R-projective, and consider the diagram of S-modules

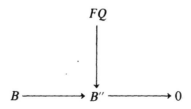

If $G = P \otimes_S$, then we have the diagram of R-modules

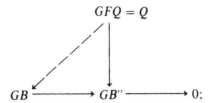

$$GFQ = Q$$

the dashed arrow exists because Q is projective. Now apply F to this diagram to obtain the desired arrow $FQ \to B$. A similar argument shows F preserves injectives. ∎

The Morita theorems go on to say, among other things, that every equivalence between module categories arises as above, that is, from a small projective generator P. We should also point out that the functor $F = \mathrm{Hom}_R(P, \ )$ in this case is (right) exact and preserves sums, so Watts' Theorem 3.33 gives $F = \mathrm{Hom}_R(P, R) \otimes_R$. An exposition of the Morita theorems may be found in DeMeyer–Ingraham [1971].

FLAT MODULES

A module P is projective if and only if $\mathrm{Hom}_R(P, \ )$ is exact, and a module E is injective if and only if $\mathrm{Hom}_R(\ , E)$ is exact. Let us now fix a variable of tensor.

Definition: A right R-module B is **flat** if the functor $B \otimes_R$ is exact.

Of course, there is a similar definition for left R-modules C and $\otimes_R C$. Since $B \otimes_R$ is always right exact, a module B is flat if and only if f monic implies $1_B \otimes f$ monic.

Theorem 3.43: R *is a flat R-module.*

Proof: Let $f : A' \to A$ be monic. If $t : 1 \to R \otimes_R$ is the natural equivalence of Theorem 1.12, then there is a commutative diagram

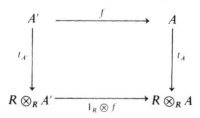

Since the vertical maps are isomorphisms, the map $1_R \otimes f$ is monic and R is flat. ∎

An example of a nonflat module is contained in Example 18 of Chapter 2: $\mathbf{Z}/2\mathbf{Z}$ is not a flat $\mathbf{Z}$-module. The next theorem generalizes Theorem 3.26 and Exercise 3.22.

Theorem 3.44: *Assume B is a bimodule ${}_SB_R$ that is R-flat and $C = {}_SC$ is injective. Then $\mathrm{Hom}_S(B, C)$ is an injective left R-module.*

Proof: First of all, $\mathrm{Hom}_S(B, C)$ is a left R-module, by Theorem 1.15.

We must prove $\mathrm{Hom}_R(\ , \mathrm{Hom}_S(B, C))$ is an exact functor. The adjoint isomorphism shows this functor is naturally equivalent to $\mathrm{Hom}_S(B \otimes_R \ , C)$, which is the composite $\mathrm{Hom}_S(\ , C) \circ (B \otimes_R \ )$. Each of these is exact, since B is R-flat and C is S-injective, and the composite of exact functors is exact. ∎

Setting $B = R$, we obtain the theorem and exercise mentioned above.

Theorem 3.45: *Let $\{B_k : k \in K\}$ be a family of right R-modules. Then $\coprod B_k$ is flat if and only if each B_k is flat.*

Proof: Note first that if $\{f_k : A'_k \to A_k\}$ is a family of maps, there is a unique map $\coprod f_k : \coprod A'_k \to \coprod A_k$ with $\sum a'_k \mapsto \sum f_k a'_k$; moreover, $\coprod f_k$ is monic if and only if each f_k is monic.

Assume $f : A' \to A$ is monic. There is a commutative diagram

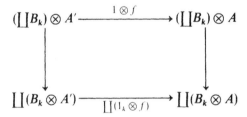

where $1_k = 1_{B_k}$ and the vertical maps are the usual isomorphisms

$$(\textstyle\sum b_k) \otimes a \mapsto \sum(b_k \otimes a).$$

Our initial remarks give $1 \otimes f$ monic if and only if each $1_k \otimes f$ is monic, i.e., $\coprod B_k$ is flat if and only if each B_k is flat. ∎

Corollary 3.46: *Every projective module is flat.*

Proof: Theorems 3.43 and 3.45 show that every free module is flat; Theorem 3.45 shows that every summand of a free module is flat. ∎

It follows that every module A has a **flat resolution**, i.e., there is an exact sequence $\cdots \to F_1 \to F_0 \to A \to 0$ in which each F_n is flat.

Theorem 3.47: *If $\{B_k, \varphi_j^k\}$ is a direct system of flat modules over a directed index set K, then $\varinjlim B_k$ is flat.*

Proof: If $f: A' \to A$ is monic, there is a commutative diagram where the

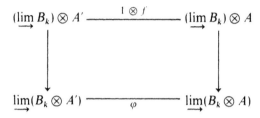

vertical maps are the isomorphisms of Corollary 2.20, and φ is the monomorphism of Theorem 2.18. It follows that $1 \otimes f$ is monic. ∎

This theorem may be false if the index set is not directed; for example, coker(flat → flat) need not be flat.

Corollary 3.48: *If R is a domain, then its quotient field Q is a flat R-module.*

Proof: We saw in Exercise 2.34 that Q is a direct limit of copies of R, indexed by a directed set. ∎

Corollary 3.49: *If every f.g. submodule of B is flat, then B is flat.*

Proof: We saw in Exercise 2.32 that every module is a direct limit of its f.g. submodules, and the index set there is directed. ∎

Let us give another proof of Corollary 3.49 avoiding direct limits. Assume $f: A' \to A$ is monic and that $x \in \ker 1 \otimes f$. As $x \in B \otimes A'$, we have $x = \sum_{i=1}^{n} b_i \otimes a_i'$. Let F be the free abelian group with basis $B \times A$, and let S be the subgroup of relations (as in Theorem 1.4) so that $F/S = B \otimes A$. To say that $\sum b_i \otimes fa_i' = 0$ in $B \otimes A$ is to say that $\sum (b_i, fa_i') \in S$. Let B' be the submodule of B generated by $b_1, \ldots, b_n$ together with the (finite number of) first coordinates of relators exhibiting $\sum (b_i, fa_i') \in S$. If $\bar{x} = \sum b_i \otimes a_i'$ in $B' \otimes A'$, then $\bar{x} \in \ker(B' \otimes A' \to B' \otimes A)$. As B' is f.g., the hypothesis gives $\bar{x} = 0$ in $B' \otimes A'$, from which it follows that $x = 0$ in $B \otimes A$ (if $j: B' \to B$ is the inclusion, then $x = (j \otimes 1_A)\bar{x}$).

We now aim toward a characterization of flat modules that links them to injectives.

Lemma 3.50: *The group $\mathbf{Q}/\mathbf{Z}$ is an injective cogenerator for $\mathbf{Ab}$.*

Proof: Since $\mathbf{Q}/\mathbf{Z}$ is divisible, being a quotient of $\mathbf{Q}$, it is injective (Theorem 3.24). Let M be an abelian group, and let $m \in M$, $m \neq 0$. If m has infinite order, define $f: \langle m \rangle \to \mathbf{Q}/\mathbf{Z}$ by $m \mapsto \frac{1}{2} + \mathbf{Z}$; if m has finite order n, define

$f: \langle m \rangle \to \mathbf{Q}/\mathbf{Z}$ by $m \mapsto 1/n + \mathbf{Z}$. In either case, $f(m) \neq 0$. Now use injectivity of $\mathbf{Q}/\mathbf{Z}$ to extend f to all of M. ∎

Definition: If B is a right R-module, its **character module** B^* is the left R-module $\mathrm{Hom}_\mathbf{Z}(B, \mathbf{Q}/\mathbf{Z})$.

Since B is a bimodule $_\mathbf{Z}B_R$, Theorem 1.15(i) gives the left R-module structure on $\mathrm{Hom}_\mathbf{Z}(B, \mathbf{Q}/\mathbf{Z})$ by $rf: b \mapsto f(br)$. We now improve Lemma 2.13.

Lemma 3.51: *A sequence of right R-modules*

$$0 \to A \xrightarrow{\alpha} B \xrightarrow{\beta} C \to 0$$

is exact if and only if the sequence of character modules

$$0 \to C^* \xrightarrow{\beta^*} B^* \xrightarrow{\alpha^*} A^* \to 0$$

is exact.

Proof: If the original sequence is exact, then the exact contravariant functor $\mathrm{Hom}_\mathbf{Z}(, \mathbf{Q}/\mathbf{Z})$ carries it into an exact sequence.

To prove the converse, we shall show that $\ker \alpha^* = \mathrm{im}\,\beta^*$ implies $\ker \beta = \mathrm{im}\,\alpha$ without assuming β^* monic or α^* epic; the result will then follow.

Is $\mathrm{im}\,\alpha \subset \ker \beta$? Suppose $a \in A$ and $\alpha a \notin \ker \beta$, i.e., $\beta \alpha a \neq 0$. Since $\mathbf{Q}/\mathbf{Z}$ is a cogenerator, there is a map $f: C \to \mathbf{Q}/\mathbf{Z}$ with $f \beta \alpha a \neq 0$. Thus, $f \in C^*$ and $f \beta \alpha \neq 0$, i.e., $\alpha^* \beta^*(f) \neq 0$. This contradicts $\alpha^* \beta^* = 0$.

Is $\ker \beta \subset \mathrm{im}\,\alpha$? If not, there is an element $b \in \ker \beta$ with $b \notin \mathrm{im}\,\alpha$. Thus, $b + \mathrm{im}\,\alpha$ is a nonzero element of $B/\mathrm{im}\,\alpha$, so there is a map $g: B/\mathrm{im}\,\alpha \to \mathbf{Q}/\mathbf{Z}$ with $g(b + \mathrm{im}\,\alpha) \neq 0$. If $\pi: B \to B/\mathrm{im}\,\alpha$ is the natural map, then $f = g\pi: B \to \mathbf{Q}/\mathbf{Z}$ is such that $fb \neq 0$ and $f(\mathrm{im}\,\alpha) = 0$. Therefore $0 = f\alpha = \alpha^*(f)$, so $f \in \ker \alpha^* = \mathrm{im}\,\beta^*$. Thus $f = \beta^*(h) = h\beta$ for some $h \in C^*$. In particular, $fb = h\beta b$. This is a contradiction, for $fb \neq 0$ while $b \in \ker \beta$. ∎

Theorem 3.52: *A right R-module B is flat if and only if its character module B^* is an injective left R-module.*

Proof: As every right R-module, B is a bimodule $B = {}_\mathbf{Z}B_R$. Since $\mathbf{Q}/\mathbf{Z}$ is an injective $\mathbf{Z}$-module, Theorem 3.44 shows $B^* = \mathrm{Hom}_\mathbf{Z}(B, \mathbf{Q}/\mathbf{Z})$ is injective.

For the converse, assume B^* is R-injective and $f: A' \to A$ is monic. There is a commutative diagram with vertical isomorphisms

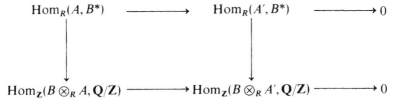

so that exactness of the top row gives exactness of the bottom. In character module notation, $(B \otimes A)^* \to (B \otimes A')^* \to 0$ is exact, and Lemma 3.51 applies to give exactness of $0 \to B \otimes A' \to B \otimes A$. Therefore, B is flat. ∎

This theorem may be exploited by coupling it with Baer's criterion.

Theorem 3.53: *If B is a right R-module such that $0 \to B \otimes I \to B \otimes R$ is exact for every f.g. left ideal I of R, then B is flat.*

Proof: Since every left ideal is a direct limit (with directed index set) of f.g. left ideals, Theorem 2.18 implies that $0 \to B \otimes I \to B \otimes R$ is exact for every left ideal I of R. This gives exactness of $(B \otimes R)^* \to (B \otimes I)^* \to 0$, which, in turn, gives exactness of

$$\operatorname{Hom}_R(R, B^*) \to \operatorname{Hom}_R(I, B^*) \to 0$$

as in the proof of Theorem 3.52. By Baer's criterion (Theorem 3.20), we have B^* injective, and this shows B is flat. ∎

Of course, the converse of Theorem 3.53 is true.

Theorem 3.54: *If B_R is flat and I is a left ideal, then the map $B \otimes_R I \to BI$ given by $b \otimes i \mapsto bi$ is an isomorphism.*

Proof: Recall that $BI = \{\sum b_j i_j : b_j \in B, i_j \in I\}$ is a subgroup of B. The composite $B \otimes_R I \to B \otimes_R R \to B$, the second map being the usual isomorphism, has image BI and is monic, by Theorem 3.53. ∎

In general, neither quotients nor submodules of flats are flat (the same is true if we replace "flat" by "projective" or "injective"), but we must wait until we examine specific rings before examples can be given. The following instance, however, guarantees that a quotient of a flat is flat if the kernel is "nicely" imbedded; the situation is related to the notion of "purity" in abelian groups.

Theorem 3.55: *Let F be flat and $0 \to K \to F \xrightarrow{\beta} B \to 0$ be an exact sequence of right R-modules. The following conditions are equivalent:*

  (i)   *B is flat;*
  (ii) *$K \cap FI = KI$ for every left ideal I;*
  (iii) *$K \cap FI = KI$ for every f.g. left ideal I.*

Proof: In conditions (ii) and (iii), note that $KI \subset K \cap FI$, so only the reverse inclusion is significant. We give a preliminary discussion before proving the implications.

Tensoring the original exact sequence by a left ideal I gives exactness of

$$K \otimes I \to F \otimes I \xrightarrow{\beta \otimes 1} B \otimes I \to 0.$$

By Corollary 3.54, we may identify $F \otimes I$ with FI via $f \otimes i \mapsto fi$; this identifies $\operatorname{im}(K \otimes I \to F \otimes I)$ with KI. There results an isomorphism

$$\gamma : FI/KI \xrightarrow{\sim} B \otimes I$$

given by $fi + KI \mapsto \beta f \otimes i$. Next,

$$\begin{aligned} BI &= \{\textstyle\sum (\beta f_j) i_j : f_j \in F, i_j \in I\} && \text{since } \beta \text{ is epic} \\ &= \{\beta(\textstyle\sum f_j i_j)\} = \beta(FI). \end{aligned}$$

The first isomorphism theorem provides an isomorphism

$$\delta : BI = \beta(FI) \to FI/FI \cap K$$

given by $bi \mapsto fi + FI \cap K$, where $\beta f = b$. We assemble these maps to obtain the composite σ:

$$FI/KI \xrightarrow{\gamma} B \otimes I \xrightarrow{\theta} BI \xrightarrow{\delta} FI/FI \cap K,$$

where $\theta : b \otimes i \mapsto bi$. Explicitly, $\sigma : fi + KI \mapsto fi + FI \cap K$. Since $KI \subset FI \cap K$, σ is an isomorphism if and only if $KI = FI \cap K$. Moreover, since the flanking maps γ and δ are isomorphisms, σ is an isomorphism if and only if θ is.

(i) $\Rightarrow$ (ii)   If B is flat, Corollary 3.54 says θ is an isomorphism. Thus σ is an isomorphism and $KI = FI \cap K$.

(ii) $\Rightarrow$ (iii)   Trivial.

(iii) $\Rightarrow$ (i)   If $KI = FI \cap K$ for every f.g. left ideal I, then θ is an isomorphism for every such I, and Theorem 3.53 gives B flat.   ∎

Another technical lemma and we will be able to relate flat modules to projectives.

Lemma 3.56:   *Let F be free with basis $\{x_j : j \in J\}$, and let $0 \to K \to F \to B \to 0$ be an exact sequence of right R-modules. If $v \in F$ is written*

$$v = x_{j_1} r_1 + \cdots + x_{j_t} r_t,$$

define $I(v)$ as the left ideal of R generated by the "coordinates" $r_1, \ldots, r_t$. Then B is flat if and only if $v \in KI(v)$ for every $v \in K$.

Proof: If B is flat and $v \in K$, then $v \in K \cap FI(v) = KI(v)$, by Theorem 3.55. Conversely, let I be any left ideal and let $v \in K \cap FI$. Then $I(v) \subset I$; by hypothesis, $v \in KI(v) \subset KI$. Therefore, $K \cap FI \subset KI$. As the reverse inclusion always holds, Theorem 3.55 gives B flat.   ∎

Theorem 3.57 **(Villamayor):** *Let F be free and let $0 \to K \to F \to B \to 0$ be an exact sequence of right R-modules. The following are equivalent.*

(i) *B is flat;*

(ii) *for every $v \in K$, there is a map $\theta : F \to K$ with $\theta v = v$;*

(iii) *for every $v_1, \dots, v_n \in K$, there is a map $\theta : F \to K$ with $\theta v_i = v_i$, $i = 1, \dots, n$.*

Proof: (i) $\Rightarrow$ (ii) Suppose B is flat and $v \in K$; choose a basis $\{x_j : j \in J\}$ of F. Let $v = x_{j_1} r_1 + \cdots + x_{j_t} r_t$ and let $I(v)$ be the left ideal generated by $r_1, \dots, r_t$. Since B is flat, Lemma 3.56 gives $v \in KI(v)$. Hence $v = \sum k_\lambda s_\lambda$, where $k_\lambda \in K$, $s_\lambda \in I(v)$; as $s_\lambda \in I(v)$, $s_\lambda = \sum a_{\lambda i} r_i$, where $a_{\lambda i} \in R$. Thus $v = \sum k'_i r_i$, where $k'_i = \sum k_\lambda a_{\lambda i} \in K$. Define $\theta : F \to K$ by $x_{j_i} \mapsto k'_i$ and the other basis elements going to 0. Clearly $\theta v = v$.

(ii) $\Rightarrow$ (i) Choose a basis $\{x_j : j \in J\}$ of F, and let $v = x_{j_1} r_1 + \cdots + x_{j_t} r_t \in K$; let $\theta : F \to K$ satisfy $\theta v = v$. Then $v = \theta v = \theta(x_{j_1}) r_1 + \cdots + \theta(x_{j_t}) r_t \in KI(v)$. By Lemma 3.56, B is flat.

Since (iii) trivially implies (ii), it only remains to show (ii) $\Rightarrow$ (iii). We do an induction on n, the case $n = 1$ being our hypothesis (ii). Suppose $n > 1$ and $v_1, \dots, v_n \in K$. By (ii), there is a map $\theta_n : F \to K$ with $\theta_n v_n = v_n$. Define $v'_i \in K$, $i = 1, \dots, n - 1$, by $v'_i = v_i - \theta_n v_i$. By induction, there is a map $\theta' : F \to K$ with $\theta' v'_i = v'_i$, $i = 1, \dots, n - 1$. Finally, define $\theta : F \to K$ by

$$\theta v = \theta_n v + \theta'(v - \theta_n v).$$

It is a routine check that $\theta v_i = v_i$, $i = 1, \dots, n$. ∎

Definition: A right R-module B is **finitely related** (or **finitely presented**) if there is an exact sequence

$$F_1 \to F_0 \to B \to 0$$

where F_1 and F_0 are f.g. free.

It is easy to see that B is finitely related if and only if there is an exact sequence $0 \to K \to F \to B \to 0$ in which F is free and both F and K are f.g. Note that every f.g. projective module B is finitely related: choose a f.g. free module F mapping onto B, say, with kernel K; this sequence must split, $F \cong B \oplus K$, and this shows that K is f.g. If B is flat, then the converse is true.

Corollary 3.58: *Every finitely related flat R-module is projective.*

Proof: There is an exact sequence

$$0 \to K \xrightarrow{\lambda} F \to B \to 0,$$

where F is free and both F and K are f.g. If $K = \langle v_1, \dots, v_n \rangle$, then Theorem

3.57 provides a map $\theta: F \to K$ with $\theta v_i = v_i$, $i = 1, \ldots, n$ (we are assuming, without loss of generality, that λ is an inclusion). Therefore $\theta\lambda = 1_K$, so the sequence splits and B is projective. ∎

There is a quicker way to prove this last corollary.

Lemma 3.59: *Let R and S be rings and consider the situation $({}_RA, {}_RB_S, C_S)$. If A is a f.g. projective R-module, then there is a natural isomorphism*

$$\text{Hom}_S(B, C) \otimes_R A \cong \text{Hom}_S(\text{Hom}_R(A, B), C).$$

Proof: Observe that B being a bimodule allows us to regard $\text{Hom}_S(B, C)$ and $\text{Hom}_R(A, B)$ as modules, so the above terms do make sense.
Define $\sigma_A: \text{Hom}_S(B, C) \otimes_R A \to \text{Hom}_S(\text{Hom}_R(A, B), C)$ by

$$f \otimes a \mapsto \sigma_A(f \otimes a),$$

where, for $g \in \text{Hom}_R(A, B)$,

$$\sigma_A(f \otimes a): g \mapsto f(g(a)) \in C.$$

It is a routine check, with no hypothesis on A other than $A = {}_RA$, that σ is a homomorphism, natural in A. Moreover, if $A = R$, then σ_A is visibly an isomorphism; indeed, it is just as clear that σ_A is an isomorphism when A is f.g. free. Finally, we let the reader check that if σ_A is an isomorphism and A' is a summand of A, then $\sigma_{A'}$ is also an isomorphism. ∎

Let us record another isomorphism as above, even though we do not need the result for the new proof of Corollary 3.58.

Lemma 3.60: *Let R and S be rings and consider the situation $({}_RA, {}_RB_S, C_S)$. If A is finitely related and C is injective, then there is a natural isomorphism*

$$\text{Hom}_S(B, C) \otimes_R A \cong \text{Hom}_S(\text{Hom}_R(A, B), C).$$

Remark: The map σ_A of Lemma 3.59 is the desired isomorphism; we shall prove this in the midst of the next proof. ∎

Theorem 3.61: *Every finitely related flat R-module E is projective.*

Proof: Let $F_1 \to F_0 \to E \to 0$ be exact, where F_1, F_0 are f.g. free. It suffices to show that $\text{Hom}(E, \ )$ is exact, i.e., if $A \to A'' \to 0$ is exact, then $\text{Hom}(E, A) \to \text{Hom}(E, A'') \to 0$ is exact. By Lemma 3.51, it suffices to show exactness of

$$0 \to (\text{Hom}(E, A''))^* \to (\text{Hom}(E, A))^*,$$

where $(\text{Hom}(E, A))^*$ is the character module $\text{Hom}_{\mathbf{Z}}(\text{Hom}_R(E, A), \mathbf{Q}/\mathbf{Z})$.

Consider the following diagram, for any module B and for C injective:

$$\text{Hom}_{\mathbf{Z}}(B, C) \otimes F_1 \quad \to \quad \text{Hom}_{\mathbf{Z}}(B, C) \otimes F_0 \quad \to \quad \text{Hom}_{\mathbf{Z}}(B, C) \otimes E \quad \to 0$$

$$\downarrow \sigma \qquad\qquad\qquad\qquad \downarrow \sigma \qquad\qquad\qquad\qquad \downarrow \sigma$$

$$\text{Hom}_{\mathbf{Z}}(\text{Hom}_R(F_1, B), C) \to \text{Hom}_{\mathbf{Z}}(\text{Hom}_R(F_0, B), C) \to \text{Hom}_{\mathbf{Z}}(\text{Hom}_R(E, B), C) \to 0$$

By Lemma 3.59, the diagram commutes and the first two vertical maps are isomorphisms. The top row is exact, for tensor product is right exact; the bottom row is exact because $\text{Hom}_R(\ , B)$ is left exact and $\text{Hom}_{\mathbf{Z}}(\ , C)$ is exact (for C is injective). The Five Lemma gives the final vertical map an isomorphism for every B:

$$\text{Hom}_{\mathbf{Z}}(B, C) \otimes E \xrightarrow{\ \sim\ } \text{Hom}_{\mathbf{Z}}(\text{Hom}(E, B), C)$$

(we have proved Lemma 3.60).

Finally, consider the diagram

$$0 \to \quad \text{Hom}_{\mathbf{Z}}(A'', \mathbf{Q}/\mathbf{Z}) \otimes E \quad \to \quad \text{Hom}_{\mathbf{Z}}(A, \mathbf{Q}/\mathbf{Z}) \otimes E$$

$$\downarrow \sigma \qquad\qquad\qquad\qquad\qquad \downarrow \sigma$$

$$0 \to \text{Hom}_{\mathbf{Z}}(\text{Hom}(E, A''), \mathbf{Q}/\mathbf{Z}) \to \text{Hom}_{\mathbf{Z}}(\text{Hom}(E, A), \mathbf{Q}/\mathbf{Z})$$

Naturality of the maps σ gives commutativity, and we have just seen that the maps σ are isomorphisms. Recall that we began with an epimorphism $A \to A'' \to 0$, so that we have exactness of $0 \to \text{Hom}_{\mathbf{Z}}(A'', \mathbf{Q}/\mathbf{Z}) \to \text{Hom}_{\mathbf{Z}}(A, \mathbf{Q}/\mathbf{Z})$. Since E is flat, the top row is exact, and this implies exactness of the bottom row, which is precisely what we had to show.   ∎

Are there any f.g. modules that are not finitely related? Let $R = k[x_1, x_2, \ldots]$, polynomials in infinitely many variables over a field k; if I is the ideal generated by the indeterminates, then R/I is cyclic (hence f.g.), and we shall show it is not finitely related.

Theorem 3.62   (Schanuel's Lemma):   *Given exact sequences*

$$0 \to K_1 \to P_1 \to B \to 0 \qquad and \qquad 0 \to K_2 \to P_2 \to B \to 0,$$

where P_1 and P_2 are projective, then $K_1 \oplus P_2 \cong K_2 \oplus P_1$.

Proof: Consider the diagram with exact rows

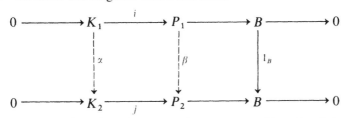

Since P_1 is projective, there is a map $\beta: P_1 \to P_2$ making the right square commute; diagram-chasing shows there exists a map $\alpha: K_1 \to K_2$ making the left square commute. There is an exact sequence

$$0 \to K_1 \xrightarrow{\theta} P_1 \oplus K_2 \xrightarrow{\psi} P_2 \to 0,$$

where $\theta: k_1 \mapsto (ik_1, \alpha k_1)$ and $\psi: (p_1, k_2) \mapsto \beta p_1 - jk_2$ (verification of exactness is routine). Since P_2 is projective, this last sequence splits, giving $P_1 \oplus K_2 \cong K_1 \oplus P_2$. ∎

Corollary 3.63:   *If B is finitely related and*

$$0 \to K \to M \xrightarrow{f} B \to 0$$

is exact, where M is f.g., then K is f.g.

Proof: Assume first that M is free. Since B is finitely related, there is an exact sequence $0 \to K_1 \to F_1 \to B \to 0$ with F_1 free and both F_1 and K_1 f.g. By Schanuel's lemma, $M \oplus K_1 \cong F_1 \oplus K$. But $M \oplus K_1$ is f.g., hence so is its summand K (summands are images).

Let us drop the assumption that M is free; choose a f.g. free F mapping onto M. There is a commutative diagram with exact rows

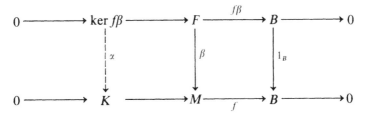

Now $\ker f\beta$ is f.g., by the first part of the proof, and the induced map α is epic, by the Five Lemma. Therefore K is f.g. ∎

It follows immediately from the Corollary that the cyclic module exhibited above is not finitely related. The real significance of Schanuel's lemma is that a strong connection exists between any two descriptions of a module by generators and relations.

Exercises: 3.36. State and prove the dual of Schanuel's lemma involving injectives.

3.37. Given two exact sequences, where the P's and Q's are projective,

$$0 \to K \to P_n \to P_{n-1} \to \cdots \to P_0 \to B \to 0$$

and

$$0 \to L \to Q_n \to Q_{n-1} \to \cdots \to Q_0 \to B \to 0,$$

then $K \oplus Q_n \oplus P_{n-1} \oplus \cdots \cong L \oplus P_n \oplus Q_{n-1} \oplus \cdots$. The dual is also true.

3.38. Let B be a bimodule $_RB_S$ and let C be a left S-module. If B is a flat R-module and C is a flat S-module, then $B \otimes_S C$ is a flat R-module. (Hint: The composite of exact functors is exact.)

PURITY

In this brief section, we examine loss of exactness from a different viewpoint. If $0 \to A' \to A \to A'' \to 0$ is exact and $0 \to B \otimes A' \to B \otimes A \to B \otimes A'' \to 0$ is not exact (of course, this means the 0 at front cannot be there), then we have made B the culprit. Good modules preserve exactness, and we have called them flat. Perhaps the fault is not in our modules but in our sequences.

Definition: An exact sequence of left R-modules

$$0 \to A' \xrightarrow{\lambda} A \to A'' \to 0$$

is **pure exact** if, for every right R-module B, we have exactness of

$$0 \to B \otimes A' \xrightarrow{1 \otimes \lambda} B \otimes A \to B \otimes A'' \to 0.$$

We say that $\lambda A'$ is a **pure submodule** of A in this case.

Of course, one need only worry whether $1_B \otimes \lambda$ is monic for every B. It is easy to see that every split short exact sequence is pure exact; every summand is a pure submodule.

Exercises: 3.39. Prove that a right R-module A is flat if and only if whenever $\sum a_i r_i = 0$ in A (where $a_i \in A$ and $r_i \in R$), there exist elements $b_j \in A$ and $s_{ij} \in R$ such that

$$a_i = \sum_j b_j s_{ij} \quad \text{and} \quad \sum_i s_{ij} r_i = 0.$$

Purity

3.40. $0 \to A' \to A \to A'' \to 0$ is pure exact if and only if it remains exact after tensoring by every finitely related module B. (Hint: To say an element lies in $\ker(B \otimes A' \to B \otimes A)$ involves only finitely many elements of B.)

3.41. If $0 \to A' \to A \to A'' \to 0$ remains exact after applying $\operatorname{Hom}(B,)$ for all B, then the original sequence is split.

Lemma 3.64: *Let B be a finitely related right R-module with generators $b_1, \ldots, b_n$ and relations $\sum b_j r_{ji}$, $i = 1, \ldots, m$. If A is a left R-module with*

$$\sum_{j=1}^{n} b_j \otimes a_j = 0 \qquad \text{in} \quad B \otimes_R A,$$

then there are elements $h_i \in A$ with $a_j = \sum r_{ji} h_i$, all j.

Proof: We have varied the definition of generators and relations a bit. We really mean there is an exact sequence

$$0 \to K \xrightarrow{\mu} F \xrightarrow{p} B \to 0$$

with F free on $x_1, \ldots, x_n$, K generated by $\sum x_j r_{ji}$, $i = 1, \ldots, m$, and $px_j = b_j$. Tensoring by A gives exactness of

$$K \otimes A \xrightarrow{\mu \otimes 1} F \otimes A \xrightarrow{p \otimes 1} B \otimes A \to 0,$$

and, by hypothesis, $\sum x_j \otimes a_j \in \ker p \otimes 1 = \operatorname{im} \mu \otimes 1$. Now every element of $K \otimes A$ has an expression of the form

$$\sum_{i,j} x_j r_{ji} \otimes h_i, \qquad \text{where} \quad h_i \in A.$$

In particular,

$$\sum x_j \otimes a_j = (\mu \otimes 1) \sum_{i,j} x_j r_{ji} \otimes h_i = \sum_j x_j \otimes \left(\sum_i r_{ji} h_i \right).$$

Since F is free on the x_j's, every element of $F \otimes A = \coprod (x_j R) \otimes A$ has a unique expression of the form $\sum x_j \otimes \alpha_j$, where $\alpha_j \in A$. It follows that $a_j = \sum r_{ji} h_i$, as desired. ∎

Theorem 3.65 (P. M. Cohn): *Let $\lambda : A' \to A$ be monic. Then $\lambda A'$ is a pure submodule of A if and only if, given any commutative diagram with F_1, F_0 f.g. free, there is a map $F_0 \to A'$ making the top triangle commute.*

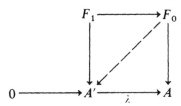

Remark: We restate the condition in terms of equations and then in terms of matrices.

1. If $a'_1, \ldots, a'_n$ in A' satisfy a system of equations $\lambda a'_j = \sum r_{ji} a_i$, where $a_i \in A$, then there exist $h'_i \in A'$ with $a'_j = \sum r_{ji} h'_i$ for all j.

2. Assume V' is a column vector in $\lambda A'$ and D is a rectangular matrix over R with $V' = DV$ for some column vector V in A; then there is a column vector H' in λA with $V' = DH'$.

Proof: Assume $\lambda A'$ is pure in A and $\lambda a'_j = \sum r_{ji} a_i$ for elements $a_i \in A$. Define a finitely related module $B = F/K$, where F is free on $x_j, j = 1, \ldots, n$, and K is generated by $\sum x_j r_{ji}$, $i = 1, \ldots, m$; let $b_j = x_j + K$. In $B \otimes A$, we have

$$\sum b_j \otimes \lambda a'_j = \sum_j b_j \otimes (\sum r_{ji} a_i) = \sum_i \left(\sum_j b_j r_{ji} \otimes a_i \right) = 0.$$

Since $1 \otimes \lambda$ is monic, by purity, we have $\sum b_j \otimes a'_j = 0$ in $B \otimes A'$. By Lemma 3.64, there are elements $h'_i \in A'$ with $a'_j = \sum r_{ji} h'_i$, all j.

For the converse, we must show $1 \otimes \lambda : B \otimes A' \to B \otimes A$ is monic for every B. By Exercise 3.40, we may assume B is finitely related, say, with generators $b_1, \ldots, b_n$ and relators $\sum b_j r_{ji}$, $i = 1, \ldots, m$. A typical element of $B \otimes A'$ can be written $\sum b_j \otimes a'_j$ for $a'_j \in A'$. If $(1 \otimes \lambda) \sum b_j \otimes a'_j = 0$ in $B \otimes A$, is $\sum b_j \otimes a'_j = 0$ in $B \otimes A'$? By Lemma 3.64, there are elements $h_i \in A$ with $\lambda a'_j = \sum r_{ji} h_i$, all j. By hypothesis, there are elements $h'_i \in A'$ with $a'_j = \sum r_{ji} h'_i$, all j. Therefore

$$\sum b_j \otimes a'_j = \sum_j b_j \otimes (\sum r_{ji} h'_i) = \sum_i \left(\sum_j b_j r_{ji} \right) \otimes h'_i = 0 \qquad \text{in} \quad B \otimes A'.$$

Hence $1 \otimes \lambda$ is monic and $\lambda A'$ is a pure submodule of A. ∎

Exercises: 3.42. If $R = \mathbf{Z}$, prove that a subgroup A' of A is pure if and only if $A' \cap nA = nA'$ for all $n \in \mathbf{Z}$.

3.43. If A and B are torsion abelian groups, prove that $A \otimes B$ is a sum of cyclic groups. (Hint: Use Kulikov's theorem [Rotman, 1973, p. 197]: if A is a torsion abelian group, there is a pure exact sequence

$$0 \to C \to A \to D \to 0,$$

where C is a sum of cyclic groups and D is divisible.)

3.44. Let A' be a subgroup of a torsion-free abelian group A. Prove that A' is pure in A if and only if A/A' is torsion-free.

3.45. If $\{A_i : i \in I\}$ is a family of pure subgroups of a torsion-free group A, then $\bigcap A_i$ is pure in A. (Hint: There is an exact sequence

$$0 \to \bigcap A_i \to A \to \prod (A/A_i).)$$

(If A is not torsion-free, an intersection of pure subgroups need not be pure.)

LOCALIZATION

A very important construction in commutative algebra is "localizing" a ring R (it is the ring-theoretic way to focus on the behavior of an algebraic variety at a neighborhood of one of its points). All rings in this section are commutative; I saw the elegant exposition below in unpublished lecture notes of M. Artin. We assume the reader accepts the existence of $R[U]$, the ring of all polynomials in (possibly infinitely many) variables U. Moreover, we assume the reader knows $R[U]$ is the free commutative R-algebra with basis U:

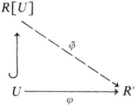

if R' is an R-algebra and $\varphi : U \to R'$ any function, there exists a unique R-algebra map $\tilde{\varphi}$ extending φ (that $\tilde{\varphi}$ is an R-algebra map means $\tilde{\varphi}$ is a ring map with $\tilde{\varphi}(rx) = r\tilde{\varphi}(x)$ for all $r \in R$ and $x \in R[U]$). This merely says R-algebra maps from $R[U]$ are completely determined by their values on the variables U.

Definition: Let S be a subset of a commutative ring R. The **localization** $S^{-1}R$ is a commutative R-algebra and an R-algebra map $\theta : R \to S^{-1}R$ such that

(i) $\theta(s)$ is invertible for every $s \in S$;
(ii) $S^{-1}R$ is universal with this property:

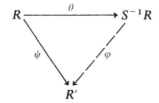

if R' is a commutative R-algebra and $\psi: R \to R'$ is an R-algebra map with $\psi(s)$ invertible for all $s \in S$, then there exists a unique R-algebra map $\varphi: S^{-1}R \to R'$ with $\varphi\theta = \psi$.

As any solution to a universal mapping problem, $S^{-1}R$ is unique if it exists.

Theorem 3.66:   *For every subset S of a commutative ring R, the localization $S^{-1}R$ exists.*

Proof: Let $R[U]$ be the polynomial ring over R in variables $U = \{u_s : s \in S\}$, a set in one–one correspondence with S. Define $S^{-1}R = R[U]/I$, where I is the ideal generated by all elements of the form $su_s - 1$, $s \in S$; define $\theta: R \to S^{-1}R$ as the composite $R \to R[U] \to S^{-1}R$, the first arrow imbedding R as the constants, the second being the natural map. It is clear that $S^{-1}R$ is a commutative R-algebra, θ is an R-algebra map, and that $\theta(s)$ is invertible for every $s \in S$.

Assume R' is a commutative R-algebra and $\psi: R \to R'$ is an R-algebra map with $\psi(s)$ invertible for all $s \in S$. Now the definition of θ is precisely

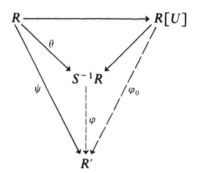

commutativity of the top triangle in the diagram. Define $\varphi_0: R[U] \to R'$ as the unique R-algebra map with $u_s \mapsto \psi(s)^{-1}$, all $s \in S$. Since $\psi(s)$ is invertible for all $s \in S$, $\ker \varphi_0 \supset I$; it follows that φ_0 induces an R-algebra map $\varphi: S^{-1}R = R[U]/I \to R'$ with $\varphi\theta = \psi$. Uniqueness of φ follows from $S^{-1}R$ being generated, as an R-algebra, by the image of $R[U]$.   ∎

We now know $S^{-1}R$ exists, but what does it look like? Here are some simple observations, whose verifications are left as exercises.

Definition: A subset S of a ring R is **multiplicatively closed** if it is a monoid: $1 \in S$; if $s_1, s_2 \in S$, then $s_1 s_2 \in S$.

If S is any subset of R, let $\bar{S}$ be the monoid in R generated by S.

Exercises: 3.46. For any subset S of R, there is an isomorphism $S^{-1}R \cong (\overline{S})^{-1}R$. (Hint: Both are solutions to the same universal mapping problem.)

3.47. If S is finite, say, $S = \{s_1, \ldots, s_n\}$, and if $s = \prod_{i=1}^n s_i$, then $S^{-1}R \cong s^{-1}R$. (Hint: If s^{-1} exists, so does $s^{-1}(s_1 \cdots \hat{s}_i \cdots s_n) = s_i^{-1}$, where $\hat{\ }$ means delete.)

Theorem 3.67:   *If S is a subset of R, then each $x \in S^{-1}R$ has a factorization*

$$x = \theta(r)\theta(s)^{-1}, \qquad s \in \overline{S}, \quad r \in R.$$

Remark: This factorization is not unique.

Proof: The set A of all $x \in S^{-1}R$ admitting such a factorization is a subalgebra containing $\operatorname{im}\theta$; in particular, $\theta(s)$ is invertible in A for every $s \in S$. If $i: A \to S^{-1}R$ is the inclusion, there is a commutative diagram

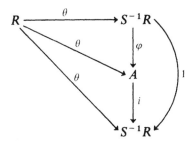

where φ exists by universality. As the identity on $S^{-1}R$ makes the large triangle commute, uniqueness gives $i\varphi = 1$. By set theory, i is onto and $A = S^{-1}R$. ∎

It follows that we may write the elements of $S^{-1}R$ as "fractions" r/s, where $r \in R$ and $s \in \overline{S}$; moreover, addition and multiplication of fractions have the usual formulas. The only difficulty is that "cross multiplication" is not the answer to the question of when $r/s = r'/s'$. This is plain in case $0 \in S$, for 0 invertible in $S^{-1}R$ forces $0 = 1$ there; in this case, $S^{-1}R = \{0\}$, a ring whose existence we admit, but whose abode we choose to be outer darkness. In particular, if we denote $\theta: R \to S^{-1}R$ by $r \mapsto r/1$, then θ need not be monic.

Theorem 3.68:   *If $\theta: R \to S^{-1}R$, then*

$$\ker\theta = \{r \in R : sr = 0 \text{ for some } s \in \overline{S}\}.$$

Proof: If $sr = 0$ for some $s \in \overline{S}$, then $\theta(r) = \theta(rs)\theta(s)^{-1} = 0$.

Conversely, assume $\theta(r) = 0$. As $S^{-1}R = R[U]/I$, we have $r \in I$, the ideal generated by all $su_s - 1$. Writing what this means gives an equation involving only finitely many u_s, say, $u_{s_1}, \ldots, u_{s_n}$. If $S_0 = \{s_1, \ldots, s_n\}$, then $r \in \ker \theta_0$, where $\theta_0 : R \to S_0^{-1} R$. By Exercise 3.47, we may assume the finite set S_0 consists of just one element, s, whence $s^{-1} R = R[u]/(su - 1)$. But $\theta_0(r) = 0$ says

$$r = (su - 1) f(u), \qquad \text{where} \quad f(u) \in R[u]$$
$$= (su - 1)(a_0 + a_1 u + \cdots + a_m u^m).$$

Expanding and equating coefficients of like powers of u give equations:

$$r = -a_0, \qquad sa_0 = a_1, \ldots, sa_{m-1} = a_m, \qquad sa_m = 0.$$

This gives $s^{m+1} r = 0$, as desired. ∎

Corollary 3.69:   (i)   *If S contains no zero divisors, then $R \to S^{-1} R$ is monic.*
     (ii)   *If R is a domain with quotient field Q, then $S^{-1} R$ is a subring of Q.*
     (iii)   *If R is a domain and $S = R - \{0\}$, then $S^{-1} R = Q$.*

Corollary 3.70:   *Let r/s and $r'/s' \in S^{-1} R$. Then $r/s = r'/s'$ if and only if there is $\sigma \in \bar{S}$ with $\sigma(rs' - r's) = 0$.*

Proof: If $r/s = r'/s'$, then $(rs' - r's)/ss' = 0$ in $S^{-1} R$. Since ss' is invertible, $rs' - r's = 0$ in $S^{-1} R$; by Theorem 3.68, $\sigma(rs' - r's) = 0$ for some $\sigma \in \bar{S}$.

     The converse is left to the reader; it is easy and does not need Theorem 3.68. ∎

     The usual construction of $S^{-1} R$ mimics the standard construction of the quotient field of a domain. Consider all ordered pairs $(r, s) \in R \times \bar{S}$ and define an equivalence relation by the, *a priori*, mysterious condition in Corollary 3.70 (the usual relation of cross multiplication does not work, for the possible presence of zero divisors in S prevents it from being an equivalence relation). The standard definitions of addition and multiplication of equivalence classes are given, and the myriad easy calculations (the operations are well defined and all the ring axioms hold) are left as a tedious exercise.

     Usually one assumes S is multiplicatively closed ($S = \bar{S}$) when constructing $S^{-1} R$; by Exercise 3.46, the same localizations arise. The two most popular examples of multiplicatively closed subsets S of R are: (i) $S = R - P$, where P is a prime ideal of R; (ii) $S = \{r^n : n \geq 0\}$ for some fixed $r \in R$.

     The ring $S^{-1} R$ is an R-module (if we forget some multiplication). One of its most useful features, and the reason localization appears in this chapter, is that $S^{-1} R$ is flat as an R-module. We now prove this.

Definition: If S is a subset of a commutative ring R and if M is an R-module, then

$$S^{-1} M = S^{-1} R \otimes_R M.$$

There is an obvious map $\theta_M : M \to S^{-1}M$ given by $m \mapsto 1 \otimes m$. The next result generalizes Theorem 3.68, which describes $\ker \theta_R$ (we wrote θ for θ_R then).

Theorem 3.71: $\ker \theta_M = \{m \in M : \sigma m = 0 \text{ for some } \sigma \in \bar{S}\}$.

Proof: If $\sigma m = 0$ for some $\sigma \in \bar{S}$, then $\theta_M(m) = 1/\sigma\theta_M(\sigma m) = 0$. The proof of the reverse inclusion parallels that of Theorem 3.68. As there, we may reduce to the case $S = \{s\}$, so there is an exact sequence

$$0 \to (su - 1) \xrightarrow{j} R[u] \xrightarrow{\pi} S^{-1}R \to 0.$$

There is a commutative diagram

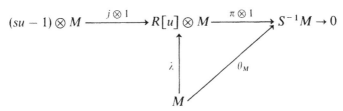

where the top row is exact because of right exactness of $\otimes_R M$, and $\lambda : m \mapsto 1 \otimes m$ is monic (because, as R-modules, R is a summand of $R[u]$). It follows that

$$\ker \theta_M = \{m \in M : 1 \otimes m \in \ker \pi \otimes 1\} = \{m \in M : 1 \otimes m \in \operatorname{im} j \otimes 1\}.$$

Now $R[u]$ is a free R-module with basis $\{1, u, u^2, \ldots\}$. Therefore $R[u] \otimes M \cong \coprod (Ru^i \otimes M)$, so that every $x \in R[u] \otimes M$ has a unique expression

$$x = \sum_{i=0}^{n} u^i \otimes m_i, \qquad \text{where} \quad m_i \in M.$$

Hence,

$$\ker \theta_M = \{m \in M : 1 \otimes m = (su - 1) \sum_{i=0}^{n} u^i \otimes m_i\}.$$

We now proceed as in Theorem 3.68, expanding and equating coefficients. There are equations:

$$1 \otimes m = -1 \otimes m_0; \qquad u \otimes sm_0 = u \otimes m_1; \ldots;$$
$$u^n \otimes sm_{n-1} = u^n \otimes m_n; \qquad u^{n+1} \otimes sm_n = 0.$$

These give $u^{n+1} \otimes s^{n+1}m = 0$ in $Ru^{n+1} \otimes M \cong M$, whence $s^{n+1}m = 0$, as desired. ∎

Corollary 3.72:   *Every element of $S^{-1}M$ has the form $s^{-1} \otimes m$, where $s \in \bar{S}$. Moreover, $s_1^{-1} \otimes m_1 = s_2^{-1} \otimes m_2$ if and only if there is some $\sigma \in \bar{S}$ with $\sigma(s_2 m_1 - s_1 m_2) = 0$.*

Proof: A typical element of $S^{-1}M = S^{-1}R \otimes_R M$ has the form $\sum x_i \otimes m_i$, where $x_i \in S^{-1}R$ and $m_i \in M$. But $x_i = r_i/s_i$, where $s_i \in \bar{S}$, so that

$$\sum x_i \otimes m_i = \sum r_i/s_i \otimes m_i = \sum 1/s_i \otimes r_i m_i = \sum \sigma_i/s \otimes r_i m_i$$

(where $s = \prod s_j$ and $\sigma_i = \prod_{j \neq i} s_j$)

$$= 1/s \otimes \sum \sigma_i r_i m_i = 1/s \otimes m.$$

If $s_1^{-1} \otimes m_1 = s_2^{-1} \otimes m_2$, then $(1/s_1 s_2) \otimes (s_2 m_1 - s_1 m_2) = 0$ and hence $1 \otimes (s_2 m_1 - s_1 m_2) = 0$. By Theorem 3.71, there is $\sigma \in \bar{S}$ with $\sigma(s_2 m_1 - s_1 m_2) = 0$.   ∎

As a consequence of Corollary 3.72, one writes the elements of $S^{-1}M$ as m/s, where $m \in M$ and $s \in \bar{S}$, with addition and scalar multiplication as suggested by the notation.

Theorem 3.73:   *For any subset S of R, the localization $S^{-1}R$ is flat as an R-module.*

Proof: Assume $f:M' \to M$ is monic; we must show $1 \otimes f:S^{-1}M' \to S^{-1}M$ is monic. Let $(1 \otimes f)(s^{-1} \otimes m') = s^{-1} \otimes fm' = 0$ in $S^{-1}M$. Then multiplying by s gives $1 \otimes fm' = 0$. By Theorem 3.71, there is $\sigma \in \bar{S}$ with $\sigma fm' = 0$. But $\sigma fm' = f(\sigma m')$, so that f monic implies $\sigma m' = 0$. Since σ is a unit in $S^{-1}R$, however, $0 = s^{-1} \otimes \sigma m' = \sigma(s^{-1} \otimes m')$ implies $s^{-1} \otimes m' = 0$.   ∎

Corollary 3.74: $S^{-1}:_R\mathfrak{M} \to {}_{S^{-1}R}\mathfrak{M}$ *is an exact functor.*

Proof: S^{-1} is, by definition, the functor $S^{-1}R \otimes_R$; to say this functor is exact is precisely to say that $S^{-1}R$ is a flat R-module.   ∎

We remark that if $f:M \to M'$, then $S^{-1}f:S^{-1}M \to S^{-1}M'$ is given by $m/s \mapsto (fm)/s$.

Exercises: 3.48.   If $S_1 \subset S \subset R$, then $S^{-1}(S_1^{-1}A) = S^{-1}A = S_1^{-1}(S^{-1}A)$ for every R-module A.

   3.49.   If A_1 and A_2 are submodules of B, then $S^{-1}(A_1 \cap A_2) = S^{-1}A_1 \cap S^{-1}A_2$.

   3.50.   Every R-map $f:B \to B'$, where B and B' are $S^{-1}R$ modules, is automatically an $S^{-1}R$ map. (Hint: Consider $sf(m/s)$.)

   3.51.   The maps $\theta_M:M \to S^{-1}M$ define a natural transformation $\theta:1 \to S^{-1}$.

   3.52.   Find a functor G so that (S^{-1}, G) is an adjoint pair.

3.53. Prove that $S^{-1}R \otimes_R S^{-1}R \cong S^{-1}R$ as $S^{-1}R$-modules. (Hint: Consider multiplication on $S^{-1}R$.)

Lemma 3.75: *If B is an $S^{-1}R$-module, then $B \cong S^{-1}B$; moreover, if $f:B \to B'$ is an $S^{-1}R$-map between $S^{-1}R$-modules, then $f = S^{-1}f$.*

Proof: Define $\varphi:B \to S^{-1}B$ by $b \mapsto 1 \otimes b$. If $\varphi(b) = 1 \otimes b = 0$, then $\sigma b = 0$ for some $\sigma \in \bar{S}$; since σ is a unit in $S^{-1}R$ and B is an $S^{-1}R$-module, we have $0 = \sigma^{-1}(\sigma b) = b$. To see φ is epic, note that $s^{-1} \otimes b = \varphi(s^{-1}b)$.

The second statement is obvious from the identification $s^{-1} \otimes b = b/s$. Without identification, it says we have commutativity of

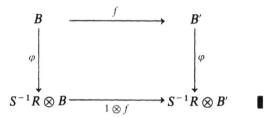

The next theorem says S^{-1} preserves most nice modules.

Theorem 3.76: *Let S be a subset of a commutative ring R, and let A be an R-module.*

(i)   *If A is R-free, then $S^{-1}A$ is $S^{-1}R$-free.*
(ii) *If A is f.g., then $S^{-1}A$ is f.g.*
(iii) *If A is finitely related, then $S^{-1}A$ is finitely related.*
(iv) *If A is R-projective, then $S^{-1}A$ is $S^{-1}R$-projective.*

Proof: (i) S^{-1} preserves sums.
(ii) If $F \to A \to 0$ is exact with F f.g. free, then $S^{-1}F \to S^{-1}A \to 0$ is exact, with $S^{-1}F$ f.g. free.
(iii) If $F_1 \to F_0 \to A \to 0$ is exact with F_1 and F_0 f.g. free, then $S^{-1}F_1 \to S^{-1}F_0 \to S^{-1}A \to 0$ has the same properties.
(iv) S^{-1} preserves summands. ∎

It is true that S^{-1} preserves flatness, but this requires a short lemma.

Lemma 3.77: *For R-modules B and A, there is a natural isomorphism*

$$S^{-1}(B \otimes_R A) \cong S^{-1}B \otimes_{S^{-1}R} S^{-1}A.$$

Proof: An elementary argument is to define an isomorphism by bringing an element of $S^{-1}B \otimes_{S^{-1}R} S^{-1} A$ into a normal form:

$$\sum b_i/s_i \otimes a_i/\sigma_i = \sum(b_i/s_i\sigma_i) \otimes a_i \qquad \text{(slip } 1/\sigma_i \text{ across tensor sign)}$$
$$= \sum(1/s_i\sigma_i)(b_i \otimes a_i).$$

A fancier argument is to show $S^{-1}(B \otimes A)$ satisfies the universal mapping problem for $S^{-1}R$-bilinear maps f:

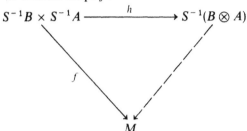

Define $h:(b/s, a/\sigma) \mapsto (1/s\sigma)(b \otimes a)$. Details of either version are left to the reader. ∎

Theorem 3.78:   *If B is a flat R-module, then $S^{-1}B$ is a flat $S^{-1}R$-module.*

Proof: Let $f:A \to A'$ be monic in $_{S^{-1}R}\mathfrak{M}$; is $1 \otimes f: S^{-1}B \otimes_{S^{-1}R} A \to S^{-1}B \otimes_{S^{-1}R} A'$ monic? By Lemma 3.75, we may ask whether

$$1 \otimes S^{-1}f: S^{-1}B \otimes_{S^{-1}R} S^{-1}A \to S^{-1}B \otimes_{S^{-1}R} S^{-1}A'$$

is monic; by Lemma 3.77, we may even ask whether

$$S^{-1}(1 \otimes f): S^{-1}(B \otimes_R A) \to S^{-1}(B \otimes_R A')$$

is monic. The affirmative answer is now apparent: $1 \otimes f$ is monic because B is R-flat, and $S^{-1}(1 \otimes f)$ is monic because S^{-1} is an exact functor. ∎

This theorem may also be proved using Exercise 3.38.

Corollary 3.79:   *Let $S_1 \subset S$ be subsets of R. For any R-module A,*

$$S^{-1}A = S^{-1}(S_1^{-1}A) = S_1^{-1}A \otimes_{S_1^{-1}R} S^{-1}R.$$

Proof: In Exercise 3.48, we observed that $S^{-1}A = S^{-1}(S_1^{-1}A) = S_1^{-1}(S^{-1}A)$ (for all satisfy the same universal mapping problem). But

$$S_1^{-1}(S^{-1}A) = S_1^{-1}(A \otimes_R S^{-1}R) = S_1^{-1}A \otimes_{S_1^{-1}R} S_1^{-1}(S^{-1}R), \text{ by Lemma 3.77}.$$

Since $S_1^{-1}(S^{-1}R) = S^{-1}R$, we have the desired formula. ∎

The proper attitude is that $S^{-1}R$ is a less complicated ring than R. For example, if R is a domain, its quotient field is a localization (thus, Theorem 3.73 is a vast generalization of Corollary 3.48). More generally, if P is a prime ideal in R and $S = R - P$ ($\{0\}$ is a prime ideal when R is a domain),

$S^{-1}R$ is a ring having only one maximal ideal. In this very important special case, one writes R_P, M_P, and f_P instead of $S^{-1}R$, $S^{-1}M$, and $S^{-1}f$, where M is an R-module and f is an R-map. The next theorem illustrates how one may use localization to reduce a problem to a simpler setting.

Theorem 3.80: (i) *If M is an R-module with $M_P = 0$ for every maximal ideal P of R, then $M = 0$.*

(ii) *If $f: M \to N$ has $f_P: M_P \to N_P$ monic (or epic) for every maximal ideal P, then f is monic (or epic).*

Proof: (i) Assume $M \neq 0$, and choose $m \in M$ with $m \neq 0$. If $I = \{r \in R : rm = 0\}$, then I is a proper ideal of R, hence is contained in some maximal ideal P. Now $m/1 \in M_P = 0$, so Theorem 3.71 provides $\sigma \in R - P$, i.e., $\sigma \notin P$, with $\sigma m = 0$. But $\sigma \in I \subset P$, a contradiction.

(ii) Consider the exact sequence $0 \to K \to M \xrightarrow{f} N$, where $K = \ker f$. For each maximal ideal P, Corollary 3.74 gives an exact sequence

$$0 \to K_P \to M_P \xrightarrow{f_P} N_P.$$

By hypothesis, each f_P is monic, whence $K_P = 0$ for all P, and so $K = 0$, by part (i). Therefore f is monic.

If each f_P is epic, there is a similar argument using coker f. ∎

A variation on the proof just given is useful.

Theorem 3.81: *Let M be f.g. and $S \subset R$. Then $S^{-1}M = 0$ if and only if there exists $\sigma \in \bar{S}$ with $\sigma M = 0$.*

Proof: Let $M = \langle x_1, \ldots, x_n \rangle$. If $S^{-1}M = 0$, then each $x_i/1 = 0$ in $S^{-1}M$. By Theorem 3.71, there is $\sigma_i \in \bar{S}$ with $\sigma_i x_i = 0$. Note that $\sigma = \prod \sigma_i \in \bar{S}$ and that $\sigma x_i = 0$ for all i. It follows that $\sigma M = 0$.

For the converse, $\sigma M = 0$ implies $\sigma(S^{-1}M) = 0$. On the other hand, $S^{-1}M$ is an $S^{-1}R$-module, and σ is invertible in $S^{-1}R$. Therefore, multiplication by σ is an automorphism of $S^{-1}M$ (whose inverse is multiplication by $1/\sigma$). It follows that $M = 0$. ∎

Our final desire in this chapter is to obtain an analogue of Lemma 3.77 for Hom in place of $\otimes$: we wish to show $S^{-1} \operatorname{Hom}_R(N, M) \cong \operatorname{Hom}_{S^{-1}R}(S^{-1}N, S^{-1}M)$, but, in contrast to Lemma 3.77, some restriction must be imposed on N. To see that the formula is not generally true, take $R = \mathbf{Z}$ and $S^{-1}R = \mathbf{Q}$. If $N = \mathbf{Q}$ and $M = \mathbf{Z}$, then $\operatorname{Hom}_{\mathbf{Z}}(\mathbf{Q}, \mathbf{Z}) = 0$, hence its localization is 0; on the other hand, $\operatorname{Hom}_{\mathbf{Q}}(\mathbf{Q} \otimes \mathbf{Q}, \mathbf{Q} \otimes \mathbf{Z}) = \operatorname{Hom}_{\mathbf{Q}}(\mathbf{Q}, \mathbf{Q}) = \mathbf{Q}$.

We first give two very general results.

Lemma 3.82: *Let A be an R-algebra, and N a finitely related R-module. For* *every A-module M, there is a natural isomorphism*

$$\theta: \text{Hom}_R(N, M) \xrightarrow{\sim} \text{Hom}_A(N \otimes_R A, M)$$

given by $f \mapsto \tilde{f}$, *where* $\tilde{f}(x \otimes 1) = f(x)$ *for all* $x \in N$.

Proof: We first establish an isomorphism when N is f.g. free. If N has a basis $\{e_1, \ldots, e_n\}$, then $N \otimes_R A$ has a basis $\{e_1 \otimes 1, \ldots, e_n \otimes 1\}$. Define $\theta: \text{Hom}_R(N, M) \to \text{Hom}_A(N \otimes A, M)$ by $f \mapsto \tilde{f}$, where $\tilde{f}(e_i \otimes 1) = f(e_i)$, $i = 1, \ldots, n$. That θ is a well-defined isomorphism is a restatement of the definition of basis.

Now assume N is finitely related, so there is an exact sequence

$$R^r \to R^s \to N \to 0.$$

There is a commutative diagram with exact rows:

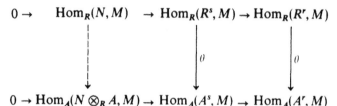

As the vertical maps θ are isomorphisms, the (dashed) induced map is also an isomorphism, and it has the desired formula. We leave the check of naturality to the reader. ∎

Lemma 3.83: *Let B be flat and N finitely related. For any module M, there is a natural isomorphism*

$$\psi: B \otimes_R \text{Hom}_R(N, M) \xrightarrow{\sim} \text{Hom}_R(N, M \otimes_R B)$$

given by $b \otimes g \mapsto g_b$, *where* $g_b(n) = g(n) \otimes b$.

Proof: It is clear that ψ arises from the R-bilinear function sending $(b, g) \mapsto g_b$, it is natural (in N), and it is an isomorphism when N is f.g. free. There is an exact sequence when N is finitely related:

$$R^r \to R^s \to N \to 0.$$

Since B is flat, there is a commutative diagram with exact rows:

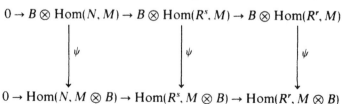

Since the second two vertical maps are isomorphisms, so is the first. ∎

Theorem 3.84: *Let $S \subset R$, and let N be a finitely related R-module. There is a natural isomorphism, for every R-module M,*

$$\varphi : S^{-1} \operatorname{Hom}_R(N, M) \overset{\sim}{\to} \operatorname{Hom}_{S^{-1}R}(S^{-1}N, S^{-1}M)$$

given by $g/1 \mapsto \bar{g}$, where $\bar{g}(x/1) = g(x) \otimes 1$ for all $x \in N$.

Proof: By definition, $S^{-1} \operatorname{Hom}_R(N, M) = S^{-1}R \otimes_R \operatorname{Hom}_R(N, M)$. Since $S^{-1}R$ is flat (Theorem 3.73), Lemma 3.83 gives an isomorphism

$$S^{-1} \operatorname{Hom}_R(N, M) \cong \operatorname{Hom}_R(N, S^{-1}M);$$

since $S^{-1}R$ is an R-algebra, Lemma 3.82 gives an isomorphism

$$\operatorname{Hom}_R(N, S^{-1}M) \cong \operatorname{Hom}_{S^{-1}R}(S^{-1}N, S^{-1}M).$$

The composite is an isomorphism having the given formula. ∎

Theorem 3.85: Assume R is a ring in which every ideal is finitely related. If $S \subset R$ and A is an injective R-module, then $S^{-1}A$ is an injective $S^{-1}R$-module.

Proof: Note first that if $I = \langle r_1/s_1, \ldots, r_n/s_n : r_i \in R, s_i \in \bar{S} \rangle$ is an ideal in $S^{-1}R$, then there is an ideal J in R, namely, $J = \langle r_1, \ldots, r_n \rangle$, with $S^{-1}J = I$. By Baer's criterion, it suffices to show

$$i^* : \operatorname{Hom}_{S^{-1}R}(S^{-1}R, S^{-1}A) \to \operatorname{Hom}_{S^{-1}R}(I, S^{-1}A)$$

is epic for every ideal I, where $i : I \to S^{-1}R$ is the inclusion. But Theorem 3.84 gives a commutative diagram

$$
\begin{array}{ccc}
S^{-1} \operatorname{Hom}_R(R, A) & \longrightarrow & S^{-1} \operatorname{Hom}_R(J, A) \\
\downarrow & & \downarrow \\
\operatorname{Hom}_{S^{-1}R}(S^{-1}R, S^{-1}A) & \to & \operatorname{Hom}_{S^{-1}R}(S^{-1}J, S^{-1}A).
\end{array}
$$

Injectivity of A implies $\operatorname{Hom}_R(R, A) \to \operatorname{Hom}_R(J, A)$ is epic, so that (right) exactness of S^{-1} shows the top arrow is epic. Since the vertical arrows are isomorphisms, the bottom arrow is epic, too. This completes the proof because $S^{-1}J = I$. ∎

Remarks: 1. If every ideal of R is f.g. (i.e., R is noetherian), then Corollary 4.2 shows that every ideal of R must be finitely related.

2. Dade [1981] shows that if k is a commutative ring and $R = k[X]$, where X is an uncountable set of indeterminates, then there exists a subset S of R and an injective R-module E such that $S^{-1}E$ is not an injective $S^{-1}R$-module. If, however, X is countable and k is noetherian, he shows E an injective R-module implies $S^{-1}E$ is an injective $S^{-1}R$-module for every subset S of R.

4 Specific Rings

The time for examples has come. We pose twin problems: if conditions are imposed on projective, injective, or flat modules, how does this affect the ring of scalars R; if we put conditions on R, how does this affect these special modules? We deviate from earlier style in that a number of results are stated without proof (though references are given). There are almost no exercises, for they would be more ring-theoretic than homological. We do, however, include a little ring theory to give the reader some feeling for the theorems, thereby removing some of their mystery. Of course, what is included and excluded is a matter of taste.

NOETHERIAN RINGS

Let us begin with a class of rings that arose several times in Chapter 3.

Definition: A ring R is **left noetherian** if every left ideal is f.g.

Examples: 1. Every principal ideal ring (possibly with zero divisors, e.g., $\mathbf{Z}/n\mathbf{Z}$, possibly noncommutative, e.g., division rings).
2. If R is left noetherian, so is $R[x]$, polynomials in which x commutes with all constants in R **(Hilbert Basis Theorem)** [Lambek, 1966, p. 70]. By induction on n, R left noetherian implies $R[x_1, \ldots, x_n]$ left noetherian.

3. If R is commutative noetherian, so is any localization $S^{-1}R$ [Kaplansky, 1970, p. 57].

4. Every quotient ring of a left noetherian ring is left noetherian.

5. If R is the ring of all matrices of the form $\begin{pmatrix} a & b \\ 0 & c \end{pmatrix}$, where $a \in \mathbf{Z}$ and b, $c \in \mathbf{Q}$, then R is right noetherian but not left noetherian [Small, 1966]. (Dieudonné gave an example of this phenomenon in the 1950s [Cartan–Eilenberg, 1956, p. 16].)

Theorem 4.1: *R is left noetherian if and only if every submodule of a f.g. left R-module M is also f.g.*

Proof: Assume $M = \langle x_1, \ldots, x_n \rangle$, and S is a submodule of M. We do an induction on n that S is f.g. If $n = 1$, then M is cyclic, hence $M \cong R/I$ for some left ideal I (Theorem 2.2). It follows that $S \cong J/I$, where J is some left ideal containing I. Since R is left noetherian, J, hence its image $J/I \cong S$, is f.g.

Suppose $n > 1$. If $M' = Rx_n$, there is an exact sequence

$$0 \to M' \to M \to M/M' \to 0,$$

and M/M' can be generated by $n - 1$ elements. There is another exact sequence

$$0 \to S \cap M' \to S \to S/(S \cap M') \to 0.$$

Now $S \cap M'$ is f.g., being a submodule of the cyclic module M', while $S/(S \cap M') \cong (S + M')/M' \subset M/M'$ is f.g. by induction. It follows that S is f.g.

For the converse, the submodules of R, namely, the left ideals, are f.g., hence R is left noetherian. ∎

Corollary 4.2: *If R is left noetherian, then every f.g. module is finitely related.*

Proof: If $0 \to K \to F \to B \to 0$ is exact, where F is f.g. free, then K is f.g. by Theorem 4.1. ∎

Corollary 4.3: *If R is left noetherian, every f.g. flat module is projective.*

Proof: Couple Corollary 4.2 with Corollary 3.58. ∎

Note that we have identified the rings in the hypothesis of Corollary 3.34.

Corollary 4.4: *If R is left noetherian, every f.g. module M has a free resolution*

$$\cdots \to F_n \to F_{n-1} \to \cdots \to F_1 \to F_0 \to M \to 0$$

in which each F_n is f.g.

Proof: Since M is f.g., we may choose a f.g. free module F_0 and an exact sequence $0 \to K_0 \to F_0 \to M \to 0$. Since R is left noetherian, K_0 is f.g., so we may choose a f.g. free module F_1 mapping onto K_0 and obtain an exact sequence $0 \to K_1 \to F_1 \to K_0 \to 0$. Moreover, K_1 is f.g., so we may iterate this process as in Theorem 3.8. ∎

Definition: A module M (over any ring) has **ACC (ascending chain condition)** if every ascending chain of submodules

$$M_1 \subset M_2 \subset M_3 \subset \cdots$$

stops, i.e., there is an integer n with $M_n = M_{n+1} = M_{n+2} = \cdots$.

Theorem 4.5: *A module M (over any ring) has ACC if and only if every submodule of M is f.g.*

Proof: Assume every submodule is f.g. and let $M_1 \subset M_2 \subset \cdots$ be an ascending chain of submodules of M. If $M^* = \bigcup_{i=1}^{\infty} M_i$, then $M^* = \langle x_1, \ldots, x_k \rangle$, by hypothesis. Each $x_j \in M_{i(j)}$; let $n = \max i(j)$. Then $M^* \subset M_n \subset M^*$, so that the chain stops.

Assume S is a submodule of M that is not f.g. Choose $s_1 \in S$ and define $M_1 = Rs_1$. Assume, by induction, that we have chosen $s_1, \ldots, s_n \in S$ so that if $M_i = \langle s_1, \ldots, s_i \rangle$, then $M_1 \subsetneqq M_2 \subsetneqq \cdots \subsetneqq M_n$. Since S is not f.g., M_n is a proper submodule of S; we may thus choose $s_{n+1} \in S$ with $s_{n+1} \notin M_n$. Define $M_{n+1} = M_n + Rs_{n+1}$, so that $M_n \subsetneqq M_{n+1}$. By the axiom of choice,[1] there is an ascending sequence of submodules of M, namely, $M_1 \subset M_2 \subset \ldots$, that plainly does not stop. This contradicts ACC. ∎

A module with ACC is often called "noetherian" because of this theorem.

Corollary 4.6: *A ring R is left noetherian if and only if it has ACC on left ideals.*

Proof: If we consider R as a module, its submodules are the left ideals. ∎

Corollary 4.7: *If R is left noetherian, every f.g. module has ACC.*

Proof: Theorems 4.1 and 4.5. ∎

Lemma 4.8: *Let A be a module (over any ring) and $\varphi : A \to A$ an epimorphism. If A has ACC, then φ is an isomorphism.*

---

[1] This is a subtle logical point. Induction allows us to choose a suitable s_n for each integer n. The axiom of choice allows us to make all these choices simultaneously. Of course, one could phrase this more formally.

Proof: Let $K_1 = \ker \varphi$ and let $K_2 = \ker(\varphi^2)$, where $\varphi^2 = \varphi\varphi$. Clearly $K_1 \subset K_2$ and φ^2 is epic. There is a commutative diagram

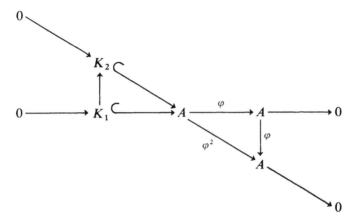

The version of the Third Isomorphism Theorem in Exercise 2.10 shows that if $K_1 \neq 0$, then K_1 is a proper submodule of K_2. This procedure may then be iterated, by induction, to obtain $\ker(\varphi^n) \subsetneq \ker(\varphi^{n+1})$. Since A has ACC, it must be that $K_1 = \ker \varphi = 0$, so that φ is an isomorphism.   ∎

Theorem 4.9:   *If R is left noetherian, then R has* **IBN**. *Moreover, if A is free of rank n, then any generating set with n elements is a basis.*

Proof: Assume A is free and $A \cong R^n \cong R^m$, where $m \geq n$ (R^m is the sum of m copies of R). If $m > n$, there is an epimorphism $\varphi: A \to A$ having a nonzero kernel. But A has ACC, by Corollary 4.7. Therefore φ is an isomorphism, by Lemma 4.8, contradicting $\ker \varphi \neq 0$.

Let A have a basis $\{x_1, \ldots, x_n\}$ and let $A = \langle a_1, \ldots, a_n \rangle$. Define an epimorphism $\varphi: A \to A$ by $x_i \mapsto a_i$. If $\ker \varphi \neq 0$, we reach a contradiction as above. The result now follows from Exercise 3.3.   ∎

Theorem 4.10:   *The following are equivalent for a ring R.*

(i)   *R is left noetherian;*
(ii)   *every direct limit (directed index set) of injective modules is injective;*
(iii)   *every sum of injective modules is injective.*

Proof: (i) $\Rightarrow$ (ii)   Let $\{E_i, \varphi_j^i\}$ be a direct system of injective modules over a directed index set. By Theorem 2.17, $\varinjlim E_i$ consists of elements $[e_i]$, $e_i \in E_i$, where $[e_i] = \lambda_i e_i + S$ and $\lambda_i: E_i \to \coprod E_i$ is the ith injection.

Consider the diagram

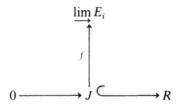

where J is a left ideal. Since R is left noetherian, $J = \langle a_1, \ldots, a_n \rangle$. If F is free on $\{x_1, \ldots, x_n\}$ and we map $F \to J$ by $x_i \mapsto a_i$, there is an exact sequence $0 \to K \to F \to J \to 0$. Moreover, Theorem 4.1 gives $K = \langle y_1, \ldots, y_t \rangle$; each $y_p = \sum r_{pj} x_j$, where $r_{pj} \in R$.

The idea is to force f to take values in some one E_i. In detail, $fa_j = [e_{i(j)}]$. Since the index set is directed, there is an index $k \geq i(j)$, all j. Define $e^j \in E_k$, $j = 1, \ldots, n$, by $e^j = \varphi_k^{i(j)} e_{i(j)}$, so that $fa_j = [e_{i(j)}] = [e^j]$. Each y_p in K yields $[\sum r_{pj} e^j] = 0$. By Theorem 2.17, there is an index $q(p) \geq k$ with $\varphi_{q(p)}^k \sum r_{pj} e^j = 0$. Finally, choose an index $m \geq q(p)$, all p, and define $b^j \in E_m$ by $b^j = \varphi_m^k e^j$. The end result of these choices is that $b^j \in E_m$, $j = 1, \ldots, n$, $fa_j = [b^j]$, and $\sum r_{pj} b^j = 0$, all p.

Define $f' : J \to E_m$ by $a_j \mapsto b^j$. Note that f' is well defined, for we have taken care that all the relations are sent to 0. Since E_m is injective, there is a map $g' : R \to E_m$ extending f'. Finally, define $g : R \to \varinjlim E_i$ by $r \mapsto [g'r]$. The map g does extend f, for

$$ga_j = [g'a_j] = [f'a_j] = [b^j] = fa_j.$$

Therefore, $\varinjlim E_i$ is injective.

(ii) $\Rightarrow$ (iii)   A finite sum of injectives is injective and, by Exercise 2.33, every sum is a direct limit, with directed index set, of its finite partial sums. Thus $\coprod E_i$ is injective.

(iii) $\Rightarrow$ (i)   (This implication is due to Bass.) We show that if R is not noetherian, there is an ideal I and a map from I to a sum of injectives that cannot be extended to R. If R is not noetherian, there is a strictly increasing sequence of left ideals $I_1 \subsetneqq I_2 \subsetneqq \cdots$. Let $I = \bigcup_{n=1}^{\infty} I_n$; note that $I/I_n \neq 0$ for all n. Imbed I/I_n in an injective module E_n. We claim that $\coprod E_n$ is not injective.

Let $\pi_n : I \to I/I_n$ be the natural map. For each $a \in I$, $\pi_n a = 0$ for large n, so the map $f : I \to \prod E_n$ given by

$$a \mapsto (\pi_1 a, \pi_2 a, \ldots, \pi_n a, \ldots) = (\pi_n a)$$

does have its image in $\coprod E_n$. We regard f as a map $I \to \coprod E_n$. Suppose there were a map $g : R \to \coprod E_n$ extending f. Write $g(1) = (x_n)$. Choose m and choose

$a \in I$, $a \notin I_m$. Now $\pi_m a \neq 0$, so that $ga = fa$ has nonzero mth coordinate $\pi_m a$. But $g(a) = ag(1) = a(x_n) = (ax_n)$. Therefore $\pi_m a = ax_m$, so that $x_m \neq 0$. Since m is arbitrary, we have contradicted the fact that almost all the coordinates of an element in a sum are zero. ∎

Let us give a direct proof that R left noetherian implies $\coprod E_i$ is injective when each E_i is injective. Consider the diagram

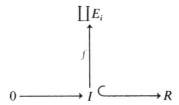

where I is a left ideal. Since R is left noetherian, $I = \langle a_1, \ldots, a_n \rangle$. For each j, the "vector" fa_j in $\coprod E_i$ has only finitely many nonzero coordinates; since there are only finitely many a_j, $\{fa_1, \ldots, fa_n\}$ collectively involve only finitely many E_i, say, $E_{i_1}, \ldots, E_{i_m}$. It follows that $\operatorname{im} f \subset E_{i_1} \oplus \cdots \oplus E_{i_m}$, which is injective, being a finite sum of injectives. There is thus a map $g: R \to E_{i_1} \oplus \cdots \oplus E_{i_m}$ extending f, which we may regard as having its image in the larger module $\coprod E_i$.

We quote a theorem of Chase analogous to Theorem 4.10.

Definition: A ring R is **left coherent** if every f.g. left ideal is finitely related.

Examples: 6. Every left noetherian ring is left coherent.
7. If k is a field, then the ring of polynomials over k in infinitely many variables is coherent (but not noetherian).

Theorem A (Chase): *A ring R is left coherent if and only if every product of flat left R-modules is flat.*

Proof: [Anderson–Fuller, 1974, p. 229].

Another important class of rings is dual (in the lattice-theoretic sense) to noetherian rings.

Definition: A ring R is **left artinian** if it has **DCC (descending chain condition)** on left ideals (i.e., every descending chain of left ideals $I_1 \supset I_2 \supset \cdots$ stops).

Examples: 8. If k is a field, every finite-dimensional k-algebra R is left (and right) artinian.

9.   A product of finitely many left artinian rings is left artinian (in a product of rings $R = \prod R_i$, each R_i is a two-sided ideal of R).

10.   Every finite ring is left (and right) artinian.

11.   There exist rings that are right artinian but not left artinian. Small observed that $\{\begin{smallmatrix} a & b \\ 0 & c \end{smallmatrix}\}: a \in \mathbf{Q}, b, c \in \mathbf{R}\}$ is such a ring.

12.   Every left artinian ring is left noetherian (**Hopkins–Levitzki Theorem**) [Anderson–Fuller, 1974, p. 172].

Definition: A ring R is **left perfect** if it satisfies DCC on principal right ideals.

(This curious mixing of left and right is explained by the theorems to come.)

Examples: 13.   If R is left or right artinian, then R is left perfect [Anderson–Fuller, 1974, p. 318].

14.   A ring may be left perfect but not right perfect [Anderson–Fuller, 1974, p. 322].

Theorem B   (Bass):   *A ring R is left perfect if and only if every flat left R-module is projective.*

Proof: [Anderson–Fuller, 1974, p. 315].

Theorem C   (Chase):   *Every product of projective left R-modules is projective if and only if R is left perfect and right coherent.*

Proof: [Chase, 1960].

Theorem D   (Chase):   *If R is commutative, every product of projective R-modules is projective if and only if R is artinian.*

Proof: [Chase, 1960].

Theorem E   (Bass):   *A ring R is left perfect if and only if every left R-module has a projective cover.*

Proof: [Anderson–Fuller, 1974, p. 315].

We shall return to projective covers later in this chapter (Theorems 4.46 and 4.50).

Exercises: 4.1.   If R is a division ring, every left R-module is projective and injective.

4.2.   If R is a domain but not a field, an R-module that is simultaneously projective and injective must be 0.

4.3.   If R is left artinian, then R contains a **minimal left ideal** S, i.e., $S \neq 0$ and S contains no smaller nonzero left ideal.

SEMISIMPLE RINGS

We are going to characterize those rings for which every module is projective.

Definition: A module A is **simple** (or **irreducible**) if $A \neq 0$ and A has no proper submodules; A is **semisimple** if it is a sum of (possibly infinitely many) simple modules.

Theorem 4.11:   *A module A is semisimple if and only if every submodule of A is a summand.*

Proof: Suppose A is semisimple: $A = \coprod_{k \in K} S_k$, where S_k is simple. Given a subset $I \subset K$, define $S_I = \coprod_{k \in I} S_k$. If B is a submodule of A, choose, by Zorn's lemma, a subset I of K maximal with $S_I \cap B = 0$. We claim $A = B \oplus S_I$; to prove this, it suffices to show $S_k \subset S_I + B$ for every $k \in K$. Clearly this is so for $k \in I$. If $k \notin I$, then maximality gives $(S_I + S_k) \cap B \neq 0$. Writing what this means in terms of elements, one sees $(S_I + B) \cap S_k \neq 0$. Since S_k is simple, $(S_I + B) \cap S_k = S_k$, i.e., $S_k \subset S_I + B$.

Suppose, conversely, that every submodule of A is a summand. We first show each nonzero submodule B contains a simple submodule. If $b \in B$, $b \neq 0$, Zorn's lemma gives a submodule C of B maximal with $b \notin C$. Since every submodule of B is a summand of A, hence of B (Exercise 2.23), $B = C \oplus D$ for some submodule D. We claim D is simple, for if D' is a proper submodule of D, then $D = D' \oplus D''$ and $B = C \oplus D = C \oplus D' \oplus D''$. Either $C \oplus D'$ or $C \oplus D''$ does not contain b, contradicting the maximality of C. Second, we show A is a sum of simple modules. By Zorn's lemma, there is a family of simple submodules of A, $\{S_k : k \in K\}$, maximal such that the submodule M they generate is their sum $\coprod S_k$. By hypothesis, $A = M \oplus B$ for some B. If $B = 0$, we are done. If $B \neq 0$, then $B = S \oplus B'$ for some simple S, by our first argument. The family $\{S, S_k : k \in K\}$ violates the maximality of $\{S_k : k \in K\}$, whence A is semisimple.   ∎

Corollary 4.12:   *Every submodule and every quotient of a semisimple module A is semisimple.*

Proof: If B is a submodule of A, Exercise 2.23 shows the criterion of Theorem 4.11 applies to B. If $A/B = C$, then B is a summand of A, so that C is isomorphic to a submodule of A, hence is semisimple. ∎

Definition: A ring R is **left semisimple** if, as a left R-module, it is semisimple.

Since the submodules of R are its left ideals, R is a sum of simple left ideals. Moreover, since R has a unit, it is easy to see R is a sum of a finite number of simple left ideals. Of course, a simple left ideal is just a minimal left ideal.

Example: 15. **Wedderburn's theorem** asserts that every semisimple ring R is a finite product of full matrix rings over division rings: $R \cong \prod_{i=1}^{r} \text{End}_{\Delta_i}(V_i)$, where $\text{End}_{\Delta_i}(V_i) = \text{Hom}_{\Delta_i}(V_i, V_i)$, and V_i is a left vector space over a division ring Δ_i (say, of dimension n_i). Moreover, the numbers r and n_i and the division rings Δ_i are uniquely determined by R.

It follows from Wedderburn's theorem that R is left semisimple if and only if it is right semisimple. Proofs of this theorem abound in the literature. In particular, it is shown in [Anderson–Fuller, 1974, p. 154] that R is semisimple if and only if R is a finite product of simple artinian rings (where "simple" means no nontrivial two-sided ideals). Here is a proof, due to Rieffel,[2] that R simple left artinian implies $R \cong \text{End}_\Delta(V)$ for a division ring Δ and a left vector space V over Δ. If V is a minimal left ideal of R (which exists, by Exercise 4.3), then $\Delta = \text{End}_R(V)$ is a division ring (an elementary fact called Schur's lemma). Now V is a left Δ-module if we write endomorphisms on the left, and V is finite-dimensional (since R is left artinian). Define $\varphi: R \to \text{End}_\Delta(V)$ by

$$\varphi_r(v) = rv, \qquad r \in R, \quad v \in V.$$

One checks easily that φ is a ring map. Since φ is not identically 0 (it preserves the identity), $\ker \varphi \neq R$; since R is simple, φ is one–one. To see that φ is onto, we first show that $\varphi(V)$ is a left ideal in $\text{End}_\Delta(V)$: indeed, we show $f\varphi_v = \varphi_{f(v)}$ for all $f \in \text{End}_\Delta(V)$ and all $v \in V$. If $u \in V$, then $v \mapsto vu \in \text{End}_R(V) = \Delta$, so that f a Δ-map says

$$f(vu) = (f(v))u$$

(i.e., f commutes with the right multiplication by u). This last equation may be restated as $(f\varphi_v)u = \varphi_{f(v)}u$, all $u \in V$, as claimed.

Now VR is a nonzero two-sided ideal in R (for $1 \in R$), so that $R = VR$. Hence $\varphi(R) = \varphi(V)\varphi(R)$ is a left ideal in $\text{End}_\Delta(V)$, which must be the whole

---

[2] I am grateful to I. Kaplansky for calling my attention to this proof.

ring since $\varphi(R)$ contains the identity of $\text{End}_\Delta(R)$. Therefore $\varphi: R \to \text{End}_\Delta(V)$ is an isomorphism of rings.

Definition: Let G be a (multiplicative) group and k a commutative ring. The **group ring** kG is the k-algebra whose additive structure is that of a free k-module with basis the elements of G, and whose multiplication is given by the multiplication in G and the distributive law.

Examples: 16. **Maschke's theorem** [Curtis–Reiner, 1962, p. 41] says that if G is a finite group and k is a field whose characteristic does not divide the order of G, then kG is semisimple. (See Theorem 10.28.)

17. If k is a field of characteristic 0, then $R = k[x]/(x^n - 1)$ is semisimple (for R is just kG, where G is cyclic of order n).

18. $R = \mathbf{Z}/n\mathbf{Z}$ is semisimple if and only if n is square free.

Here is the reason we have been discussing semisimple rings.

Theorem 4.13: *The following are equivalent for a ring R:*

  (i)   *R is semisimple;*
 (ii) *every left R-module is semisimple;*
(iii) *every left R-module is injective;*
(iv) *every short exact sequence of left R-modules splits;*
 (v) *every left R-module is projective.*

Proof: (i) $\Rightarrow$ (ii) Since R is semisimple as a module, every free module is semisimple. By Corollary 4.12, every quotient of a free module is semisimple. But this says every module is semisimple.

(ii) $\Rightarrow$ (iii) If M is a module, then M may be imbedded in an injective module E. By Theorem 4.11, M is a summand of E, and hence is injective.

(iii) $\Rightarrow$ (iv) Theorem 3.19.

(iv) $\Rightarrow$ (v) If M is a module, there is a short exact sequence $0 \to K \to F \to M \to 0$ with F free. As this sequence splits, M is a summand of F, hence is projective.

(v) $\Rightarrow$ (i) If I is a left ideal of R, then the module R/I is projective, by hypothesis. By Corollary 3.13, I is a summand of R. By Theorem 4.11, R is semisimple as an R-module, whence R is a semisimple ring. ∎

Remark: It now follows from Corollary 3.42 that if R is semisimple, so is R_n, the ring of all $n \times n$ matrices over R. This may also be proved directly from Wedderburn's theorem.

This is an appropriate place to show that projective modules really do occur naturally. The next theorem describes separability of field extensions in such a way that Galois theory may be done for commutative rings, not just for fields.

Theorem 4.14:   *Let L and k be fields, with L a finite separable extension of k. Then L is a projective $L \otimes_k L$-module.*

Proof: First of all, Exercise 1.13 shows that $L \otimes_k L$ is a k-algebra, where multiplication is given by $(l_1 \otimes l_2)(l'_1 \otimes l'_2) = l_1 l'_1 \otimes l_2 l'_2$. It is obvious that L is an $(L - L)$-bimodule, so that Exercise 1.15 shows that L is a module over $L \otimes_k L$ (note that the exercise does apply, for L commutative allows us to identify L with L^{op}). To prove the theorem, it suffices to prove $L \otimes_k L$ is a product of fields, for then it is semisimple and every module is projective.

Since L is a finite separable extension of k, the theorem of the primitive element [Lang, 1965, p. 185] provides an element $\alpha \in L$ with $L = k(\alpha)$. Let $f(x) \in k[x]$ be the irreducible polynomial of α. There is an exact sequence of k-modules

$$0 \to (f) \to k[x] \overset{\pi}{\to} L \to 0,$$

where (f) is the principal ideal generated by $f(x)$; moreover, π is a k-algebra map. Since k is a field, every k-module is flat, so we have exactness of

$$0 \to (f) \otimes_k L \to k[x] \otimes_k L \xrightarrow{\pi \otimes 1} L \otimes_k L \to 0.$$

Now $\pi \otimes 1$ is a ring map, even a k-algebra map, so that $(f) \otimes_k L$ is an ideal in $k[x] \otimes_k L$. Therefore, there is a k-algebra isomorphism

$$L \otimes_k L \cong k[x] \otimes L/(f) \otimes L.$$

If $L[y]$ is the polynomial ring with indeterminate y, then there is a k-algebra isomorphism $\theta : k[x] \otimes_k L \to L[y]$ given by

$$g(x) \otimes l \mapsto lg(y)$$

(we identify $x \otimes 1$ with y); moreover θ takes $(f) \otimes L$ onto $(f(y))$. Therefore,

$$L \otimes_k L \cong L[y]/(f(y)).$$

Now f, though irreducible over k, may factor over L. However, separability tells us there are no repeated factors:

$$f(y) = \prod p_i(y) \qquad \text{in} \quad L[y],$$

where the $p_i(y)$ are distinct irreducible polynomials. The principal ideals $(p_i(y))$ are thus distinct maximal ideals in the principal ideal domain $L[y]$, so the Chinese Remainder Theorem[3] gives a k-algebra isomorphism

$$L[y]/(f(y)) \cong \prod_i (L[y]/(p_i(y))).$$

This concludes the proof, for each $L[y]/(p_i)$ is a field. ∎

The converse of Theorem 4.14 is true [DeMeyer–Ingraham, 1971, p. 49], so finite separable field extensions may be characterized in terms of projectivity. Notice that the characterization "L is a projective $L \otimes_k L$-module" makes sense for k any commutative ring and L any commutative k-algebra.

VON NEUMANN REGULAR RINGS

The next class of rings deals with flatness.

Definition: A ring R is **von Neumann regular** if, for each $a \in R$, there is an element $a' \in R$ with $aa'a = a$.

Examples: 19. **Boolean rings** ($r^2 = r$ for every $r \in R$).
20. $R = \text{End}_k(V)$, where V is a (possibly infinite-dimensional) vector space over a field k.
21. Semisimple rings are von Neumann regular (this follows easily from Example 20 and the Wedderburn theorem).

Lemma 4.15: *If R is von Neumann regular, every f.g. left ideal is principal, generated by an* **idempotent** *(an element e with $e^2 = e$).*

Proof: First of all, let us show every principal left ideal Ra is generated by an idempotent. There is an element $a' \in R$ with $aa'a = a$. It follows that $e = a'a$ is idempotent. Moreover, $Ra = Re$, for $a \in Re$ and $e \in Ra$.

To show that an arbitrary f.g. left ideal is principal, it suffices, by induction, to prove $Ra + Rb$ is principal. There is an idempotent e with $Ra = Re$. Also, $Ra + Rb = Re + Rb(1 - e)$: both a and b are in the right side; both e and $b(1 - e)$ are in the left side. As in the first paragraph, there is an idempotent f with $Rb(1 - e) = Rf$; moreover, $f = rb(1 - e)$ for some $r \in R$. It follows that $fe = 0$. We do not know whether $ef = 0$, so we adjust f.

---

[3] If R is a ring and $I_1, \ldots, I_n$ are "pairwise comaximal" two-sided ideals, i.e., $I_j + I_k = R$ for all $j \neq k$, then $R/\bigcap I_j \cong \prod(R/I_j)$ as rings. In our case, it is easy to see $\bigcap_i(p_i(y)) = (f(y))$.

Define $g = (1 - e)f$. Then

$$g^2 = (1 - e)f(1 - e)f = (1 - e)(f - fe)f = (1 - e)f^2 = (1 - e)f = g,$$
$$ge = 0 = eg, \quad \text{and} \quad Rg = Rf.$$

Therefore, $Ra + Rb = Re + Rg$. We claim that $Re + Rg = R(e + g)$. Clearly $Re + Rg \supset R(e + g)$. For the reverse inclusion, note that $re + sg = (re + sg)(e + g) \in R(e + g)$ (since $eg = 0 = ge$). ∎

Theorem 4.16: *A ring R is von Neumann regular if and only if every right R-module is flat.*

Proof: Assume R is von Neumann regular, and let B be a right R-module; let $0 \to K \to F \to B \to 0$ be exact, where F is free. By Theorem 3.55, it suffices to prove $KI = K \cap FI$ for every f.g. left ideal I. By Lemma 4.15, $I = Ra$ for some $a \in R$. We need only show that if $k \in K$ and $k = fa \in Fa$, then $k \in Ka$. But

$$k = fa = faa'a = ka'a \in Ka.$$

Therefore B is flat.

For the converse, assume $a \in R$. By hypothesis, the cyclic right R-module R/aR is flat. Theorem 3.55 applied to the exact sequence $0 \to aR \to R \to R/aR \to 0$ gives

$$(aR)I = aR \cap RI = aR \cap I$$

for every left ideal I. In particular, if $I = Ra$, then $aRa = aR \cap Ra$ and

$$a \in aR \cap Ra = aRa.$$

There is thus some $a' \in R$ with $a = aa'a$, and R is von Neumann regular. ∎

Here is a stupid argument that R semisimple implies R von Neumann regular. If R is semisimple, every module is projective, hence flat, and so R is von Neumann regular.

HEREDITARY AND DEDEKIND RINGS

We have seen how the assumption that every module is "special" forces constraints on R. Let us now assume every ideal (or every f.g. ideal) is special.

Definition: A ring R is **left hereditary** if every left ideal is projective. A **Dedekind ring** is a hereditary domain.[4]

---

[4] Recall that "domain" means "commutative integral domain".

Examples: 22. Every semisimple ring is left hereditary.

23. Every principal ideal domain R is hereditary, hence is a Dedekind ring (for every nonzero ideal is isomorphic to R).

24. Small's ring of triangular matrices (Example 5) is right hereditary but not left hereditary. For the first example of this phenomenon, see [Kaplansky, 1958b].

There are more examples after Corollary 4.26.

Theorem 4.17 (Kaplansky) *If R is left hereditary, then every submodule of a free module is isomorphic to a sum of left ideals.*

Proof: Let F be a free module with basis $\{x_k : k \in K\}$, and suppose the index set K is well ordered. For each $k \in K$, define

$$F_k = \coprod_{i < k} Rx_i;$$

therefore $F_0 = 0$ and $F_{k+1} = \coprod_{i \le k} Rx_i$. Now let A be a submodule of F. Each $a \in A \cap F_{k+1}$ has a unique expression $a = b + rx_k$, where $b \in F_k$ and $r \in R$. If $\varphi_k : A \cap F_{k+1} \to R$ is defined by $a \mapsto r$, then there is an exact sequence

$$0 \to A \cap F_k \to A \cap F_{k+1} \xrightarrow{\varphi_k} I_k \to 0,$$

where $I_k = \operatorname{im} \varphi_k$ is an ideal of R. Since I_k is projective, this sequence splits: $A \cap F_{k+1} = (A \cap F_k) \oplus C_k$, where $C_k \cong I_k$. We claim that $A = \coprod_{k \in K} C_k$, which will complete the proof.

(i) $A = \langle \bigcup C_k \rangle$. Since $F = \bigcup F_{k+1}$, each $a \in A$ (as any element of F) lies in some F_{k+1}. Let $\mu(a)$ be the least k with $a \in F_{k+1}$. Clearly $C = \langle \bigcup C_k \rangle \subset A$. Suppose $C \ne A$, and consider $\{\mu(a) : a \in A, a \notin C\}$. Let j be the least such index, and choose $y \in A$ with $y \notin C$ and $\mu(y) = j$. Now $\mu(y) = j$ says $y \in A \cap F_{j+1}$, so that $y = b + c$, where $b \in A \cap F_j$ and $c \in C_j$. Therefore $b = y - c \in A$, $b \notin C$ (lest $y \in C$), and $\mu(b) < j$, a contradiction. Hence $A = C = \langle \bigcup C_k \rangle$.

(ii) Uniqueness of expression. Suppose $c_1 + \cdots + c_n = 0$, where $c_i \in C_{k_i}$, and $k_1 < k_2 < \cdots < k_n$. Then

$$c_1 + \cdots + c_{n-1} = -c_n \in (A \cap F_{k_n}) \cap C_{k_n} = 0.$$

It follows that $c_n = 0$. Induction on n now gives $c_i = 0$ for all i. ∎

Corollary 4.18: *If R is left hereditary, every submodule of a projective module is projective.*

Corollary 4.19: *Let R be a principal ideal domain.*

(i) *If A is a submodule of a free module F, then A is free and*

$$\operatorname{rank} A \le \operatorname{rank} F.$$

(ii)   *If B' is a submodule of a f.g. module $B = \langle b_1, \ldots, b_n \rangle$, then B' is f.g. and B' can be generated by $\leq n$ elements.*

Proof: (i)   Every nonzero ideal is isomorphic to R. In the notation of the proof of Theorem 4.17, we have $A = \coprod_{k \in K} C_k$. Since $C_k = 0$ or $C_k \cong R$, we have A free and rank $A \leq \operatorname{card}(K) = \operatorname{rank} F$.

(ii)   Let F be free with basis $\{x_1, \ldots, x_n\}$, and define $\varphi: F \to B$ by $x_i \mapsto b_i$. If $A = \varphi^{-1}(B')$, then $\varphi \mid A: A \to B'$ is epic. By (i), A is free with some basis $\{y_1, \ldots, y_m\}$ and $m \leq n$. It follows that B' is generated by $\{\varphi(y_1), \ldots, \varphi(y_m)\}$. ∎

Corollary 4.20:   *If R is a principal ideal domain, every projective module is free.*

If R is Dedekind and A is a f.g. projective R-module, then Theorem 4.17 shows that A is a sum of finitely many ideals:

$$A = I_1 \oplus \cdots \oplus I_m.$$

This decomposition is not unique: $A = J \oplus F$, where J is an ideal and F is free of rank $m - 1$ (indeed, J is the product ideal $I_1 I_2 \cdots I_m = \{\sum_i a_{i1} a_{i2} \cdots a_{im} : a_{ij} \in I_j\}$). This latter decomposition is unique to isomorphism, and this theorem is due to Steinitz [Milnor, 1971, p. 11].

Although Theorem C of Chase, cited earlier, tells us that products of projectives may not be projective, let us actually see an example.

Theorem 4.21   (Baer):   *If $\{\mathbf{Z}e_i \cong \mathbf{Z} : i = 1, 2, \ldots\}$ is a family of infinite cyclic groups, then $G = \prod \mathbf{Z}e_i$ is not free (hence not projective).*

Proof: By Corollary 4.19, it suffices to exhibit a subgroup of G that is not free. Choose a prime p; define a subgroup S by

$$S = \{(m_i e_i) \in G : \text{for each } k \geq 1, \text{ we have } p^k \mid m_i \text{ for almost all } i\}.$$

For example, $(a_i p^i e_i) \in S$ for every $a_i \in \mathbf{Z}$; it follows that card $S = 2^{\aleph_0}$. Were S free, then its rank would be uncountable (for a countable sum of copies of $\mathbf{Z}$ is countable). It would then follow that dim S/pS, as a vector space over $\mathbf{Z}/p\mathbf{Z}$, is uncountable. We finish the proof by showing that dim $S/pS \leq \aleph_0$. Let us identify e_i with $(0, \ldots, 0, e_i, 0, \ldots) \in S$. We claim the cosets $\{e_i + pS, i = 1, 2, \ldots\}$ span S/pS, which will suffice. If $s = (m_i e_i) \in S$, then almost all its coordinates m_i are divisible by p. There is thus an integer N so that

$$s - \sum_{i=1}^{N} m_i e_i = ps'$$

for some $s' \in S$. In S/pS, the coset of s is thus a finite linear combination of cosets of e_i. ∎

Using the continuum hypothesis, Specker proved that

$$B = \{(m_i e_i) \in G : \text{there exists } N \text{ with } |m_i| \leq N \text{ for all } i\},$$

the subgroup of all bounded sequences, is free. Nöbeling [Fuchs, 1973, p. 173] was able to generalize this result in two ways; first, the continuum hypothesis need not be assumed; second, the index set of the product need not be countable. The cited proof is an elegant one due to Bergman.

Corollary 4.18 characterizes left hereditary rings. In order to see this, we first prove a lemma.

Lemma 4.22: *A module P is projective if and only if every diagram with Q*

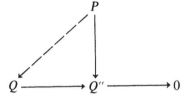

injective can be completed to a commutative diagram. The dual is also true.

Proof: If P is projective, the diagram can be completed with no hypothesis on Q.

For the converse, consider the diagram

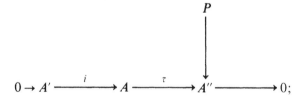

we want a map $P \to A$ making the diagram commute. There is an injective Q and a monic $\sigma : A \to Q$. Imbed the diagram in

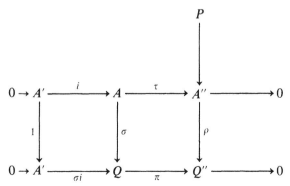

where $Q'' = \operatorname{coker} \sigma i$, π is the natural map, and ρ exists by diagram chasing. By hypothesis, there is a map $\gamma : P \to Q$ giving commutativity. Another diagram chase gives $\operatorname{im} \gamma \subset \operatorname{im} \sigma$. ∎

Theorem 4.23   (Cartan–Eilenberg): *The following are equivalent for a ring R:*

   (i)   *R is left hereditary;*
  (ii)   *every submodule of a projective module is projective;*
 (iii)   *every quotient of an injective module is injective.*

Proof: (i) $\Rightarrow$ (ii)   Corollary 4.18.

(ii) $\Rightarrow$ (i) R itself is projective, being free, and hence its submodules, the left ideals, are projective.

(iii) $\Rightarrow$ (ii)   Consider the diagram with exact rows

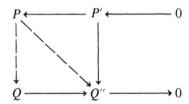

where P is projective and Q is injective. By Lemma 4.22, it suffices to find a map $P' \to Q$ making the diagram commute. Now Q'' is injective, by hypothesis, so there exists a map $P \to Q''$ giving commutativity. Since P is projective, there is a map $P \to Q$ giving commutativity. The composite $P' \to P \to Q$ is the desired map.

(ii) $\Rightarrow$ (iii)   Dualize the argument just given, using the dual of Lemma 4.22. ∎

Recall Theorem 3.23: every injective module is divisible; recall Exercise 3.14: every quotient of a divisible module is divisible. We conclude that divisible modules may not be injective, i.e., the converse of Theorem 3.23 is false. If R is not left hereditary, there exists an injective having a non-injective quotient.

Another consequence of Theorem 4.23, using Corollary 3.42: if R is left hereditary, so is the matrix ring R_n. (This result may be proved without the categorical Corollary 3.42.)

Let us show that the definition of Dedekind ring above coincides with a more classical definition.

Definition: Let R be a domain with quotient field Q. An ideal I in R is

invertible if there are elements $a_1, \ldots, a_n \in I, q_1, \ldots, q_n \in Q$ with

(i) $q_i I \subset R, i = 1, \ldots, n$;
(ii) $1 = \sum_{i=1}^{n} q_i a_i$.

Remarks: 1.   Every nonzero principal ideal Ra is invertible: choose $a_1 = a$ and $q_1 = 1/a$.
  2.   If I is invertible, then I is generated by $a_1, \ldots, a_n$.
  3.   Define I^{-1} to be the R-submodule of Q generated by $q_1, \ldots, q_n$; then if I is invertible, $II^{-1} = R = I^{-1}I$, where $II^{-1} = \{\sum \alpha_j \gamma_j : \alpha_j \in I, \gamma_j \in I^{-1}\}$.
  4.   A **fractional ideal** is a f.g. nonzero R-submodule of Q. All the fractional ideals form a commutative monoid under multiplication, with unit R. We shall see, in Corollary 4.25, that every nonzero ideal in a Dedekind ring R is invertible. It follows that the family of all fractional ideals forms an abelian group (which turns out to be free with basis all prime ideals). One defines the **class group** of R as the quotient group of this group by the subgroup of all principal ideals.
  5.   One can show $I^{-1} \cong \operatorname{Hom}_R(I, R)$.

Theorem 4.24:   *If R is a domain, a nonzero ideal I is projective if and only if it is invertible.*

Proof: If I is projective, it has a projective basis (Theorem 3.15): there are elements $\{a_k : k \in K\} \subset I$ and maps $\varphi_k : I \to R$ such that

(1)   if $a \in I$, almost all $\varphi_k a = 0$;
(2)   if $a \in I$, then $a = \sum (\varphi_k a) a_k$.

If $b \in I$ and $b \neq 0$, define $q_k \in Q$ by

$$q_k = \varphi_k(b)/b.$$

Note that q_k does not depend on the choice of b: if $b' \in I$ and $b' \neq 0$, then

$$b' \varphi_k(b) = \varphi_k(b'b) = \varphi_k(bb') = b\varphi_k(b'),$$

so that

$$\varphi_k(b)/b = \varphi_k(b')/b'.$$

It follows that $q_k I \subset R$ for all k: if $b \in I$ and $b \neq 0$, then $q_k b = [\varphi_k(b)/b]b = \varphi_k(b) \in R$. By condition (1), if $b \in I$ and $b \neq 0$, then almost all $\varphi_k(b) = 0$. Since $q_k = \varphi_k(b)/b$, there are only finitely many nonzero q_k. Finally, condition (2) gives, for $b \in I$,

$$b = \sum (\varphi_k b) a_k = \sum (q_k b) a_k = b(\sum q_k a_k).$$

If we discard all a_k for which $q_k = 0$, there remain finitely many $a_k \in I$. Moreover, if $b \neq 0$, we may cancel b from both sides of the above equation to obtain $1 = \sum q_k a_k$. Therefore, I is invertible.

Conversely, assume I is invertible, and let $a_1, \ldots, a_n \in I$, $q_1, \ldots, q_n \in Q$ be as in the definition. Define $\varphi_k : I \to R$ by $a \mapsto q_k a$ (which lies in R because $q_k I \subset R$). If $a \in I$, then

$$\sum (\varphi_k a) a_k = \sum q_k a a_k = a \sum q_k a_k = a.$$

Therefore, I has a projective basis, hence is a projective module.   ∎

Corollary 4.25:   *A domain is a Dedekind ring if and only if every nonzero ideal is invertible.*

This corollary is the link to the classical notion of Dedekind ring [Zariski–Samuel, 1958, p. 275].

Corollary 4.26:   *Every Dedekind ring is noetherian.*

Proof: Invertible ideals are always f.g.   ∎

Remarks: 1.   It can be shown [Zariski–Samuel, 1958, p. 279] that every ideal in a Dedekind ring can be generated by two elements.

2.   The ring of integers in an algebraic number field is a Dedekind ring [Zariski–Samuel, 1958, p. 283]. In particular, there are Dedekind rings that are not principal ideal domains; the easiest example is $R = \{a + b\sqrt{-5} : a, b \in \mathbf{Z}\}$.

3.   The polynomial ring R in noncommuting variables over a field is left and right hereditary; it can be shown [Cohn, 1971, p. 80] that every ideal in R is free. There are thus hereditary rings that are not noetherian.

We now generalize Theorem 3.24.

Theorem 4.27:   *A domain R is a Dedekind ring if and only if every divisible module is injective.*

Proof: If every divisible module is injective, then every quotient of an injective is divisible, hence injective. Theorem 4.22 shows that R is hereditary, and it is Dedekind since it is a domain.

Assume R is Dedekind. If D is divisible, we show that, for every nonzero ideal I and every map $f : I \to D$, there exists an extension g of f to R. Since I is invertible, there are elements $a_1, \ldots, a_n \in I$, $q_1, \ldots, q_n \in Q$ with $\sum q_i a_i = 1$ and $q_i I \subset R$. Since D is divisible, there are elements $d_i \in D$ with $f a_i = a_i d_i$. If $a \in I$,

$$f a = f(\sum q_i a_i a) = \sum (q_i a) f a_i = \sum (q_i a) a_i d_i = a \sum (q_i a_i) d_i.$$

If we set $d = \sum(q_i a_i)d_i$, then $d \in D$ and $fa = ad$ for all $a \in I$. Define $g: R \to D$ by $r \mapsto rd$. Therefore D is injective, by Theorem 3.20. $\blacksquare$

As long as we are discussing projective ideals, let us note the following fact.

Theorem 4.28: *If R is a unique factorization domain* (UFD $=$ *factorial ring*), *an ideal I is projective if and only if it is principal.*

Proof: In a domain R, every nonzero principal ideal is isomorphic to R, hence is projective. Conversely, assume I is projective and nonzero, hence invertible. Choose elements $a_1, \ldots, a_n \in I, q_1, \ldots, q_n \in Q$ with $1 = \sum q_i a_i$ and $q_i I \subset R$. Write $q_i = b_i/c_i$, and assume, by unique factorization, that b_i and c_i have no nonunit factors in common. Since $(b_i/c_i)a_j \in R$, we must have c_i dividing a_j, all i, j. Let $c = \operatorname{lcm}\{c_i\}$. We claim $I = Rc$. First,

$$c = c\sum b_i a_i/c_i = \sum (b_i c/c_i)a_i \in I,$$

for $b_i c/c_i \in R$; therefore, $Rc \subset I$. For the reverse inclusion, c_i divides a_j for all i, j implies c divides a_j, all j, whence $a_j \in Rc$ for all j. $\blacksquare$

It follows that a Dedekind ring is a UFD if and only if it is a principal ideal domain.

SEMIHEREDITARY AND PRÜFER RINGS

Definition: A ring R is **left semihereditary** if every f.g. left ideal is projective. A semihereditary domain is called a **Prüfer ring**.[5]

Examples: 25. Every left hereditary ring R is left semihereditary (of course, if R is left noetherian, the distinction vanishes).

26. Every **valuation ring** (i.e., a domain such that, given any two elements, one divides the other) is a Prüfer ring. (In such a ring, every f.g. ideal is principal; these special Prüfer rings are called **Bézout rings**.)

27. Let X be a noncompact Riemann surface, and let R be the ring of all complex-valued analytic functions on X. Helmer [1940] proved that R is a Bézout ring.

28. Every von Neumann regular ring R is left and right semihereditary (this follows quickly from Lemma 4.15).

29. The ring of all algebraic integers is a Bézout ring [Kaplansky, 1970, p. 72].

30. Chase, [1961] gives an example of a left semihereditary ring that is not right semihereditary.

---

[5] Recall that "domain" means "commutative integral domain".

We remark that left semihereditary rings are left coherent, for f.g. projectives are finitely related.

Theorem 4.29:   *If R is left semihereditary, then every f.g. submodule A of a free module F is a sum of finitely many f.g. left ideals.*

Proof: Let F have basis $\{x_k : k \in K\}$. Since A is f.g. and each generator of A is a finite linear combination of x_k's, we see that A is contained in a free summand of F generated by finitely many x_k's. We may, therefore, assume F is free with basis $\{x_1, \ldots, x_n\}$. We do an induction on n that A is a finite sum of f.g. ideals.

If $n = 1$, then A is isomorphic to a f.g. ideal. If $n > 1$, define $B = A \cap (Rx_1 \oplus \cdots \oplus Rx_{n-1})$. Each $a \in A$ has a unique expression $a = b + rx_n$, where $b \in B$ and $r \in R$. If $\varphi : A \to R$ is defined by $a \mapsto r$, then there is an exact sequence

$$0 \to B \to A \xrightarrow{\varphi} I \to 0,$$

where $I = \operatorname{im} \varphi$ is a f.g. ideal of R. Since I is projective, this sequence splits and $A \cong B \oplus I$. Since B is contained in $Rx_1 \oplus \cdots \oplus Rx_{n-1}$ the inductive hypothesis gives B, hence A, a finite sum of f.g. ideals.   ∎

The reader has undoubtedly observed that the proof just given is Theorem 4.17 stripped of its transfinite apparel. Albrecht [1961] proves that every (perhaps not f.g.) projective module over a left semihereditary ring is a sum of f.g. left ideals.

Theorem 4.30:   *A ring R is left semihereditary if and only if every f.g. submodule of a projective module is projective.*

Proof: Let A be a f.g. submodule of a projective module P. Now P is a submodule of a free module F (even a summand). By Theorem 4.29, A is a sum of f.g. ideals, each of which is projective since R is semihereditary. Therefore, A is projective.

For the converse, R itself is projective, hence its f.g. submodules, the f.g. left ideals, are projective. Therefore, R is left semihereditary.   ∎

Let us now say a few words about Prüfer rings. Note first that if R is any domain and M an R-module, then its **torsion submodule** is defined by

$$tM = \{m \in M : rm = 0 \text{ for some } r \in R, r \neq 0\}.$$

Because R is a domain, tM is a submodule. Moreover, M/tM is **torsion-free**, i.e., its torsion submodule is 0.

Lemma 4.31: *If R is a domain with quotient field Q, then every torsion-free R-module A can be imbedded in a vector space over Q. If A is torsion free and f.g., then A can be imbedded in a f.g. free R-module.*

Proof: Imbed A in an injective module E. Since A is torsion free, A is also imbedded in E/tE, which is torsion free and divisible. By Exercise 3.20, E/tE is a vector space over Q.

Assume A is f.g. If one chooses a basis of E/tE, then each of the generators of A is a linear combination of only finitely many basis vectors. Therefore, we may assume A imbedded in a finite-dimensional vector space V, namely, the subspace of E/tE spanned by the finitely many basis vectors just mentioned.

Let V have basis $\{v_1, \ldots, v_m\}$. If A is generated by $\{a_1, \ldots, a_n\}$, then each $a_i = \sum_j (r_{ij}/s_{ij})v_j$, where $r_{ij}, s_{ij} \in R$. If s is the product of all s_{ij}, then $\{s^{-1}v_1, \ldots, s^{-1}v_m\}$ is independent in V, and the R-submodule B of V generated by this set is free. Clearly A is imbedded in B. ∎

Theorem 4.32: *A domain R is a Prüfer ring if and only if every f.g torsion-free module A is projective.*

Proof: If R is a Prüfer ring, it is a domain, and so Lemma 4.31 shows A may be imbedded in a free module. By Theorem 4.30, A is projective.

Conversely, every submodule of R is torsion free, so every ideal is torsion free and every f.g. ideal is projective. ∎

Theorem 4.33: *If R is a Prüfer ring, then a module B is flat if and only if it is torsion free.*

Proof: Suppose B is flat, and $0 \to K \to F \xrightarrow{\varphi} B \to 0$ is exact, where F is free. By Theorem 3.55, $K \cap FI = KI$ for every ideal I. In particular, this holds when I is principal. Suppose $b \in B$ and $rb = 0$ for some $r \neq 0$. If $x \in F$ and $\varphi x = b$, then $rx \in K \cap rF = rK$; hence $rx = rk$ for some $k \in K$. But F is torsion free and $r(x - k) = 0$, so that $x = k \in K$. Therefore $b = \varphi x = 0$, and B is torsion-free.

Suppose B is torsion-free. By Corollary 3.49, it suffices to show every f.g. submodule A of B is flat. Since R is Prüfer, Theorem 4.32 shows A is projective, so it is surely flat. ∎

Remark: The first half of the proof shows that flat modules are torsion-free for every domain R. Thus, projective generalizes free, injective generalizes divisible, and flat generalizes torsion-free.

Instead of assuming ideals are projective, what happens if we assume they are flat? Flatness gives us nothing new if we assume rings are noetherian.

Theorem 4.34:   *A left noetherian ring R is left hereditary if and only if every left ideal is flat.*

Proof: If R is left hereditary, every left ideal is projective, hence flat. Conversely, every left ideal is f.g. since R is left noetherian. By Corollary 4.3, f.g. flat modules are projective, so every left ideal is projective and R is left hereditary. ∎

In Theorems 9.24 and 9.25, we shall classify all (not necessarily noetherian) rings in which every ideal is flat. It will be seen that this class contains all left semihereditary rings and all right semihereditary rings.

QUASI-FROBENIUS RINGS

If we assume every left ideal of R is injective, it is easy to see that R is semisimple. Let us now consider R itself as a module. It is no restriction at all to assume R projective or flat; it is obviously a stringent condition to assume R is injective (= **self-injective**).

Definition: A ring R is **quasi-Frobenius** if it is left and right noetherian and R is an injective left R-module.

It can be shown that the asymmetry of the definition is only virtual: R must be injective as a right R-module [Jans, 1964, p. 78]. It is also true [Jans, 1964, p. 80] that quasi-Frobenius rings are left and right artinian.

Clearly semisimple rings are quasi-Frobenius; in particular, kG is quasi-Frobenius when G is a finite group and k is a field whose characteristic does not divide the order of G. Although there are other examples, as we shall see, the most important example of a quasi-Frobenius ring is kG with no restriction on characteristic k. This ring is important in the theory of modular group representations.

Theorem 4.35:   *If R is a principal ideal domain and I = Ra is a nonzero ideal, then R/I is quasi-Frobenius.*

Proof: As an R-module, R/I is torsion and cyclic, say, with generator x. An ideal of R/I has the form J/I, where J is an ideal containing I. If $J = Rb$, then $bc = a$ for some $c \in R$ (since $Ra = I \subset J$); moreover, bx is a generator of $J/I \subset R/I$ (for $x = 1 + Ra$). Consider the diagram

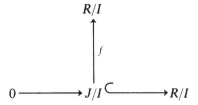

Now $f(bx) = sx$ for some $s \in R$. Since $cbx = ax = 0$, we have

$$0 = cf(bx) = csx,$$

so that $cs \in Ra$. Therefore $cs = ra = rbc$, for some $r \in R$; cancelling c gives $s = rb$, so that $f(bx) = sx = rbx$. Define $g: R/I \to R/I$ to be multiplication by r. Our calculation shows that g extends f, whence R/I is self-injective. Finally, R/I is quasi-Frobenius, for it is clearly noetherian. ∎

Remark: Exercise 9.24 generalizes this by allowing R Dedekind.

Corollary 4.36: *The following rings are quasi-Frobenius:*

(i) $\mathbf{Z}/n\mathbf{Z}$, *where* $n \neq 0$;

(ii) $k[x]/I$, *where* k *is a field and* I *is a nonzero ideal.*

Remark: Compare this Corollary with Examples 17 and 18.

Theorem 4.37: *Let R be left and right noetherian. Then R is quasi-Frobenius if and only if every projective module is injective.*

Proof: Assume R is quasi-Frobenius, and P is projective. Now P is a summand of a free module $F = \coprod R$. Since R is noetherian, Theorem 4.10 shows F is injective (since R itself is injective); thus P is injective.

For the converse, R itself is projective and hence injective, by hypothesis. Thus R is quasi-Frobenius, since we are assuming R is noetherian. ∎

Definition: A module is **indecomposable** if it is not the sum of two proper submodules.

Definition: If $R = \coprod L_i$, where each L_i is indecomposable, then the L_i are called **principal indecomposable modules**.

It follows easily from the definition that every left artinian ring R has such a decomposition, and hence has principal indecomposable modules.

Theorem 4.38: *If R is quasi-Frobenius, there is a bijection between minimal left ideals and principal indecomposable modules.*

Proof: First of all, we know that R is left artinian, so R does have a decomposition into indecomposables.

If I is a minimal left ideal, let $E(I)$ be its injective envelope in the injective module R. Being injective, $E(I)$ is a summand of R. If $E(I) = A \oplus B$, then minimality shows either $I \cap A = 0$ or $I \cap B = 0$, which contradicts $E(I)$ being an essential extension of I. Therefore $E(I)$ is a principal indecomposable module.

If E is a principal indecomposable module, then E contains a minimal left ideal I (Exercise 4.3). Now E is injective, being a summand of R. Let $E(I)$ be the injective envelope of I contained in E. Since E is a summand of $E(I)$ and $E(I)$ is indecomposable, we have $E = E(I)$. Therefore, E is an essential extension of I. If I' is another minimal left ideal in E, then $I \cap I' = 0$, contradicting E being essential. We conclude that $I \mapsto E(I)$ is a bijection. ∎

This result takes on added interest when we observe that every simple module over a quasi-Frobenius ring is isomorphic to a minimal left ideal [Curtis–Reiner, 1962, p. 401]. This fact can be used to show that f.g. injective modules E over a quasi-Frobenius ring are projective (it is true that all injectives here are projectives). The module E is a sum of indecomposables E_i, each of which contains a simple submodule. One may now repeat the argument above to show each E_i is isomorphic to a principal indecomposable module, hence is projective.

Definition: Let R be a finite-dimensional algebra over a field k. Then R is a **Frobenius algebra** if $R \cong \text{Hom}_k(R_R, k)$ as left R-modules.

Observe that the dual space $\text{Hom}_k(R_R, k)$ is a left R-module as in Theorem 1.15(i).

Theorem 4.39: *Every Frobenius algebra R is quasi-Frobenius.*

Proof: Every left or right ideal of R is a vector space over k, so finite-dimensionality of R shows R is left and right noetherian. Theorem 3.44 applies to show $\text{Hom}_k(R, k) \cong {}_R R$ is injective, so R is quasi-Frobenius. ∎

Lemma 4.40: *Let R be a finite-dimensional k-algebra over a field k. If there is a k-map $f : R \to k$ whose kernel contains no nonzero left ideals, then R is a Frobenius algebra.*

Proof: Define $\theta : R \to \text{Hom}_k(R, k)$ by $\theta_r(x) = f(xr)$. It is easy to check that each θ_r is a k-map and that θ is an R-map: $r'\theta_r = \theta_{r'r}$. Next, we claim θ is monic. If $\theta_r = 0$, then

$$0 = \theta_r(x) = f(xr) \qquad \text{for all} \quad x \in R,$$

and $Rr \subset \ker f$. By hypothesis, $r = 0$. Finally, if $\dim_k R = n$, then $\dim_k \operatorname{Hom}(R, k) = n$, being the dual space. Therefore, θ must be epic, hence an isomorphism. ∎

We are now able to give the important example.

Theorem 4.41: *If k is a field and G a finite group, then kG is a Frobenius algebra, hence is quasi-Frobenius.*

Remark: There is no restriction on the characteristic of k.

Proof: By Lemma 4.40, it suffices to exhibit a linear functional $f : R \to k$ whose kernel contains no nonzero left ideals. Each $r \in R$ has a unique expression

$$r = \sum_{x \in G} m_x x, \qquad \text{where} \quad m_x \in k.$$

Define $f : R \to k$ by $r \mapsto m_1$, the coefficient of 1. If $r = \sum m_x x$, then $f(x^{-1}r) = m_x$, so that $Rr \subset \ker f$ gives $m_x = 0$ for all $x \in G$, whence $r = 0$. ∎

LOCAL RINGS AND ARTINIAN RINGS

Definition: A ring R is **local** if it has a unique maximal left ideal.

One must be aware of the context when reading about local rings, for many authors assume local rings are commutative and noetherian as well.

Examples: 31. Every division ring is local.

32. If p is prime, then $\mathbf{Z}/p^n\mathbf{Z}$ is local, for any n.

33. If k is commutative local, the ring of formal power series $k[[x]]$ over k is local.

34. If M is an injective indecomposable R-module, then $\operatorname{End}_R(M)$ is a local ring [Lambek, 1966, exercise on p. 104].

35. If p is a prime and $\mathbf{Q}$ is the rationals, then $\mathbf{Z}_p = \{a/b \in \mathbf{Q} : (b, p) = 1\}$ is a local ring.

36. If R is a commutative ring and P is a prime ideal in R, then the localization $R_P = S^{-1}R$ (where $S = R - P$) is a local ring [Kaplansky, 1970, p. 24]. Note that Example 35 is a special case of this example: take $R = \mathbf{Z}$ and $P = (p)$.

Lemma 4.42: *Let R be a local ring with maximal left ideal J.*

   (i)   *$r \in R$ is invertible (i.e., is a unit) if and only if $r \notin J$;*
   (ii) *if $r \in J$, then $1 - r$ is invertible;*
  (iii) *if A is a f.g. R-module with $JA = A$, then $A = 0$;*
  (iv) *J is a two-sided ideal;*
   (v) *R/J is a division ring.*

Proof: (i)[6] In any ring R, Zorn's lemma shows that every proper left ideal is contained in a maximal left ideal; in particular, if $x \in R$ is not invertible, then $x \in Rx$ is contained in the unique maximal left ideal J. The converse is trivial: if $x \in J$, then x cannot be invertible, for J is a proper ideal.

(ii) If $r \in J$, then $1 - r \notin J$ and hence is invertible.

(iii) Let $\{a_1, \ldots, a_n\}$ be a minimal generating set for A in the sense that no a_i may be deleted leaving a generating set. Since $JA = A$, we have

$$a_1 = \sum_{i=1}^{n} r_i a_i, \qquad r_i \in J, \quad \text{all } i.$$

Since $1 - r_1$ is invertible, a_1 is a linear combination of the remaining a_i, violating minimality.

(iv) Since J is a left ideal, JR is a two-sided ideal. As $J \subset JR$ (for $1 \in R$) and J is a maximal left ideal, either $J = JR$ (and we are done) or $JR = R$. This last equation violates (iii).

(v) R/J is a ring because J is a two-sided ideal. Since every r outside J is already invertible in R, by (i), every nonzero element in R/J is invertible, i.e., R/J is a division ring. $\blacksquare$

Remark: Lemma 4.42(iii) is a special case of Nakayama's lemma (see Theorem 4.47 for a better version). It is essential that the R-module A be f.g. For example, let $R = \mathbf{Z}_p$ (Example 35) and let $A = \mathbf{Q}$. Here $J = (p)$, the principal ideal generated by p, and it is easy to see $p\mathbf{Q} = \mathbf{Q}$.

Lemma 4.43: *Let R be a local ring with maximal left ideal J; let A be an R-module with minimal generating set $\{a_1, \ldots, a_n\}$. If F is free on $\{x_1, \ldots, x_n\}$, $\varphi: F \to A$ is defined by $\varphi x_i = a_i$, and $K = \ker \varphi$, then $K \subset JF$.*

Proof: If $K \not\subset JF$, there is an element $\sum r_i x_i \in K$ not in JF. By Lemma 4.42(i), one of the coefficients, say, r_1, must be invertible. But $\sum r_i a_i = 0$, so that

$$a_1 = -r_1^{-1}\left(\sum_{i=2}^{n} r_i a_i\right),$$

---

[6] We show only that $r \notin J$ implies r has a left inverse s; there is a standard argument [Anderson and Fuller, 1974, pp. 165–166] that s is a two-sided inverse.

contradicting the minimality of $\{a_1, \ldots, a_n\}$. ∎

Theorem 4.44: *If R is a local ring, every f.g. projective module A is free.*

Proof: Construct an exact sequence

$$0 \to K \to F \to A \to 0$$

as in Lemma 4.43, so that $K \subset JF$. Now projectivity of A gives $F = K \oplus A'$, where $A' \cong A$. Hence $JF = JK \oplus JA'$. Now $JK \subset K \subset JF$, so Exercise 2.23 gives

$$K = JK \oplus (K \cap JA').$$

Since $K \cap JA' \subset K \cap A' = 0$, we have $K = JK$; since K is f.g., being a summand (hence an image) of the f.g. module F, we have $K = 0$ (Lemma 4.42(iii)). Therefore, A is free. ∎

It follows from Exercise 3.4 and Lemma 4.42(v) that local rings have IBN, i.e., the rank of a free module is well defined.

Endo [1961] proves that every f.g. flat module over a local ring is free; Endo [1962] proves that every f.g. flat module over a (commutative) domain is projective (he attributes this result to Cartier). Kaplansky [1958a] proves that every summand of $\coprod M_\alpha$, where each M_α is a countably generated module over an arbitrary ring, is again of the same form ("countably generated" means there is a set of generators of cardinality $\leq \aleph_0$). It follows immediately that, for any ring, every projective module is a sum of countably generated projectives. If a ring R has every countably generated projective actually free, therefore, then every projective R-module is free. In particular, Kaplansky proves in the paper cited above that all R-projectives are free when R is local or when R is Bézout.

We now show that f.g. modules over a local ring R have projective covers. This is not true for arbitrary R-modules, for Bass's Theorem E above would imply R left perfect, contradicting the existence of local rings that do not have DCC on principal right ideals (Example 35). Rings for which every f.g. module has a projective cover are called **semiperfect**.

Lemma 4.45: *If R is local with maximal left ideal J and if A is a f.g. R-module, then JA is a superfluous submodule.*

Remark: See Exercise 3.31.

Proof: If B is a submodule of A with $B + JA = A$, we must show that $B = A$. Now $A/B = (B + JA)/B = J(A/B)$. Since A/B is f.g., Lemma 4.42(iii) gives $A/B = 0$, whence $B = A$. ∎

Theorem 4.46:   *If R is a local ring, every f.g. module A has a projective cover.*

Proof: Let us call an epimorphism **essential** if its kernel is superfluous. We must find a projective P and an essential epimorphism $\varphi: P \to A$. This is precisely what is done in Lemma 4.43, for $\ker \varphi \subset JP$ and Lemma 4.45 shows that JP, hence any submodule of JP, is a superfluous submodule of P. ■

We now give a second, more complicated, proof of Theorem 4.46, because this second proof will allow a generalization for artinian rings.

Second proof: By Lemma 4.45, the natural map $\pi: A \to A/JA$ is an essential epimorphism. Now A/JA is a f.g. R/J-module, i.e., a finite-dimensional left vector space over the division ring R/J; hence $A/J \cong (R/J)^n$, the sum of n copies of R/J. Consider the diagram

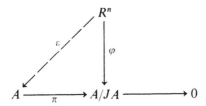

where $\varphi: R^n \to (R/J)^n$ is the natural map on each summand. Each coordinate map of φ is an essential epimorphism (Lemma 4.45 applied to R itself), so Exercise 3.33 shows that $\varphi: R^n \to A/JA$ is a projective cover. Since R^n is projective (even free), there is a map $\varepsilon: R^n \to A$ making the diagram commute: $\pi\varepsilon = \varphi$. Now $\ker \varepsilon \subset \ker \varphi$, so $\ker \varepsilon$ is a superfluous submodule of R^n (since $\ker \varphi$ is). Finally, if $a \in A$, there exists $x \in R^n$ with $\pi a = \varphi x = \pi\varepsilon x$; thus, $\varepsilon x - a \in \ker \pi = JA$. Hence $A = \operatorname{im} \varepsilon + JA$, so that ε is epic because JA is a superfluous submodule. Therefore $\varepsilon: R^n \to A$ is an essential epimorphism, i.e., a projective cover. ■

One can axiomatize this proof if one can find an ideal to play the role of J. Once we do this, we will be able to prove the existence of projective covers when R is left artinian.

Definition: If R is a ring, its **(Jacobson) radical** J is the intersection of all its maximal left ideals.

In a local ring, the radical is the unique maximal left ideal.

Theorem F: *Let R be a ring with radical J.*

(i)   *if $r \in J$, then $1 - r$ is invertible;*
(ii) *J is a two-sided ideal in R.*

If R is left artinian, then

(iii) *R/J is a semisimple ring;*
(iv)   *idempotents can be lifted: if $\bar{e} \in R/J$ is an idempotent, then there is an idempotent $e \in R$ with $e + J = \bar{e}$.*

Proof: All proofs may be found in [Anderson–Fuller 1974]: (i) and (ii) on p. 166; (iii) on p. 170; (iv) on p. 301 (in order to apply the result there, one needs to know J is nil; even more is true: it is nilpotent, as proved on p. 172).   ∎

Theorem 4.47   (Nakayama's Lemma):   *If R is a ring with radical J and A is a f.g. module with $JA = A$, then $A = 0$.*

Proof: Exactly as the proof of Lemma 4.42(iii).   ∎

Corollary 4.48:   *If R is a ring with radical J and A is a f.g. module, then JA is superfluous.*

Proof: Exactly as the proof of Lemma 4.44.   ∎

Lemma 4.49:   *Let R be a left artinian ring with radical J. If $\bar{e}$ is an idempotent in R/J, there is a projective R-module P and a projective cover $\varepsilon : P \to (R/J)\bar{e}$.*

Proof: By Theorem F, we may lift the idempotent $\bar{e}$ to an idempotent $e \in R$. There is thus an exact sequence

$$0 \to Je \to Re \overset{\pi}{\to} (R/J)\bar{e} \to 0,$$

where $\pi : re \mapsto r\bar{e}$ (one readily checks that $\ker \pi = Je$). Setting $A = Re$ in Corollary 4.48, we see that Je is a superfluous submodule of Re. Finally, e idempotent in R gives Re a summand of R, hence Re is R-projective.   ∎

Theorem 4.50:   *If R is left artinian, every f.g. module A has a projective cover.*

Proof: The proof proceeds exactly as that for local rings. Let J be the radical of R, and note that the natural map $\pi : A \to A/JA$ is an essential epimorphism (Corollary 4.48). Since R/J is semisimple, the R/J-module A/J has a decomposition into simple modules: $A/J \cong \coprod S_i$, and S_i is isomorphic to a minimal left ideal of R/J. Every ideal of R/J is a summand, and there is an idempotent $\bar{e}_i \in R/J$ with $S_i \cong (R/J)\bar{e}_i$. By Lemma 4.49, there are projective covers $\varepsilon_i : P_i \to S_i$, hence a projective cover $\varepsilon : \coprod P_i \to A/JA$ (Exercise 3.33). The proof is completed as in Theorem 4.46. ∎

Theorems 4.46 and 4.50 show that local rings and artinian rings are semiperfect.

POLYNOMIAL RINGS

In the mid-1950s, Serre proved that if $A = R[t_1, \ldots, t_k]$, where R is a field, then every f.g. projective A-module P has a f.g. free complement: there is a f.g. free module F with $P \oplus F$ free. Serre posed the problem: are projective A-modules free? If $k = 1$, then $A = R[t]$ is a principal ideal domain, so projectives are free, by Corollary 4.20. In 1958, Seshadri proved that if B is a principal ideal domain, then f.g. projective $B[x]$-modules are free; in particular, setting $B = R[t]$ gives an affirmative answer to Serre's problem when $k = 2$. There was much interest in this problem for $k > 2$; indeed, it was one of the main reasons for the development of algebraic K-theory. Remarkably, the problem was solved simultaneously, in January 1976, by Quillen in the United States and Suslin in the Soviet Union. We first present Vaserstein's version of Suslin's proof; afterwards, we shall outline Quillen's solution. We remark that Bass [1963] proved that "big projectives" (i.e., non-f.g. projectives) are free over these rings.

Definition: Let A be a commutative ring and let A^n be the free A-module of rank n. A **unimodular column** is an element $\alpha = (a_1, \ldots, a_n) \in A^n$ such that there exist $b_i \in A$, $i = 1, \ldots, n$, with $\sum_{i=1}^{n} a_i b_i = 1$.

Definition: A commutative ring A has the **unimodular column property** if, for every n, every unimodular column α is the first column of some $n \times n$ invertible matrix over A.

This definition may be rephrased. Let $GL(n, A)$ denote the multiplicative group of invertible matrices over A, and let ε_1 denote the column vector having first coordinate 1 and 0's elsewhere. Then α is the first column of an invertible matrix over A if and only if

$$\alpha = M\varepsilon_1$$

for some $M \in GL(n, A)$.

That a column $\alpha = (a_1, \ldots, a_n)$ be unimodular is a necessary condition that it be the first column of some $M \in GL(n, A)$, for M is invertible if and only if its determinant, det M, is a unit in A, and Laplace expansion shows det M is a linear combination of the entries in the first column of M.

For the remainder of this chapter, all rings are commutative.

Theorem 4.51: *If every f.g. projective A-module is free, then A has the unimodular column property.*

Proof: If $\alpha = (a_1, \ldots, a_n) \in A^n$ is a unimodular column, there exist $b_i \in A$ with $\sum a_i b_i = 1$. Define $\varphi: A^n \to A$ by $(r_1, \ldots, r_n) \mapsto \sum r_i b_i$. Since $\varphi(\alpha) = 1$, there is an exact sequence

$$0 \to K \to A^n \xrightarrow{\varphi} A \to 0,$$

where $K = \ker \varphi$. As A is projective, this sequence splits, and

$$A^n = K \oplus \langle \alpha \rangle.$$

By hypothesis, K is free (of rank $n - 1$). If $\{\alpha_2, \ldots, \alpha_n\}$ is a basis of K, then adjoining α gives a basis of A^n. Let $\{\varepsilon_1, \ldots, \varepsilon_n\}$ be the **standard basis** of A^n, i.e., ε_i has all coordinates 0 except for 1 in the ith place (this agrees with our earlier notation for ε_1). The transformation $T: A^n \to A^n$ with $T\varepsilon_1 = \alpha$ and $T\varepsilon_i = \alpha_i$ for $i \geq 2$ is invertible, and the matrix of T with respect to the standard basis has first column α. ∎

In general, the converse of Theorem 4.51 is false. For example, the reader may show that $A = \mathbf{Z}/6\mathbf{Z}$ has the unimodular column property, yet we know A has nonfree projectives.

Definition: A f.g. module P is **stably free** if there exists a f.g. free module F such that $P \oplus F$ is free.

Remarks: 1. Every stably free module is projective, being a summand of a free module.

2. With this terminology, Serre's theorem states that f.g. projective A-modules are stably free when $A = R[t_1, \ldots, t_k]$ with R a field.

3. Kaplansky exhibited a stably free module that is not free over the ring of all continuous real-valued functions on the 2-sphere $\{(x, y, z) \in \mathbf{R}^3 : x^2 + y^2 + z^2 = 1\}$ (see [Swan, 1962]). This example may be modified to exhibit the same phenomenon over $A = \mathbf{Z}[x, y, z]/(x^2 + y^2 + z^2 - 1)$. The first algebraic proof of this is given by Kong [1977].

Eilenberg's observation (Exercise 3.13) shows that we must insist that the complement F be f.g. The next result shows that the notion of stably free is of interest only when all modules in sight are f.g.

Theorem 4.52 (Gabel): *Assume $P \oplus A^m = F$, where F is free. If P is not f.g., then P is free.*

Proof: Let $\{e_i : i \in I\}$ be a basis of F. Since P is an image of F, the index set I must be infinite. Consider the exact sequence

$$0 \to P \to F \xrightarrow{\varphi} A^m \to 0.$$

If $\{\varepsilon_1, \ldots, \varepsilon_m\}$ is the standard basis of A^m, choose $x_j \in F$ with $\varphi(x_j) = \varepsilon_j$, $1 \leq j \leq m$. As each x_j has only finitely many nonzero coordinates, they collectively involve only a finite subset $J \subset I$. If we define $F' = \coprod_{i \in J} Ae_i$, then $\varphi' = \varphi | F'$ is epic and $F = P + F'$. Setting $P' = P \cap F'$, there are exact sequences

$$0 \to P' \to F' \xrightarrow{\varphi'} A^m \to 0 \quad \text{and} \quad 0 \to P' \to P \to P/P' \to 0.$$

Clearly the first sequence splits: $F' = P' \oplus A^m$. The second sequence also splits, for

$$P/P' = P/P \cap F' \cong P + F'/F' = F/F' \cong \coprod_{I-J} Ae_i.$$

Thus, $P \cong P' \oplus P/P'$. Since $I - J$ is infinite, there is a free module F'' with $P/P' \cong A^m \oplus F''$. Therefore

$$P \cong P' \oplus P/P' \cong P' \oplus (A^m \oplus F'') \cong (P' \oplus A^m) \oplus F'' \cong F' \oplus F'',$$

and P is free. ∎

Theorem 4.53: *If A has the unimodular column property, then every stably free A-module P is free.*

Proof: By induction on the rank of a free complement of P, it suffices to prove P is free assuming $P \oplus A$ is free: $P \oplus A = A^n$. Let π be the projection of A^n on A having kernel P, and let $\{\varepsilon_1, \ldots, \varepsilon_n\}$ be the standard basis of A^n. There exists $\alpha = (a_1, \ldots, a_n) \in A^n$ with $1 = \pi(\alpha) = \sum a_i \pi(\varepsilon_i)$; thus, α is a unimodular column. By hypothesis, $\alpha = M\varepsilon_1$ for some $M \in GL(n, A)$. Let $T: A^n \to A^n$ be the corresponding A-homomorphism: $T\varepsilon_1 = \alpha$ and $T\varepsilon_i = \alpha_i$ for $i \geq 2$ (where $\alpha_2, \ldots, \alpha_n$ are the other columns of M). If $T\varepsilon_i \in P$ for all $i \geq 2$, we claim $T' = T | \langle \varepsilon_2, \ldots, \varepsilon_n \rangle : \langle \varepsilon_2, \ldots, \varepsilon_n \rangle \to P$ is an isomorphism, proving P is free. Visibly, T' is monic. If $\beta \in A^n$, then $\beta = T\gamma$ for some $\gamma = \sum r_i \varepsilon_i$; hence $\beta = T(r_1 \varepsilon_1 + \delta)$, where $\delta = \sum_{i=2}^n r_i \varepsilon_i$. In particular, if $\beta \in P$, then

$$\beta - T\delta = r_1 T\varepsilon_1 = r_1 \alpha \in P \cap \langle \alpha \rangle = 0,$$

whence $\beta \in \operatorname{im} T'$ and T' is epic.

It thus suffices to show $T\varepsilon_i \in P$ for $i \geq 2$; in matrix terms, it suffices to show the columns $\alpha_2, \ldots, \alpha_n$ may be chosen to lie in $P = \ker \pi$. Suppose

$\pi(\alpha_i) = \lambda_i$ for $i \geq 2$. By elementary column operations, for each $i \geq 2$, replace α_i by $\alpha_i' = \alpha_i - \lambda_i \alpha$. The new matrix is invertible, has first column α, and, if $i \geq 2$, then $\pi(\alpha_i') = \pi(\alpha_i) - \lambda_i \pi(\alpha) = 0$ and $\alpha_i' \in P$. ∎

Theorem **(Serre):** *If R is a field, then f.g. projective $R[t_1, \ldots, t_k]$-modules are stably free.*

Proof: Theorem 9.45. ∎

Corollary 4.54: *If $A = R[t_1, \ldots, t_k]$, where R is a field, has the unimodular column property, then f.g. projective A-modules are free.*

Proof: Serre's theorem and Theorem 4.53. ∎

Definition: If B is a commutative ring, a polynomial $f(y) \in B[y]$ has **virtual degree** s if

$$f(y) = b_0 y^s + b_1 y^{s-1} + \cdots + b_s, \qquad b_i \in B.$$

Since we do not demand that the leading coefficient $b_0 \neq 0$, the degree of $f(y)$ may be smaller than its virtual degree (of course, $f(y)$ has many virtual degrees).

Theorem 4.55 **(Suslin Lemma):** *Assume B is a commutative ring. Let $s \geq 1$ and consider polynomials in $B[y]$:*

$$f(y) = y^s + a_1 y^{s-1} + \cdots + a_s;.$$
$$g(y) = \qquad b_1 y^{s-1} + \cdots + b_s.$$

Then, for each j, $1 \leq j \leq s$, the ideal $(f(y), g(y))$ in $B[y]$ contains a polynomial of virtual degree $s - 1$ and leading coefficient b_j.

Proof: Define

$$I = \{\text{leading coefficients of those } h(y) \in (f, g)$$
$$\text{having virtual degree } s - 1\};$$

it is clear that I is an ideal in B containing b_1. We prove by induction on j that I contains $b_1, \ldots, b_j, j \leq s$. Define $g'(y) \in (f, g)$ by

$$g'(y) = yg(y) - b_1 f(y) = \sum_{i=1}^{s} (b_{i+1} - b_1 a_i) y^{s-i}.$$

By induction, I contains the first $j - 1$ coefficients of $g'(y)$, the last of which is $b_j - b_1 a_{j-1}$. It follows that $b_j \in I$. ∎

Observe that performing elementary row operations on a matrix $M \in \mathrm{GL}(n, A)$ corresponds to left multiplication by a matrix $N \in \mathrm{GL}(n, A)$.

Thus, if a column α can, after suitable row operations, be completed to an invertible matrix, then α itself can be so completed: $N\alpha = M\varepsilon_1$ implies $\alpha = N^{-1}M\varepsilon_1$.

Theorem 4.56 (Horrocks): *Let B be local, and let $\alpha = (a_1, \ldots, a_n)$ be a unimodular column over $A = B[\,y\,]$. If some a_i is monic, i.e., has leading coefficient 1, then α is the first column of an invertible matrix over A.*

Proof **(Suslin):** If $n = 1$ or 2, the theorem holds for any commutative ring A; therefore, we may assume $n \geq 3$. We do an induction on s, the degree of the monic polynomial a_i, noting that the case $s = 0$ is trivial. By our preceding remark about row operations, we may assume a_1 is monic of degree $s > 0$ and the other polynomials $a_2, \ldots, a_n$ have degrees $\leq s - 1$.

Let $\mathfrak{m}$ be the maximal ideal in B. Thus $\mathfrak{m}A$ consists of those polynomials each of whose coefficients lies in $\mathfrak{m}$. The column $\bar{\alpha} \in A^n/\mathfrak{m}A^n$ is unimodular over $(B/\mathfrak{m})[\,y\,]$, so that not all a_i, $i \geq 2$, lie in $\mathfrak{m}A$; assume $a_2 \notin \mathfrak{m}A$. Thus, $a_2 = b_1 y^{s-1} + \cdots + b_s$, and some $b_j \notin \mathfrak{m}$; since B is local, b_j is a unit. By Suslin's Lemma 4.55, the ideal (a_1, a_2) in A contains a monic polynomial of degree $s - 1$. Therefore, the elementary row operation of adding a linear combination of a_1 and a_2 to a_3 produces a monic polynomial in position 3 of degree $s - 1$. One may now apply the inductive hypothesis. ∎

Notation: If $A = B[\,y\,]$, where B is a commutative ring, and if $\alpha(y) = (a_i(y))$ is an n-rowed column over A, then

$$\mathrm{GL}(n, A)\alpha(y) = \{M\alpha(y): M \in \mathrm{GL}(n, A)\}.$$

As the entries of $M \in \mathrm{GL}(n, A)$ lie in $A = B[\,y\,]$, they are polynomials in y; one may thus write $M = M(y)$.

Theorem 4.57: *Let B be a domain, $A = B[\,y\,]$, and $\alpha(y) = (a_i(y))$ a unimodular column over A, one of whose coordinates is monic. Then*

$$\alpha(y) = M(y)\beta,$$

where $M(y) \in \mathrm{GL}(n, A)$ and β is a unimodular column over B.

Proof: Define

$$I = \{b \in B : \mathrm{GL}(n, A)\alpha(w + bx) = \mathrm{GL}(n, A)\alpha(w), \text{ all } w, x \in A\}.$$

It is easy to see that I is an ideal in B: in particular, if $b, b' \in I$, then

$$\mathrm{GL}(n, A)\alpha(w + bx + b'x) = \mathrm{GL}(n, A)\alpha(w + bx) = \mathrm{GL}(n, A)\alpha(w),$$

whence $b + b' \in I$.

Suppose $I = B$, so that $1 \in I$. Set $w = y$, $b = 1$, and $x = -y$ to obtain

$$GL(n, A)\alpha(y) = GL(n, A)\alpha(0).$$

Thus, $\alpha(y) = M\alpha(0)$ for some $M \in GL(n, A)$; since $\beta = \alpha(0)$ is a unimodular column over B, the theorem would be proved.

We may therefore assume I is a proper ideal of B, so that $I \subset J$ for some maximal ideal J. Since B is a domain, B is contained in the localization B_J. As B_J is a local ring and $\alpha(y)$ is a unimodular column over $B_J[y]$ having a monic coordinate, Horrocks' Lemma 4.56 applies to give

$$\alpha(y) = M(y)\varepsilon_1$$

for some $M(y) = (m_{ij}(y)) \in GL(n, B_J[y])$. Adjoin a new indeterminate z to $B_J[y]$ and define a matrix

$$N(y, z) = M(y)[M(y + z)]^{-1} \in GL(n, B_J[y, z])$$

(the matrix $M(y + z)$ is obtained from $M(y)$ by replacing each of its polynomial entries $m_{ij}(y)$ by $m_{ij}(y + z)$; if $M(y)^{-1} = (h_{ij}(y))$, then it is easy to see that $(h_{ij}(y + z))$ is the inverse of $M(y + z)$). Observe that the definition of $N(y, z)$ gives

$$N(y, 0) = 1_n, \qquad \text{the } n \times n \text{ identity matrix.}$$

Since $\alpha(y) = M(y)\varepsilon_1$, it follows that $\alpha(y + z) = M(y + z)\varepsilon_1$. Therefore

$$(*) \qquad N(y, z)\alpha(y + z) = N(y, z)M(y + z)\varepsilon_1 = M(y)\varepsilon_1 = \alpha(y).$$

Each entry of $N(y, z)$ is a polynomial in $B_J[y, z]$, hence may be written as $f(y) + g(y, z)$, where each monomial in $g(y, z)$ involves a positive power of z. Since $N(y, 0) = 1_n$, we must have $f(y) = 0$ or 1, and we conclude that the entries of $N(y, z)$ are polynomials in $B_J[y, z]$ containing no nonzero monomials of the form λy^i with $i > 0$ and $\lambda \in B_J$. Let b be the product of all denominators occurring in coefficients of the polynomial entries of $N(y, z)$; by definition of B_J, we have $b \notin J$ and hence $b \notin I$. Further, $N(y, bz) \in GL(n, B[y, z])$ for we have just seen that replacing z by bz eliminates all denominators. Equation $(*)$ gives

$$GL(n, B[y, z])\alpha(y + bz) = GL(n, B[y, z])\alpha(y).$$

For fixed $w, x \in A = B[y]$, define a B-algebra map $\varphi : B[y, z] \to A$ by $\varphi(y) = w$ and $\varphi(z) = x$. Applying φ to the equation above gives

$$GL(n, A)\alpha(w + bx) = GL(n, A)\alpha(w),$$

and this contradicts $b \notin I$. ∎

The reader surely appreciates how ingenious this elementary argument is!

The coefficient ring is seriously restricted in the next technical lemma, which is the computation at the heart of the "Noether Normalization Lemma".

Lemma 4.58 (Noether): *Let* $A = R[t_1, \ldots, t_k]$, *where R is a field, and let* $a \in A$ *have (total) degree* δ; *set* $m = \delta + 1$. *Define*

$$y = t_k$$

and, for $1 \le i \le k - 1$, *define*

$$y_i = t_i - t_k^{m^{k-i}}$$

Then $a = ra'$, *where* $r \in R$ *and* a' *is a monic polynomial over the polynomial ring* $R[y_1, \ldots, y_{k-1}]$.

Proof: First of all, $R[y_1, \ldots, y_{k-1}]$ is a polynomial ring (i.e., $\{y_1, \ldots, y_{k-1}\}$ is an independent set of transcendentals), for the defining equations for the y's give an automorphism of A (with inverse $t_k \mapsto t_k$ and $t_i \mapsto t_i + t_k^{m^{k-i}}$ for $1 \le i \le k - 1$).

Denote the k-tuple $(m^{k-1}, m^{k-2}, \ldots, m, 1)$ by μ, and denote k-tuples $(j_1, \ldots, j_k)$ with each $j_i \ge 0$ by (j). The dot product $\mu \cdot (j)$ is, of course, $m^{k-1}j_1 + \cdots + mj_{k-1} + j_k$.

The polynomial a may be written

$$a = \sum_{(j)} r_{(j)} t_1^{j_1} \cdots t_k^{j_k},$$

where $r_{(j)}$ are nonzero elements of R. Replacing each t_i as in the statement gives

$$a = \sum_{(j)} r_{(j)} (y_1 + y^{m^{k-1}})^{j_1} \cdots (y_{k-1} + y^m)^{j_{k-1}} y^{j_k}$$

Expand each monomial for fixed (j) and separate the "pure" power of y from the rest:

$$a = \sum_{(j)} r_{(j)} y^{\mu \cdot (j)} + \sum_{(j)} f_{(j)}(y_1, \ldots, y_{k-1}, y),$$

where each mixed term $f_{(j)}$ involves some y_i to a positive power; moreover, for each (j), the highest power of y is $\mu \cdot (j)$.

Since m is greater than the (total) degree of a, each indexing k-tuple $(j) = (j_1, \ldots, j_k)$ has $0 \le j_i < m$ for all i. It follows that $(j) \ne (j')$ implies $\mu \cdot (j) \ne \mu \cdot (j')$, for this is just the unique representation of a positive integer written in base m. All the exponents $\mu \cdot (j)$ distinct implies all the terms $r_{(j)} y^{\mu \cdot (j)}$ distinct. If d is the largest $\mu \cdot (j)$, then d is the degree of a as a polynomial in y with coefficients in $R[y_1, \ldots, y_{k-1}]$. Therefore,

$$a = ry^d + g(y_1, \ldots, y_{k-1}, y),$$

where $r \in R$ and g has degree in y strictly less than d. As $r \neq 0$ and R is a field, we may define $a' = y^d + r^{-1}g(y_1, \ldots, y_{k-1}, y)$. ∎

Theorem 4.59 (Quillen–Suslin): *If $A = R[t_1, \ldots, t_k]$, where R is a field, then every f.g. projective A-module is free.*

Proof: We do an induction on k, the case $k = 1$ holding because $R[t]$ is a principal ideal domain.

By Corollary 4.54, it suffices to prove A has the unimodular column property. Let $\alpha = (a_i(t_1, \ldots, t_k))$ be a unimodular column over A; we may assume $a_1 \neq 0$. By the Noether Lemma 4.58, $a_1 = ra'_1$, where $r \in R$ and $a'_1 \in R[y_1, \ldots, y_{k-1}]$ is a monic polynomial (y_i defined as in the Noether lemma). Since r is a unit (lest $a_1 = 0$), there is no loss in generality in assuming $a_1 = a'_1$, i.e., a_1 is monic. Theorem 4.57 thus applies to give

$$\alpha = M\beta,$$

where $M \in \mathrm{GL}(n, A)$ and β is a unimodular column over $B = R[y_1, \ldots, y_{k-1}]$. By induction, B has the unimodular column property, so that

$$\beta = N\varepsilon_1$$

for some N in $\mathrm{GL}(n, B)$. Hence $MN \in \mathrm{GL}(n, A)$ and

$$\alpha = MN\varepsilon_1. \quad ∎$$

Generalization of Noether's lemma to more general rings enables one to relax the hypotheses on R.

Quillen's approach is more module-theoretic, and makes use of an elementary idea used by Seshadri in his solution of Serre's problem for $k = 2$.

Definition: If A is an R-algebra, an A-module P is **extended** from R if there is an R-module P_0 and an isomorphism $P \cong P_0 \otimes_R A$.

Examples: 37. Every free A-module is extended from R.

38. If an A-module P is extended from R and S is a subset of A, then $S^{-1}P$ is extended from R (if $P \cong P_0 \otimes_R A$, then $S^{-1}P = P \otimes_A S^{-1}A$, and associativity gives

$$S^{-1}P \cong (P_0 \otimes_R A) \otimes_A S^{-1}A \cong P_0 \otimes_R S^{-1}A.$$

39. If $P \cong P_0 \otimes_R A$, the R-module P_0 may not be unique (if $R = \mathbf{Z}$ and $A = \mathbf{Q}$, then the $\mathbf{Q}$-module $\mathbf{Q}$ is extended from $\mathbf{Z}$: $\mathbf{Q} \cong \mathbf{Z} \otimes_{\mathbf{Z}} \mathbf{Q}$. However, for any torsion group T, we also have $\mathbf{Q} = (\mathbf{Z} \oplus T) \otimes_{\mathbf{Z}} \mathbf{Q}$.

40. There exist A-modules that are not extended (if R is a field and A is an R-algebra, then every R-module is free (being a vector space) and hence every extended A-module must also be free.)

The uniqueness question: "If P is extended from R, is the module P_0 unique?", is easily settled for polynomial rings.

Definition: An R-algebra A (not necessarily commutative) is **augmented** if there is an R-algebra map $\varepsilon: A \to R$.

Since $\varepsilon(1) = 1$, the augmentation ε must be epic. The kernel of ε, called the **augmentation ideal**, is a two-sided ideal in A.

Examples: 41. If $A = R[U]$, polynomials in a set of variables U, then A is augmented with augmentation $\varepsilon: A \to R$ given by $f \mapsto$ constant term of f.

42. If G is a group, the group ring RG has an augmentation: $\varepsilon(\sum r_g g) = \sum r_g$.

Lemma 4.60: *Let A be a commutative augmented R-algebra with augmentation ideal I; let P be an A-module.*

(i)   *If $P \cong P_0 \otimes_R A$, then $P_0 \cong P/IP$.*

(ii) *If, in addition, P is projective or f.g., then so is P_0.*

Proof: (i)   Tensor the exact sequence $0 \to I \to A \to R \to 0$ by P_0 to obtain an exact sequence

$$P_0 \otimes_R I \to P_0 \otimes_R A \to P_0 \otimes_R R \to 0.$$

The last term is P_0, the middle term, by hypothesis, is P, and the image of the first term is IP. We conclude that $P_0 \cong P/IP$.

(ii)   If P is f.g., the formula $P/IP \cong P_0$ shows P_0 is also f.g.; if P is A-projective, then Exercise 3.11 applies with $R \cong A/I$ to show that P_0 is R-projective. ∎

The existence question: "Given an A-module P, when is P extended from R?", is quite difficult. In his solution of Serre's problem, Quillen discovered the following criterion.

Theorem (Quillen): *Let R be a commutative ring, and let P be a finitely related $R[t]$-module. If the $R_{\mathfrak{m}}[t]$-module $P_{\mathfrak{m}} = P \otimes R_{\mathfrak{m}}[t]$ is extended from $R_{\mathfrak{m}}$ for every maximal ideal $\mathfrak{m}$ of R, then P is extended from R.*

Proof: [Roitman, 1979].

As in Suslin's solution, one prefers polynomials with coefficients in a local ring.

Definition: If S is the set of all monic polynomials in $R[t]$, write

$$R(t) = S^{-1}R[t].$$

The elements of $R(t)$ may be regarded as rational functions

$$(a_m t^m + a_{m-1}t^{m-1} + \cdots + a_0)/(t^n + b_{n-1}t^{n-1} + \cdots + b_0).$$

Definition: If P is an $R[t]$-module, define

$$P(t) = S^{-1}P = P \otimes_{R[t]} R(t).$$

There is an analogue of Horrocks' lemma.

Theorem H: *If R is local and P is a f.g. projective $R[t]$-module such that $P(t)$ is a free $R(t)$-module, then P is free.*

Proof: [Lam, 1978, pp. 114–116].   ∎

Assuming Quillen's theorem and Theorem H, let us see how Quillen's solution reads.

Theorem 4.61   (Quillen):   *Let $\mathfrak{A}$ be a class of rings such that*

(i)   *if $R \in \mathfrak{A}$, then $R(t) \in \mathfrak{A}$;*

(ii)   *If $R \in \mathfrak{A}$ and $\mathfrak{m}$ is a maximal ideal of R, then $R_{\mathfrak{m}}[t]$-projectives are free.*

Then, for all $k \geq 1$ and for all $R \in \mathfrak{A}$, f.g. projective $R[t_1, \ldots, t_k]$-modules are extended from R.

Proof: We do an induction on k.

Assume $k = 1$ and P is f.g. $R[t]$-projective. By condition (ii), $R_{\mathfrak{m}}[t]$-projectives are free for every maximal ideal $\mathfrak{m}$ of R. In particular, $P_{\mathfrak{m}} = P \otimes R_{\mathfrak{m}}[t]$ is free, hence is extended from $R_{\mathfrak{m}}$ (Example 37). By Quillen's theorem, P is extended from R.

Assume $k > 1$ and P is f.g. $R[t_1, \ldots, t_k]$-projective. We establish notation:

$$P_0 = P/(t_1, \ldots, t_k)P; \qquad P_1 = P/(t_2, \ldots, t_k)P;$$
$$S_1 = \{\text{all monic polynomials in } R[t_1]\},$$

which we view as a (multiplicatively closed) subset of $R[t_1, \ldots, t_k]$. Now define

$$P' = S_1^{-1}P = P \otimes_{R[t_1, \ldots, t_k]} S_1^{-1}R[t_1, \ldots, t_k]$$
$$= P \otimes_{R[t_1, \ldots, t_k]} R(t)[t_2, \ldots, t_k],$$

where we have written t in place of t_1. Since $R(t) \in \mathfrak{A}$, by (i), induction says the projective $R(t)[t_2, \ldots, t_k]$-module P' is extended from $R(t)$. As polynomial rings are augmented,

$$P' \cong P'/(t_2, \ldots, t_k)P' \otimes_{R(t)} R(t)[t_2, \ldots, t_k].$$

Using Exercise 1.12 (with $I = (t_2, \ldots, t_k)$) and the definition $P' = P \otimes R(t)[t_2, \ldots, t_k]$, we have

$$P'/(t_2, \ldots, t_k)P' \cong P_1 \otimes_{R[t]} R(t).$$

From the initial step $k = 1$, we have

$$P_1 \cong P_0 \otimes_R R[t].$$

Combining these isomorphisms gives

(*) $P' \cong P_0 \otimes_R R(t)[t_2, \ldots, t_k].$

Now define $B = R[t_2, \ldots, t_k]$ and $S = \{$all monic polynomials in $B[t]\}$; note that $S^{-1}B[t] = B(t)$. Since $S_1 \subset S$, Exercise 3.48 gives

$$P(t) = S^{-1}P = S^{-1}(S_1^{-1}P) = S^{-1}P'.$$

But, by definition of S^{-1},

$$\begin{aligned}
P(t) = S^{-1}P' &= P' \otimes_{R(t)[t_2, \ldots, t_k]} S^{-1}R(t)[t_2, \ldots, t_k] \\
&= P' \otimes_{R(t)[t_2, \ldots, t_k]} B(t) \\
&= P_0 \otimes_R B(t), \qquad \text{by equation (*)}, \\
&= (P_0 \otimes_R B) \otimes_B B(t).
\end{aligned}$$

Let $\mathfrak{m}$ be a maximal ideal in B. Defining $P_{\mathfrak{m}}(t) = P(t) \otimes_B B_{\mathfrak{m}}$, we apply Lemma 3.77 to obtain

$$P_{\mathfrak{m}}(t) = (P_0 \otimes_R B)_{\mathfrak{m}} \otimes_{B_{\mathfrak{m}}} B_{\mathfrak{m}}(t).$$

Now $(P_0 \otimes_R B)_{\mathfrak{m}}$ is $B_{\mathfrak{m}}$-projective because P_0 is R-projective; it is free because $B_{\mathfrak{m}}$ is local; therefore, $P_{\mathfrak{m}}(t)$ is free. By Theorem H, $P_{\mathfrak{m}}$ is free, and so $P_{\mathfrak{m}}$ is extended from $B_{\mathfrak{m}}$ for every $\mathfrak{m}$ (Example 37). Quillen's theorem shows P is extended from $B = R[t_2, \ldots, t_k]$, and induction shows P is extended from R. ∎

Theorem 4.62 (Quillen–Suslin): *If R is a field, then every f.g. projective $R[t_1, \ldots, t_k]$-module is free.*

Proof: Let $\mathfrak{A}$ be the class of all fields. If $R \in \mathfrak{A}$, then it is easily seen that $R(t)$ is the quotient field of $R[t]$ consisting of all rational functions (for the leading coefficient of a nonzero polynomial is a unit). To verify condition (ii) of Theorem 4.61, note that $R = R_{\mathfrak{m}}$ and that $R[t]$ is a principal ideal domain. Theorem 4.61 says that f.g. projective $R[t_1, \ldots, t_k]$-modules are extended from R. As every R-module is a vector space, hence free, extended modules are free. ∎

Theorem 4.63 (**Quillen–Suslin**): *If R is a principal ideal domain, then every f.g. projective $R[t_1, \ldots, t_k]$-module is free.*

Proof: Let $\mathfrak{A}$ be the class of all principal ideal domains. Condition (i) of Theorem 4.61 needs some standard commutative algebra: one shows that $R(t) \in \mathfrak{A}$ by showing it is a unique factorization domain in which nonzero prime ideals are maximal. Condition (ii) is not standard. Assume P is $R_m[t]$-projective, where m is a maximal ideal in R. Now $P(t) = P \otimes_{R_m[t]} R_m(t)$ is $R_m(t)$-projective, hence $R_m(t)$-free, for $R_m \in \mathfrak{A}$ (standard commutative algebra) and $R_m(t)$ is a principal ideal domain (condition (i)). Since R_m is local, Theorem H gives P free. Now apply Theorem 4.61 to conclude that projective $R[t_1, \ldots, t_k]$-modules are extended from R; since $R[t_1, \ldots, t_k]$ is augmented, such modules are extended from projective, hence free, R-modules, and so are themselves free. ∎

Remarks: 1. Seshadri's theorem would also prove condition (ii) in Theorem 4.63.

2. Theorem 4.61 applies to the class $\mathfrak{A}$ of all Dedekind rings. In particular, our remarks after Corollary 4.20 show that if R is Dedekind, then every f.g. projective $R[t_1, \ldots, t_k]$-module has the form $A \oplus B$, where A is free and B is extended from an ideal of R.

3. Brewer and Costa [1978] have proved that the class $\mathfrak{A}$ of all Bézout rings in which nonzero prime ideals are maximal satisfies the hypotheses of Theorem 4.61 (the ring of all algebraic integers is such a ring), thereby proving f.g. $R[t_1, \ldots, t_k]$-projectives are free for these rings.

4. Quillen applied Theorem 4.61 to the class $\mathfrak{A}$ of all regular rings (i.e., rings of finite global dimension; see Chapter 9 for the definition) of Krull dimension ≤ 2 (i.e., the longest chain of prime ideals has two steps: $P_0 \subset P_1 \subset P_2$) [Lam, 1978, p. 138].

5. Other examples of rings R for which projective $R[t_1, \ldots, t_k]$-modules are free are formal power series rings $K[[x_1, \ldots, x_n]]$, where K is a field [Lam, 1978, pp. 150–162]. This result was proved, independently, by Mohan–Kumar and by Lindel and Lütkebohmert.

6. Ojanguren and Sridharan [1971] exhibit a division ring D and a nonfree projective $D[x, y]$-module. Thus, the noncommutative version of the Quillen–Suslin theorem is false.

5   Extensions of Groups

With this chapter we begin the proper study of homological algebra. The group theoretic computations that we shall do will yield constructions analogous to those for the singular homology and cohomology groups of a topological space that we sketched in Chapter 1.

Let E be a (possibly nonabelian) group, A an abelian normal subgroup, and $G = E/A$. We write E and its subgroup A additively, but we write the quotient group G multiplicatively.

Definition: An **extension** of A by G is an exact sequence

$$0 \to A \to E \to G \to 1.$$

Remark: The definition of extension makes perfectly good sense for any group A, but we assume A abelian throughout.

Lemma 5.1:   *An extension of A by G determines a homomorphism $\theta: G \to$ $\mathrm{Aut}(A)$, where $\mathrm{Aut}(A)$ is the group of all automorphisms of A.*

Proof: If $x \in G$, let $\lambda x \in E$ be a **lifting** of x (i.e., $\lambda x \mapsto x$). Define $\theta_x : A \to A$ by $\theta_x(a) = \lambda x + a - \lambda x$, noting that $\operatorname{im} \theta_x \subset A$ since A is normal in E. The function $\theta: G \to \mathrm{Aut}(A)$ sending $x \mapsto \theta_x$ is well defined; this follows from a short computation using the facts: (i) A is abelian; (ii) if $\lambda' x$ is a second lifting of x, then $\lambda' x - \lambda x \in A$. Finally, θ is a homomorphism, for since the value of θ is independent of choice of liftings, we may choose $\lambda(xy) = \lambda x + \lambda y$. ∎

We recall a definition from the previous chapter.

Definition: If G is a multiplicative group (perhaps infinite), the **integral group ring** $\mathbf{Z}G$ is the ring whose additive group is the free abelian group with basis G, and whose multiplication is determined by the multiplication of G and the distributive laws.

A typical element of $\mathbf{Z}G$ is $\sum_{x \in G} m_x x$, where $m_x \in \mathbf{Z}$ and almost all $m_x = 0$.

Theorem 5.2: *If* $0 \to A \to E \to G \to 1$ *is an extension, then* A *is a left* $\mathbf{Z}G$-*module.*

Proof: In brief, G acts on A by conjugation. In detail, if $x \in G$ and $a \in A$, define $xa = \theta_x(a) = \lambda x + a - \lambda x$. Since θ is a homomorphism, $1a = a$ and $(xy)a = x(ya)$. We now define the action of an arbitrary element of $\mathbf{Z}G$ by $(\sum m_x x)a = \sum m_x(xa)$. ∎

It should now be clear why we write A additively and G multiplicatively.

There is a second way of writing conjugation. Suppose we had defined $\theta'_x(a) = -\lambda x + a + \lambda x$. The reader may check that $\theta': G \to \operatorname{Aut}(A)$ given by $x \mapsto \theta'_x$ is an antihomomorphism: $\theta'(xy) = \theta'(y)\theta'(x)$. This version of conjugation makes A into a right $\mathbf{Z}G$-module, where $ax = \theta'_x(a)$.

Exercises: 5.1. If $x \in G$, then $\theta'_x = \theta_{x^{-1}}$.

5.2. Let A be a left $\mathbf{Z}G$-module. Show that A can be made into a right $\mathbf{Z}G$-module by defining $ax = x^{-1}a$. (If G is not abelian, A need not be a $(\mathbf{Z}G - \mathbf{Z}G)$ bimodule.) Conclude that $\mathbf{Z}G$-modules can, at our pleasure, be considered as either left or right modules. It follows that if A and B are $\mathbf{Z}G$-modules, we can always arrange matters so that $A \otimes_{\mathbf{Z}G} B$ makes sense.

5.3. Give an example in which G is abelian (so that $\mathbf{Z}G$ is commutative) and a $\mathbf{Z}G$-module A in which $xa \neq ax$ (i.e., the left and right structures do not coincide). [If R is a commutative ring, one usually makes a left R-module A into a right module by defining $ar = ra$; if A happened to be a right R-module, however, there is no reason to expect that these two right actions of R should be the same.]

5.4. The integral group ring functor is adjoint to the unit group functor: there is a natural bijection

$$\operatorname{Hom}_{\mathbf{Rings}}(\mathbf{Z}G, R) \cong \operatorname{Hom}_{\mathbf{Groups}}(G, U(R)),$$

where G is a group, R a ring, and $U(R)$ the group of units ($=$ invertible elements) in R.

This last exercise is the formal reason A is a left $\mathbf{Z}G$-module in Theorem 5.2, for $\operatorname{Aut}(A)$ is the group of units in $\operatorname{End}(A)$.

5.5.   If G and H are groups, then there is a ring isomorphism

$$\mathbf{Z}(G \times H) \cong \mathbf{Z}G \otimes_{\mathbf{Z}} \mathbf{Z}H.$$

Definition: A left $\mathbf{Z}G$-module A is **trivial** if $xa = a$ for all $x \in G$ and all $a \in A$.

Exercise: 5.6.   The following are equivalent for the left $\mathbf{Z}G$-module A, where $0 \to A \to E \to G \to 1$:

   (i) A is trivial;
   (ii) A is contained in the center of E;
   (iii) $\theta : G \to \mathrm{Aut}(A)$ is trivial, i.e., $\theta_x = 1$ for all $x \in G$.

Suppose A is a left $\mathbf{Z}G$-module, and $0 \to A \to E \to G \to 1$ is an extension. It may happen that the extension equips A with a left $\mathbf{Z}G$-module structure distinct from its original structure.

Definition: A left $\mathbf{Z}G$-module A **realizes the operators** of an extension $0 \to A \to E \to G \to 1$ if the two left $\mathbf{Z}G$-actions on A coincide:

$$xa = \theta_x(a) = \lambda x + a - \lambda x, \qquad \text{all} \quad a \in A, \quad x \in G.$$

The **extension problem** is: Given a left $\mathbf{Z}G$-module A, determine all extensions of A by G that realize the operators.

From now on, we shall say "G-module" instead of left $\mathbf{Z}G$-module.

Definition: An extension $0 \to A \to E \xrightarrow{\pi} G \to 1$ is **split** if there is a homomorphism $\lambda : G \to E$ with $\pi\lambda = 1_G$; the middle group E is then called a **semidirect product**.

Theorem 5.3:   *An extension $0 \to A \to E \xrightarrow{\pi} G \to 1$ is split if and only if E contains a subgroup $C \cong G$ (not necessarily normal) with $A + C = E$ and $A \cap C = 0$.*

Remark: Such a subgroup C is called a **complement** of A in E.

Proof: If the extension splits, then define $C = \mathrm{im}\,\lambda$. Since $\pi\lambda = 1_G$, λ is monic, so it is an isomorphism between G and C. Now $\mathrm{im}\,\lambda$ is a **transversal** of A in E (i.e., a complete set of coset representatives), so that $A + C = E$. Finally, if $a \in A$, then $\pi a = 1$; if $a \in C$, then $a = \lambda x$ for some $x \in G$, and $\pi a = \pi\lambda x = x$. Therefore if $a \in A \cap C$, then $a = \lambda 1 = 0$.

   Conversely, each $e \in E$ can be written $e = a + c$, $a \in A$, $c \in C$; moreover, this expression is unique because $A \cap C = 0$. Note that $\pi | C : C \to G$ is an isomorphism, and define $\lambda : G \to E$ as $(\pi | C)^{-1}$.   ∎

Definition: Let $0 \to A \to E \xrightarrow{\pi} G \to 1$ be an extension. A **lifting** is a function $\lambda: G \to E$ with $\pi\lambda = 1_G$ and $\lambda 1 = 0$.

Exercises: 5.7. An extension is split if and only if it has a lifting that is a homomorphism.

5.8. Show there is a split exact sequence

$$0 \to \mathbf{Z}/3\mathbf{Z} \to S_3 \to \mathbf{Z}/2\mathbf{Z} \to 1,$$

where S_3 is the symmetric group on three letters. Show there is no projection of S_3 onto $\mathbf{Z}/3\mathbf{Z}$ (compare Exercise 2.22).

5.9. If an extension $0 \to A \to E \to G \to 1$ is split and if A is G-trivial, then E is isomorphic to the direct product of A and G.

Theorem 5.4: *An extension $0 \to A \to E \to G \to 1$ is split if and only if*

(i) there is a group E' whose elements are all ordered pairs $(a, x) \in A \times G$ with addition given by

$$(a, x) + (a', y) = (a + xa', xy);$$

(ii) there is a lifting λ so that the function $\varphi: E \to E'$ defined by $a + \lambda x \mapsto (a, x)$ is an isomorphism.

Proof: Suppose the extension splits, and $\lambda: G \to E$ is a lifting that is a homomorphism. As in Theorem 5.3, every element of E has a unique expression of the form $a + \lambda x$. We compute addition:

$$\begin{aligned} (a + \lambda x) + (a' + \lambda y) &= a + \lambda x + a' - \lambda x + \lambda x + \lambda y \\ &= a + (\lambda x + a' - \lambda x) + \lambda xy \\ &= a + xa' + \lambda xy. \end{aligned}$$

It follows that E' is a group and $\varphi: E \to E'$ is an isomorphism.

Conversely, define a function $\lambda_1: G \to E'$ by $x \mapsto (0, x)$, and note λ_1 is a homomorphism. But $\lambda_1 = \varphi\lambda$, so that $\lambda = \varphi^{-1}\lambda_1$ is a homomorphism and the extension splits. ∎

We now solve the extension problem in the sense that we construct all possible addition tables for "middle groups" E. Suppose an extension $0 \to A \to E \to G \to 1$ is given. If $\lambda: G \to E$ is a lifting (λ need not be a homomorphism!), then im λ is a transversal of A in E. Since $\lambda 1 = 0$, every element of E can be uniquely expressed as $a + \lambda x$. Both $\lambda(xy)$ and $\lambda x + \lambda y$ represent the same coset of A, so we have

(∗) $$\lambda x + \lambda y = [x, y] + \lambda(xy)$$

for some element $[x, y] \in A$.

Definition: The function $[\ ]:G \times G \to A$ defined by formula (∗) is called a **factor set**.

Further, A realizes the operators (there is no other G-structure in sight), so there is a second formula

(∗∗) $xa + \lambda x = \lambda x + a.$

Theorem 5.5: *Let A be a G-module. A function $[\ ]:G \times G \to A$ is a factor set if and only if*

(i) $[x, 1] = 0 = [1, x];$
(ii) $x[y, z] - [xy, z] + [x, yz] - [x, y] = 0,$

for all $x, y, z \in G$.

Proof: Suppose $[\ ]$ is a factor set. The first equation follows from (∗) and the assumption $\lambda 1 = 0$; the second follows from the associative law in E.

For the converse, we must construct an extension of A by G (in which A realizes the operators), and we must then choose a lifting λ so that $[\ ]$ satisfies (∗). Let E be the set of all ordered pairs $(a, x) \in A \times G$ with addition defined by

$$(a, x) + (a', y) = (a + xa' + [x, y], xy).$$

(Compare this with the addition in the semidirect product.) Formula (ii) gives associativity, the identity is $(0, 1)$, and

$$-(a, x) = (-x^{-1}a - x^{-1}[x, x^{-1}], x^{-1}),$$

so that E is a group. Define $\pi:E \to G$ by $(a, x) \mapsto x$; clearly π is onto with kernel $\{(a, 1):a \in A\}$ that we identify with A. Thus $0 \to A \to E \overset{\pi}{\to} G \to 1$ is an extension. If we define $\lambda x = (0, x)$, then it is easy to check that A does realize the operators, viz, formula (∗∗) holds, and $[\ ]$ satisfies formula (∗). ∎

Definition: $Z^2(G, A)$ is the abelian group of all factor sets under pointwise addition.

Observe that the zero of $Z^2(G, A)$ is the factor set that is identically zero, and it corresponds to the semidirect product.

The definition of factor set arose from considering an extension and it depends on a choice of lifting λ. Changing λ should still give the same extension.

Theorem 5.6: *Let $0 \to A \to E \to G \to 1$ be an extension, and let λ and $\lambda':G \to E$ be liftings. If $[\ ]$ and $(\ )$ are the corresponding factor sets, then there* *is a function* $\langle\ \rangle:G \to A$ *satisfying*

(i) $\langle 1 \rangle = 0$;
(ii) $(x, y) - [x, y] = x\langle y \rangle - \langle xy \rangle + \langle x \rangle$ *for all* $x, y \in G$.

Proof: For any $x \in G$, both λx and $\lambda' x$ lie in the same coset of A, so there is an element $\langle x \rangle \in A$ with

$$\langle x \rangle = \lambda' x - \lambda x.$$

Since $\lambda 1 = 0 = \lambda' 1$, we have $\langle 1 \rangle = 0$. The main formula is derived as follows:

$$
\begin{aligned}
\lambda' x + \lambda' y &= \langle x \rangle + \lambda x + \langle y \rangle + \lambda y \\
&= \langle x \rangle + x\langle y \rangle + \lambda x + \lambda y \quad \text{(A realizes the operators)} \\
&= \langle x \rangle + x\langle y \rangle + [x, y] + \lambda xy \\
&= \{\langle x \rangle + x\langle y \rangle + [x, y] - \langle xy \rangle\} + \lambda' xy.
\end{aligned}
$$

It follows that $(x, y) = \langle x \rangle + x\langle y \rangle + [x, y] - \langle xy \rangle$, and this is formula (ii) since each term lies in the abelian group A.   ∎

Definition: $B^2(G, A)$ is the set of all functions $f; G \times G \to A$ for which there is a function $\langle\ \rangle:G \to A$ with $\langle 1 \rangle = 0$ such that

$$f(x, y) = x\langle y \rangle - \langle xy \rangle + \langle x \rangle.$$

The elements of $B^2(G, A)$ are called **coboundaries**.

Theorem 5.7: $B^2(G, A)$ *is a subgroup of* $Z^2(G, A)$.

Proof: A computation is again left to the reader. In particular, one shows that coboundaries satisfy the two properties of Theorem 5.5 and hence are factor sets.   ∎

Definition: $e(G, A) = Z^2(G, A)/B^2(G, A)$.

We summarize our results.

Corollary 5.8:   *Two factor sets arising from an extension via two choices of liftings determine the same element of* $e(G, A)$.

We have been led to the following equivalence relation.

Definition: Two extensions $0 \to A \to E \to G \to 1$ and $0 \to A \to E' \to G \to 1$ are **equivalent** if there are factor sets $[\ ]$ and $(\ )$ of each with $[\ ] - (\ ) \in B^2(G, A)$.

A semantic point: equivalence is a relation among extensions and not merely among factor sets: if [ ]' and ( )' are two other factor sets of these extensions, then [ ]' − ( )' ∈ $B^2(G, A)$ also, as the reader should check.

Theorem 5.9: *Two extensions of A by G are equivalent if and only if they fit in a commutative diagram*[1]:

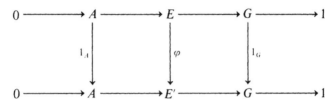

Proof: If the extensions are equivalent, then the construction of Theorem 5.5 allows us to regard E as all $a + \lambda x$ with addition

$$(a + \lambda x) + (a' + \lambda y) = a + xa' + [x, y] + \lambda xy.$$

Similarly, we may regard E' as all $a + \lambda'x$ (denote the factor set here by (x, y)). We are told that

$$[x, y] - (x, y) = x\langle y \rangle - \langle xy \rangle + \langle x \rangle.$$

Define $\varphi : E \to E'$ by $a + \lambda x \mapsto a + \langle x \rangle + \lambda'x$. It is obvious the diagram commutes, and a dull computation shows that φ is a homomorphism.

Conversely, suppose $\lambda : G \to E$ is a lifting determining a factor set [ ]. By commutativity of the first square, $[x, y] \in A$ gives $\varphi[x, y] = [x, y]$, so that $\varphi[ ] = [ ]$. By commutativity of the second square, $\varphi\lambda : G \to E'$ is also a lifting, and, applying φ to the defining equation

$$\lambda x + \lambda y = [x, y] + \lambda xy$$

shows that $\varphi[ ]$ is the factor set determined by $\varphi\lambda$. If ( ) is any factor set of E', however, we know that ( ) is equivalent to $\varphi[ ]$, by Corollary 5.8. Since $\varphi[ ] = [ ]$, the two extensions are equivalent.   ∎

Remark: Diagram chasing shows that φ must be an isomorphism.

Although the equivalence relation is defined in a natural manner, it can happen that there are inequivalent extensions of A by G with isomorphic middle groups. If p is a prime, we shall see later that $e(\mathbf{Z}/p\mathbf{Z}, \mathbf{Z}/p\mathbf{Z}) \cong \mathbf{Z}/p\mathbf{Z}$. If $0 \to \mathbf{Z}/p\mathbf{Z} \to E \to \mathbf{Z}/p\mathbf{Z} \to 1$ is an extension, then E has order p^2, hence is abelian [Rotman, 1973, p. 84]. It follows easily (one ought not invoke the

---

[1] One usually defines equivalence of extensions in terms of this diagram; we have chosen to give the definition in terms of factor sets because it arises directly from calculations.

Fundamental Theorem of Finite Abelian Groups) that $E \cong \mathbf{Z}/p^2\mathbf{Z}$ or $E \cong \mathbf{Z}/p\mathbf{Z} \oplus \mathbf{Z}/p\mathbf{Z}$. There are thus only two middle groups E, but there are p equivalence classes of extensions. Here is an explicit construction of two such inequivalent extensions. Let p be an odd prime, let A be cyclic of order p with generator a, and let E be cyclic of order p^2 with generator x. Define $i: A \to E$ by $ia = px$; define $j: A \to E$ by $ja = 2px$. Suppose these definitions lead to equivalent extensions, i.e. there is a commutative diagram

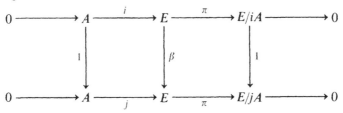

Note that $iA = jA$, so we may take the same map $\pi: E \to E/iA$ in both extensions. Now commutativity of the first square shows that β is multiplication by 2, but this is not compatible with commutativity of the second square.

Corollary 5.10:   *If $e(G, A) = 0$, then every extension*

$$0 \to A \to E \to G \to 1$$

is split, and E is a semidirect product.

Proof: By Corollary 5.8, any given extension is equivalent to the "obvious" extension $0 \to A \to E' \to G \to 1$ in which E' is a semidirect product. By Theorem 5.9, there is an isomorphism $\varphi: E \to E'$ showing that $E \cong E'$ and that the given extension is split.   ∎

The Schur–Zassenhaus lemma (Theorem 10.27) gives a condition guaranteeing that $e(G, A) = 0$: if G and A are finite groups whose orders are relatively prime. The theorem goes on to say that if C and C' are complements of A in E, then C and C' are conjugate. Let us examine this.

Definition: A **derivation** (or **crossed homomorphism**) is a function $\langle \ \rangle: G \to A$ with

$$\langle xy \rangle = x \langle y \rangle + \langle x \rangle.$$

The set of all derivations, $\text{Der}(G, A)$, is an abelian group under pointwise addition.

If A is a trivial G-module, then $\text{Der}(G, A) = \text{Hom}(G, A)$.

Exercise: 5.10.   Let A be a semidirect product of A by G, and let $\lambda\colon G \to E$ be a lifting.

   (i) $\lambda x = (\langle x \rangle, x)$ for some $\langle x \rangle \in A$;
  (ii) λ is a homomorphism if and only if $\langle\ \rangle$ is a derivation.

Definition: An automorphism φ of E **stabilizes** an extension $0 \to A \to E \to G \to 1$ if the following diagram commutes:

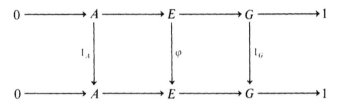

Exercise: 5.11.   Given an extension $0 \to A \to E \to G \to 1$, the set of all stabilizing automorphisms of E, denoted $s(E)$, is a subgroup of $\operatorname{Aut}(E)$. (Of course, $s(E)$ depends on the choice of extension having middle group E.)

Theorem 5.11:   *Let* $0 \to A \to E \to G \to 1$ *be an extension. There is an isomorphism* $s(E) \cong \operatorname{Der}(G, A)$.

Proof: Let $\lambda\colon G \to E$ be a lifting. Then E consists of all $a + \lambda x$, so that if $\varphi \in \operatorname{Aut}(E)$ is stabilizing, then $\varphi(a + \lambda x) = a + \langle x \rangle + \lambda x$ for some $\langle\ \rangle\colon G \to A$. A short computation (using the fact that φ is a homomorphism) shows that $\langle\ \rangle$ is a derivation.

We claim that the derivation $\langle\ \rangle$ does not depend on the choice of lifting λ. If $\lambda'\colon G \to E$ is another lifting, then $\varphi(a + \lambda'x) = a + [x] + \lambda'x$, and $[x] = \varphi\lambda'x - \lambda'x \in A$. Therefore,

$$-\langle x \rangle + [x] = \lambda x - \varphi\lambda x + \varphi\lambda'x - \lambda'x$$
$$= \lambda x + \varphi(-\lambda x + \lambda'x) - \lambda'x = 0$$

since $-\lambda x + \lambda'x \in A$, hence is fixed by φ. The function $\xi\colon\varphi \mapsto \langle\ \rangle$ is thus well defined, and is easily seen to be a homomorphism.

To construct a function inverse to ξ, associate to $\langle\ \rangle \in \operatorname{Der}(G, A)$ the map $\varphi\colon E \to E$ defined by $\varphi(a + \lambda x) = a + \langle x \rangle + \lambda x$. One checks easily that φ is a stabilizing automorphism and that φ is independent of the choice of λ. ∎

Recall that an automorphism of a group E is **inner** if it is conjugation by some element of E.

Theorem 5.12: *Let $0 \to A \to E \to G \to 1$ be a split extension, and let $\lambda: G \to E$ be a lifting that is a homomorphism. A function $\varphi: E \to E$ is an inner stabilizing automorphism if and only if*

$$\varphi(a + \lambda x) = a + a_0 - xa_0 + \lambda x$$

for some fixed $a_0 \in A$.

Proof: Suppose φ is as displayed. Then φ is conjugation by $-a_0$, for

$$-a_0 + a + \lambda x + a_0 = -a_0 + a + xa_0 + \lambda x.$$

It is clear that φ is stabilizing.

Conversely, suppose φ is stabilizing and is conjugation by $a_0 + \lambda x_0$. Then

$$
\begin{aligned}
\varphi(a + \lambda x) &= (a_0 + \lambda x_0) + a + \lambda x - (a_0 + \lambda x_0) \\
&= a_0 + x_0 a + \lambda x_0 + \lambda x - \lambda x_0 - a_0 \\
&= a_0 + x_0 a + \lambda x_0 x x_0^{-1} - a_0 \\
&= a_0 + x_0 a - x_0 x x_0^{-1} a_0 + \lambda x_0 x x_0^{-1} \\
&= a + \langle x \rangle + \lambda x,
\end{aligned}
$$

since φ is stabilizing. It follows that $x_0 x x_0^{-1} = x$, so that

$$\varphi(a + \lambda x) = a_0 + x_0 a - xa_0 + \lambda x.$$

Furthermore, when $x = 1$,

$$a = \varphi(a) = a_0 + x_0 a - a_0,$$

so that $x_0 a = a$, and we have the desired formula. ∎

Definition: A **principal derivation** (or **inner derivation**) is a function $f: G \to A$ of the form

$$f(x) = xa_0 - a_0,$$

where $a_0 \in A$ is fixed.

The set of principal derivations, $\mathrm{PDer}(G, A)$, is a subgroup of $\mathrm{Der}(G, A)$.

Definition: $\mathrm{stab}(G, A) = \mathrm{Der}(G, A)/\mathrm{PDer}(G, A)$.

Theorem 5.13: *Let E be a semidirect product of A by G, and let C and C' be complements of A in E. If $\mathrm{stab}(G, A) = 0$, then C and C' are conjugate.*

Proof: Since E is a semidirect product, there are liftings λ, λ' of G that are homomorphisms and $\mathrm{im}\,\lambda = C$, $\mathrm{im}\,\lambda' = C'$. It follows that the factor sets [ ] and [ ]' they define are identically zero. Theorem 5.6 provides a function

$\langle \rangle : G \to A$ with

$$0 = [x, y] - [x, y]' = x\langle y \rangle - \langle xy \rangle + \langle x \rangle.$$

It follows that $\langle \rangle$ is a derivation. Since $\mathrm{stab}(G, A) = 0$, $\langle \rangle$ is principal, so there is an element $a_0 \in A$ with $\langle x \rangle = xa_0 - a_0$. But $\langle x \rangle = \lambda'x - \lambda x$. Therefore,

$$\lambda'x - \lambda x = xa_0 - a_0, \qquad \text{and} \qquad \lambda x = a_0 - xa_0 + \lambda'x = a_0 + \lambda'x - a_0.$$

Since $C = \mathrm{im}\,\lambda$ and $C' = \mathrm{im}\,\lambda'$, a_0 conjugates C' into C. ∎

Let us give another consequence of the vanishing of $\mathrm{stab}(G, A)$. For a field k, we consider kG-modules A; of course, each such module is a vector space over k.

Theorem 5.14: *If* $\mathrm{stab}(G, A) = 0$ *for every kG-module A, then kG is semi-simple.*

Proof: By Theorem 4.13, it suffices to show that every exact sequence of kG-modules

$$(*) \qquad\qquad 0 \to U \overset{i}{\to} V \overset{\pi}{\to} W \to 0$$

splits, i.e., there is a kG-map $\sigma : W \to V$ with $\pi\sigma = 1_W$.
   Consider the exact sequence

$$0 \to \mathrm{Hom}_k(W, U) \overset{i_*}{\to} \mathrm{Hom}_k(W, V) \overset{\pi_*}{\to} \mathrm{Hom}_k(W, W).$$

First of all, an easy computation shows that each of these Hom's is a kG-module if we define xf, for $x \in G$, by

$$xf : w \mapsto xf(x^{-1}w).$$

Furthermore, another brief calculation shows $f \in \mathrm{Hom}_k(W, V)$ is a G-map if and only if $xf = f$ for every $x \in G$.
   Since every exact sequence of vector spaces splits, there exists a k-map $s : W \to V$ with $\pi s = 1_W$. For every $x \in G$, we claim $\pi \circ (xs - s) = 0$:

$$\begin{aligned}
[\pi \circ (xs - s)]w &= \pi[(xs)w] - (\pi s)w \\
&= \pi[x(s(x^{-1}w))] - w && (\text{since } \pi s = 1_W) \\
&= x\pi s(x^{-1}w) - w && (\text{since } \pi \text{ is a } kG\text{-map}) \\
&= w - w = 0 && (\text{since } \pi s = 1_W).
\end{aligned}$$

Therefore, $xs - s \in \ker \pi_* = \mathrm{im}\, i_*$ for every $x \in G$. Define

$$\varphi : G \to \mathrm{Hom}_k(W, U)$$

by

$$\varphi(x) = i^{-1}(xs - s).$$

Using the fact that i is a kG-map, we see that φ is a derivation:

$$x\varphi(y) + \varphi(x) = xi^{-1}(ys - s) + i^{-1}(xs - s)$$
$$= i^{-1}(xys - xs + xs - s) = \varphi(xy).$$

Since $\text{stab}(G, \text{Hom}_k(W, U)) = 0$, there exists $h \in \text{Hom}_k(W, U)$ with

$$\varphi(x) = xh - h, \quad \text{all} \quad x \in G.$$

Recalling the definition of φ, we obtain

$$i^{-1}(xs - s) = xh - h, \quad \text{all} \quad x \in G,$$

and hence

$$x(s - ih) = s - ih, \quad \text{all} \quad x \in G.$$

If we define $\sigma \in \text{Hom}_k(W, V)$ by $\sigma = s - ih$, then the equation above tells us that σ is a kG-map. Also,

$$\pi\sigma = \pi(s - ih) = \pi s - \pi ih = \pi s = 1_W,$$

so that the original sequence $(*)$ of kG-modules splits. ∎

Obviously, Theorem 5.14 is related to Maschke's theorem: if G is a finite group and k is a field whose characteristic does not divide the order of G, then kG is semisimple. Indeed, these hypotheses on G and k do imply the vanishing of stab, thereby proving Maschke's theorem (see Theorem 10.28).

Let us summarize the formulas that have arisen.

Factor set: $x[y, z] - [xy, z] + [x, yz] - [x, y] = 0;$

Coboundary: $f(x, y) = x\langle y \rangle - \langle xy \rangle + \langle x \rangle;$

Derivation: $x\langle y \rangle - \langle xy \rangle + \langle x \rangle = 0;$

Principal derivation: $f(x) = xa - a.$

We can continue this one step further. Factor sets are certain functions of two variables $G \times G \to A$; derivations are certain functions of one variable $G \to A$. Next in line are functions with no variables, i.e., constants in A; let us regard a constant in A as a function from a singleton set, with unique element $[\ ]$, into A. Now the definition of free G-module (more formally, Exercise 3.2) says that the group of all functions $G \times \cdots \times G$ (n times) $\to A$ is the same as $\text{Hom}_G(F_n, A)$, where F_n is the free G-module with basis $G \times \cdots \times G$ (n times). A moment's reflection shows that we have been applying the contravariant functor $\text{Hom}_G(\ , A)$ to

$$F_3 \overset{d_3}{\to} F_2 \overset{d_2}{\to} F_1 \overset{d_1}{\to} F_0,$$

where F_0 is the free G-module on the single generator $[\ ]$, and

$$d_1[x] = (x - 1)[\ ],$$
$$d_2[x, y] = x[y] - [xy] + [x],$$
$$d_3[x, y, z] = x[y, z] - [xy, z] + [x, yz] - [x, y].$$

After all, look at the induced sequence

$$\text{Hom}_G(F_3, A) \overset{d_3^*}{\leftarrow} \text{Hom}_G(F_2, A) \overset{d_2^*}{\leftarrow} \text{Hom}_G(F_1, A) \overset{d_1^*}{\leftarrow} \text{Hom}_G(F_0, A).$$

We see that $\ker d_3^*$ consists of all factor sets (not quite, for we must also take account of the identity $[x, 1] = 0 = [1, x]$), $\text{im } d_2^*$ consists of all coboundaries, $\ker d_2^*$ consists of all derivations, and $\text{im } d_1^*$ is all principal derivations. What is $\ker d_1^*$? Note that since $F_0 \cong \mathbf{Z}G$, we have $\text{Hom}_G(F_0, A) \cong \text{Hom}_G(\mathbf{Z}G, A) \cong A$. Thus $\ker d_1^*$ is a submodule of A. Writing these isomorphisms explicitly gives

$$\ker d_1^* = A^G = \{a \in A : xa = a, \quad \text{all} \quad x \in G\}.$$

Definition: A^G is the submodule of **fixed points** of A.

Exercises: 5.12. A^G is a G-trivial submodule of A; A^G is the (unique) maximal G-trivial submodule of A.

   5.13.   "Fixed points" is a functor on G-modules: on an object A, its value is A^G; on a G-map $f : A \to B$, its value is the restriction $f | A^G$.

   5.14.   Assume E is a semidirect product of A by G (so that we may assume G is a subgroup of E). If A is a G-module as in Theorem 5.2, then A^G is the subgroup of A consisting of all $a \in A$ that centralize G, i.e.,

$$A^G = \{a \in A : [a, x] = axa^{-1}x^{-1} = 1 \text{ for all } x \in G\}.$$

   5.15.   Let B and A be G-modules. Show that $\text{Hom}_{\mathbf{Z}}(B, A)$ is a G-module if, for $x \in G$ and $f \in \text{Hom}_{\mathbf{Z}}(B, A)$, one defines

$$(xf)(b) = xf(x^{-1}b), \quad \text{all} \quad b \in B.$$

   5.16.   With the notation of Exercise 5.15, show that

$$\text{Hom}_{\mathbf{Z}}(B, A)^G = \text{Hom}_G(B, A).$$

(This is the same computation as in the proof of Theorem 5.14.)

Theorem 5.15:   *Let $\mathbf{Z}$ be the integers considered as a G-trivial module. Then the maps $t_A : \text{Hom}_G(\mathbf{Z}, A) \to A^G$ defined by $f \mapsto f(1)$ constitute a natural equivalence of functors.*

Proof: There is a list of items to verify, and each verification is routine. We merely note that $f(1) \in A^G$, for $xf(1) = f(x1)$ since f is a G-map, and $f(x1) = f(1)$ since $\mathbf{Z}$ is G-trivial. ∎

In Example 42 of Chapter 4, we defined the **augmentation map** $\varepsilon: \mathbf{Z}G \to \mathbf{Z}$ given by $\sum m_x x \mapsto \sum m_x$; the map ε is actually a ring map, and its kernel $\mathfrak{g}$ is called the **augmentation ideal**. Of course, $\mathfrak{g}$ is a two-sided ideal in $\mathbf{Z}G$.

Theorem 5.16: *If $\mathbf{Z}$ is G-trivial, then*

$$F_3 \xrightarrow{d_3} F_2 \xrightarrow{d_2} F_1 \xrightarrow{d_1} F_0 \xrightarrow{\varepsilon} \mathbf{Z} \to 0$$

is an exact sequence of G-modules.

Proof: We leave this as a rather uninspiring exercise for the reader. ∎

It is quite tempting to continue the sequence of free modules past F_3. It is even clear how: the final result should be a free resolution of $\mathbf{Z}$. Honest things do appear. The next step $F_4 \to F_3$ yields "obstructions" which tell whether certain types of extensions with nonabelian kernel exist [MacLane, 1963, pp. 124–131]. We shall return to this resolution in a later chapter. At this point, the reader has seen that group-theoretic problems (surveying extensions and conjugacy questions) suggest constructing a G-free resolution F_* of $\mathbf{Z}$ and applying the functor $\operatorname{Hom}_G(\, , A)$. Anticipating some definitions of the next chapter, the "complex" $\operatorname{Hom}_G(F_*, A)$ has zeroth cohomology group A^G, first cohomology group $\operatorname{stab}(G, A)$, and the second cohomology group is $e(G, A)$ (this last remark requires more work since the functions $[\, , \,]: G \times G \to A$ in $\operatorname{Hom}_G(F_2, A)$ need not satisfy the identity $[x, 1] = 0 = [1, x]$).

Before leaving groups, we take the topologists' hint of applying $\otimes_{\mathbf{Z}G} A$ to the sequence

$$F_3 \xrightarrow{d_3} F_2 \xrightarrow{d_2} F_1 \xrightarrow{d_1} F_0 \xrightarrow{\varepsilon} \mathbf{Z} \to 0.$$

Observe that the free modules in the sequence are left G-modules. As we remarked earlier (Exercise 5.2), we may regard them as right G-modules by defining, for $b \in F_n$, $bx = x^{-1}b$, so that the tensor products $F_n \otimes_G A$ make sense. Thus, consider

$$F_3 \otimes_G A \xrightarrow{d_3 \otimes 1} F_2 \otimes_G A \xrightarrow{d_2 \otimes 1} F_1 \otimes_G A \xrightarrow{d_1 \otimes 1} F_0 \otimes_G A \xrightarrow{\varepsilon \otimes 1} \mathbf{Z} \otimes_G A \to 0;$$

what are the kernels, images, and quotients?

Theorem 5.17: Coker $d_1 \otimes 1 \cong \mathbf{Z} \otimes_G A$.

Proof: The sequence

$$F_1 \xrightarrow{d_1} F_0 \xrightarrow{\varepsilon} \mathbf{Z} \to 0$$

is exact, so that right exactness of $\otimes_G A$ gives exactness of

$$F_1 \otimes A \xrightarrow{d_1 \otimes 1} F_0 \otimes A \to \mathbf{Z} \otimes A \to 0,$$

whence coker $d_1 \otimes 1 = F_0 \otimes A/\mathrm{im}(d_1 \otimes 1) \cong \mathbf{Z} \otimes_G A.$ ∎

Theorem 5.18: $\mathbf{Z} \otimes_G A \cong A/\mathfrak{g}A$, where $\mathfrak{g}$ is the augmentation ideal of G.

Proof: Exactness of $0 \to \mathfrak{g} \to \mathbf{Z}G \xrightarrow{\varepsilon} \mathbf{Z} \to 0$ gives exactness of

$$\mathfrak{g} \otimes_G A \to \mathbf{Z}G \otimes_G A \to \mathbf{Z} \otimes_G A \to 0.$$

Now $\mathbf{Z}G \otimes_G A \cong A$ and, under this isomorphism, $\mathrm{im}\,\mathfrak{g} \otimes A$ goes into $\mathfrak{g}A$. ∎

Theorem 5.19:   As an abelian group, $\mathfrak{g}$ is free with basis $\{x - 1 : x \in G\}$.

Proof: An element $u = \sum m_x x \in \ker \varepsilon$ if and only if $\sum m_x = 0$. Therefore $u = u - (\sum m_x)1 = \sum m_x(x - 1)$, so that $\mathfrak{g}$ is generated by the $x - 1$.

Suppose $\sum m_x(x - 1) = 0$. Then $\sum m_x x - (\sum m_x)1 = 0$ in $\mathbf{Z}G$, which, as an abelian group, is free with basis G. Hence, each $m_x = 0$. ∎

Definition: A_G is the maximal quotient module of A that is G-trivial.

Corollary 5.20: $A_G \cong A/\mathfrak{g}A \cong \mathbf{Z} \otimes_G A.$

Proof: It is clear that $A_G = A/S$, where $S = \{xa - a : x \in G, a \in A\}$. But $xa - a = (x - 1)a$, and $S = \mathfrak{g}A$. ∎

Example: Suppose E is a semidirect product of A by G. We may consider G as a subgroup of E, and we inquire about $[G, A]$, the subgroup generated by commutators $a + x - a - x$, $a \in A$, $x \in G$. Now

$$a + x - a - x = a - xa = (1 - x)a.$$

Therefore $A_G = A/[G, A]$.

Later, we will show that if $A =$ trivial $\mathbf{Z}$, then

$$\ker d_1 \otimes 1/\mathrm{im}\, d_2 \otimes 1 \cong G/G',$$

where G' is the commutator subgroup of G. The next quotient

$$\ker d_2 \otimes 1/\mathrm{im}\, d_3 \otimes 1$$

also has group-theoretic significance: it is called the **Schur multiplier**.

Again anticipating the next chapter, tensoring the G-free resolution by a module A gives a complex whose zeroth homology group is A_G; if $A = \mathbf{Z}$, then the first homology group is G/G', and the second homology group is the Schur multiplier (we shall discuss this in more detail in Chapter 10).

Before we leave groups, let us observe that the hybrid Der(G, A) (G is a group and A is a G-module) may be viewed as an ordinary Hom between two G-modules.

Theorem 5.21: *The maps $t_A: \mathrm{Hom}_G(\mathfrak{g}, A) \to \mathrm{Der}(G, A)$ defined by $f \mapsto \langle \; \rangle$, where $\langle x \rangle = f(x - 1)$, constitute a natural equivalence of functors*

$$\mathrm{Hom}_G(\mathfrak{g}, \;) \cong \mathrm{Der}(G, \;).$$

Proof: One checks easily that $\langle \; \rangle$ defined in the statement is a derivation, so that t_A is a well-defined homomorphism. We construct its inverse. If $\langle \; \rangle \in \mathrm{Der}(G, A)$, define a function f by $f(x - 1) = \langle x \rangle \in A$. Since $\mathfrak{g}$ is a free abelian group with basis $\{x - 1 : x \in G\}$, f can be extended to a **Z**-homomorphism $\tilde{f}: \mathfrak{g} \to A$. That $\tilde{f}$ is a G-map follows from $\langle \; \rangle$ being a derivation. The reader may show the isomorphisms t_A define a natural transformation, hence constitute a natural equivalence. ∎

It is not difficult to prove directly that the functor Der($G,$) preserves inverse limits, so that Theorem 3.38 of Watts asserts the existence of some G-module B with $\mathrm{Hom}_G(B, \;) \cong \mathrm{Der}(G, \;)$. However, that general theorem does not identify B as the augmentation ideal $\mathfrak{g}$. A similar remark can be made about Theorem 5.15.

6 Homology

In this chapter we give the proper context in which to view the group-theoretic constructions of the previous chapter. Chapter 1 gives a rapid description of homology as it arises in algebraic topology; Chapter 4 illustrates the interplay between constraints on a ring R and the behavior of special R-modules (projectives, injectives, and flats). The basic idea now is to study the interplay between constraints on a ring R and the behavior of arbitrary R-modules; this is done by replacing each R-module by a resolution of it comprised of special R-modules. Happily, this idea can be made to work using the same elementary homological constructions occurring in algebraic topology.

HOMOLOGY FUNCTORS

Throughout this chapter, "module" means "left R-module", where R is a ring fixed once for all, and "map" means "R-map". Also, all functors are additive.

Definition: A **complex** (or **chain complex**) **A** is a sequence of modules and maps

$$\mathbf{A} = \cdots \to A_{n+1} \xrightarrow{d_{n+1}} A_n \xrightarrow{d_n} A_{n-1} \to \cdots, \qquad n \in \mathbf{Z},$$

with $d_n d_{n+1} = 0$ for all n. The maps d_n are called **differentiations**. If it is necessary to display the differentiations, we will write $(\mathbf{A}, d)$ instead of $\mathbf{A}$.

Note that the condition $d_n d_{n+1} = 0$ is equivalent to

$$\operatorname{im} d_{n+1} \subset \ker d_n.$$

Examples: 1. Every exact sequence is a complex, for the required inclusions $\operatorname{im} d_{n+1} \subset \ker d_n$, all n, are even equalities $\operatorname{im} d_{n+1} = \ker d_n$.

2. If X is a topological space, then

$$\mathbf{S}(X) = \cdots \to S_n(X) \overset{\partial_n}{\to} S_{n-1}(X) \to \cdots$$

constructed in Chapter 1 is a complex; $\mathbf{S}(X)$ is called the **singular complex** of X.

Recall that the singular complex has its terms $S_n(X)$ defined only for $n \geq -1$. Since the definition of complex requires a module for each index $n \in \mathbf{Z}$, one defines $S_n(X) = 0$ for all negative n (this forces $\partial_n = 0$ for all negative n). Of course, this device of addings 0's is always available.

3. Let A be a module and $k \in \mathbf{Z}$ be a fixed integer. If we regard A as the kth term and all other terms 0, then we have a complex **concentrated in degree** k.

4. If A is a module, every projective resolution $\mathbf{P}$ of A

$$\mathbf{P} = \cdots \to P_1 \to P_0 \to A \to 0$$

is a complex (add necessary 0's to the right).

5. If A is a module, every injective resolution $\mathbf{E}$ of A

$$\mathbf{E} = 0 \to A \to E^0 \to E^1 \to E^2 \to \cdots$$

is a complex (add necessary 0's to the left). We have used a convenient notation. The indices of a complex must decrease as one goes to the right; this is easily arranged by changing the sign of the index, and has been indicated by raising the index. One would conform to the original definition if he wrote

$$\mathbf{E} = 0 \to A \to E_0 \to E_{-1} \to E_{-2} \to \cdots.$$

The main reason one defines a complex so that the indices range over all $n \in \mathbf{Z}$ is that "positive" (e.g., projective resolutions) and "negative" (e.g., injective resolutions) complexes may be treated simultaneously. There are, however, some interesting complexes ("complete resolutions" arising in the homology of finite groups) which are doubly infinite [Cartan–Eilenberg, 1956, Chapter XII].

6. If **A** is a complex and F is a functor, then

$$F(\mathbf{A}) = \cdots \to F(A_n) \xrightarrow{Fd_n} F(A_{n-1}) \to \cdots$$

is also a complex. In particular, if **A** is an exact sequence, then $F(\mathbf{A})$ is a complex (which is usually not exact).

7. A similar construction is available for a complex **A** and a contravariant functor F. Reversing the direction of arrows causes two minor notational problems:

$$F(\mathbf{A}) = \cdots \to F(A_{n-1}) \xrightarrow{Fd_n} F(A_n) \to \cdots.$$

The problem of increasing indices is solved as in Example 5: change sign and raise. Thus, define $B^{-n} = F(A_n)$ and the sequence becomes

$$F(\mathbf{A}) = \cdots \to B^{-n+1} \xrightarrow{Fd_n} B^{-n} \to \cdots.$$

The second problem is that the map $B^{-n+1} \to B^{-n}$ should have index $-n + 1$. Define

$$\Delta^{-n+1} = Fd_n$$

(change sign and add 1). With these conventions, $F(\mathbf{A})$ is a complex

$$\begin{aligned} F(\mathbf{A}) &= \cdots \to F(A_{n-1}) \xrightarrow{Fd_n} F(A_n) \to \cdots \\ &= \cdots \to B^{-n+1} \xrightarrow{\Delta^{-n+1}} B^{-n} \to \cdots. \end{aligned}$$

Definition: If **A** and **A**$'$ are complexes, a **chain map** $f : \mathbf{A} \to \mathbf{A}'$ is a sequence of maps $f_n : A_n \to A'_n$, all $n \in \mathbf{Z}$, such that the following diagram commutes:

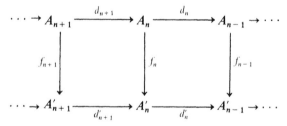

Exercises: 6.1. Construe the ordered set **Z** as a category with objects the integers $n \in \mathbf{Z}$ and exactly one morphism $n \to n + 1$ for each n (this is a special case of Chapter 1, Example 7). Prove that a complex **A** is a functor $\mathbf{Z} \to {}_R\mathfrak{M}$, and that a chain map $f : \mathbf{A} \to \mathbf{A}'$ is a natural transformation.

6.2. Define **Comp** (or R-**Comp**) as the class of all complexes and chain maps (with the obvious composition). Prove that **Comp** is a pre-additive category (the Hom's are abelian groups and distributivity laws hold).

In exercises below, it will be seen that almost every construction available in $_R\mathfrak{M}$ is also available in **Comp**. Since we are more interested in homology than in complexes, however, let us first give the important definition.

Definition: If $(\mathbf{A}, d)$ is a complex, its **nth homology module** is

$$H_n(\mathbf{A}) = \ker d_n/\operatorname{im} d_{n+1}.$$

We observed earlier that $d_n d_{n+1} = 0$ means $\operatorname{im} d_{n+1} \subset \ker d_n$, and so the quotient module does make sense. Every ingredient has a name; see Chapter 1 for the etymology.

Definition: The elements of A_n are called n-**chains,** the elements of $\ker d_n$ are called n-**cycles**, and the elements of $\operatorname{im} d_{n+1}$ are called n-**boundaries**. One writes

$$\ker d_n = Z_n(\mathbf{A}) = Z_n,$$
$$\operatorname{im} d_{n+1} = B_n(\mathbf{A}) = B_n,$$

and thus

$$H_n(\mathbf{A}) = Z_n(\mathbf{A})/B_n(\mathbf{A}).$$

As H_n is really a functor, its definition is still incomplete, for we must define its action on morphisms ($=$ chain maps).

Definition: If $f: \mathbf{A} \to \mathbf{A}'$ is a chain map, define

$$H_n(f): H_n(\mathbf{A}) \to H_n(\mathbf{A}')$$

by

$$z_n + B_n(\mathbf{A}) \mapsto f_n z_n + B_n(\mathbf{A}').$$

$H_n(f)$ is called the map **induced** by f, and it is usually denoted f_* (the subscript n being suppressed).

Theorem 6.1:   *For each n, $H_n: \mathbf{Comp} \to {}_R\mathfrak{M}$ is an additive functor.*

Proof: We only show that f_* is well defined; the other verifications are also routine and are left to the reader.

Assume $f:(\mathbf{A}, d) \to (\mathbf{A}', d')$ is a chain map; thus (omitting indices), $fd = d'f$. If z_n is an n-cycle, then $dz_n = 0$; it follows that $d'fz_n = f\,dz_n = 0$ and so fz_n is also an n-cycle. If $b_n \in B_n(\mathbf{A})$, then $b_n = da$ for some a; hence $fb_n = f\,da = d'fa \in B_n(\mathbf{A}')$. As f_* preserves cycles and boundaries, it follows easily that the formula for f_* does give a well-defined map.   ∎

Exercises: 6.3. A complex $\mathbf{A}$ is an exact sequence if and only if $H_n(\mathbf{A}) = 0$ for every $n \in \mathbf{Z}$. (For this reason, exact sequences are often called **acyclic** complexes.)

6.4. Let $T:_R\mathfrak{M} \to _R\mathfrak{M}$ be an exact (covariant) functor. For each $n \in \mathbf{Z}$ and every complex $\mathbf{A}$ of R-modules, prove that $H_n(T\mathbf{A}) \cong TH_n(\mathbf{A})$ (where $T\mathbf{A}$ is defined as in Example 6). If T is contravariant exact, prove that $H_{-n}(T\mathbf{A}) \cong TH_n(\mathbf{A})$ (where $T\mathbf{A}$ is defined as in Example 7).

6.5. Let $(\mathbf{A}, d)$ be a complex with **zero differentiation**, i.e., $d_n = 0$ for all n. Prove that $H_n(\mathbf{A}) \cong A_n$ for all n.

6.6. A chain map $f:\mathbf{A} \to \mathbf{A}'$ is an isomorphism in **Comp** if and only if each f_n is an isomorphism.

6.7. If $\{\mathbf{A}^k : k \in K\}$ is a family of complexes, where

$$\mathbf{A}^k = \cdots \to A_n^k \xrightarrow{d_n^k} A_{n-1}^k \to \cdots,$$

then the **sum** $\coprod \mathbf{A}^k$ is defined as the complex

$$\coprod \mathbf{A}^k = \cdots \to \coprod_k A_n^k \xrightarrow{\coprod d_n^k} \coprod_k A_{n-1}^k \to \cdots.$$

Prove that $H_n(\coprod \mathbf{A}^k) \cong \coprod_k H_n(\mathbf{A}^k)$ for every $n \in \mathbf{Z}$.

6.8. If K is a quasi-ordered set, define the direct limit of complexes, $\varinjlim \mathbf{A}^k$. If K is directed, prove that

$$H_n(\varinjlim \mathbf{A}^k) \cong \varinjlim H_n(\mathbf{A}^k), \qquad \text{all} \quad n \in \mathbf{Z}.$$

(Hint: Use Theorem 2.18.)

We remark that Exercise 6.8 may be false when the index set K is not directed; for example, the functors H_n need not be right exact.

Definition: A complex $(\mathbf{A}, d)$ is a **subcomplex** of $(\mathbf{A}', d')$ if each A_n is a submodule of A_n' and $d_n = d_n' | A_n$ for all n. In this case, there is a **quotient complex**

$$\mathbf{A}'/\mathbf{A} = \cdots \to A_n'/A_n \xrightarrow{\bar{d}_n'} A_{n-1}'/A_{n-1} \to \cdots,$$

where $\bar{d}_n' : a_n' + A_n \mapsto d_n' a_n' + A_{n-1}$.

One should observe that $\mathbf{A}$ is a subcomplex of $\mathbf{A}'$ precisely when the inclusions $i_n : A_n \to A_n'$ constitute a chain map. Equivalently, $\mathbf{A}$ is a subcomplex of $\mathbf{A}'$ if it is a sequence of submodules A_n of A_n' with $d_n'(A_n) \subset A_{n-1}$, all n.

Exercises: 6.9. If $f:\mathbf{A} \to \mathbf{A}'$ is a chain map, define complexes **ker** f, **im** f, and **coker** f in the obvious way: for example,

$$\ker f = \cdots \to \ker f_n \xrightarrow{d_n} \ker f_{n-1} \to \cdots,$$

where d_n is the restriction of the differentiation $A_n \to A_{n-1}$. Prove that $A/\ker f \cong \operatorname{im} f$.

6.10.   Define $\mathbf{A}' \xrightarrow{f} \mathbf{A} \xrightarrow{g} \mathbf{A}''$ to be **exact** if $\operatorname{im} f = \ker g$. Prove this sequence of complexes is exact if and only if the sequences of modules

$$A'_n \xrightarrow{f_n} A_n \xrightarrow{g_n} A''_n$$

are exact for every $n \in \mathbf{Z}$.

At this point, the reader should be comfortable with complexes, and he should regard them as generalized modules. The next natural question is what homology functors do to short exact sequences of complexes. They are, in general, neither left nor right exact (we shall see they are **half exact**: if $0 \to \mathbf{A}' \to \mathbf{A} \to \mathbf{A}'' \to 0$ is exact, then $H_n(\mathbf{A}') \to H_n(\mathbf{A}) \to H_n(\mathbf{A}'')$ is exact).

The proof of the next lemma is a routine diagram-chase, but we give the details because of the importance of the result.

Theorem 6.2   (Connecting Homomorphism):   *Let* $0 \to \mathbf{A}' \xrightarrow{i} \mathbf{A} \xrightarrow{p} \mathbf{A}'' \to 0$ *be an exact sequence of complexes. For each n, there is a homomorphism*

$$\partial_n : H_n(\mathbf{A}'') \to H_{n-1}(\mathbf{A}')$$

defined by

$$z'' + B_n(\mathbf{A}'') \mapsto i_{n-1}^{-1} d_n p_n^{-1} z'' + B_{n-1}(\mathbf{A}').$$

Proof: Consider the commutative diagram with exact rows:

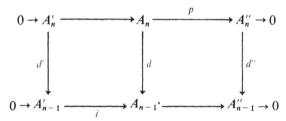

Suppose $z'' \in A''_n$ and $d''z'' = 0$. Since p is epic, we may lift z'' to $a_n \in A_n$ and then push down to $da_n \in A_{n-1}$. By commutativity,

$$da_n \in \ker(A_{n-1} \to A''_{n-1}) = \operatorname{im} i.$$

It follows that $i^{-1} da_n$ makes sense, i.e., there is a unique (i is monic) $a'_{n-1} \in A'_{n-1}$ with $ia'_{n-1} = da_n$.

Suppose we had lifted z'' to $\bar{a}_n \in A_n$. Then the construction above yields $\bar{a}'_{n-1} \in A'_{n-1}$ with $i\bar{a}'_{n-1} = d\bar{a}_n$. We also know

$$a_n - \bar{a}_n \in \ker p = \operatorname{im}(A'_n \to A_n),$$

so there is $x'_n \in A'_n$ with $a'_{n-1} - \bar{a}'_{n-1} = d'x'_n \in B_{n-1}(\mathbf{A}')$. There is thus a well defined homomorphism

$$Z_n(\mathbf{A}'') \to A'_{n-1}/B_{n-1}(\mathbf{A}').$$

It is easy to check that this map sends $B_n(\mathbf{A}'')$ into 0 and also that $i^{-1} dp^{-1} z'' = a'_{n-1}$ is a cycle. Therefore the formula does give a map $H_n(\mathbf{A}'') \to H_{n-1}(\mathbf{A}')$, as desired. $\blacksquare$

Definition: The maps $\partial_n : H_n(\mathbf{A}'') \to H_{n-1}(\mathbf{A}')$ are called **connecting homomorphisms**.

Theorem 6.3 (Long Exact Sequence): *If* $0 \to \mathbf{A}' \xrightarrow{i} \mathbf{A} \xrightarrow{p} \mathbf{A}'' \to 0$ *is an exact sequence of complexes, then there is an exact sequence of modules*

$$\cdots \to H_n(\mathbf{A}') \xrightarrow{i_*} H_n(\mathbf{A}) \xrightarrow{p_*} H_n(\mathbf{A}'') \xrightarrow{\partial} H_{n-1}(\mathbf{A}') \xrightarrow{i_*} H_{n-1}(\mathbf{A}) \to \cdots.$$

Proof: Again the argument is routine and again we supply details. There are six inclusions to verify; the notation is self-explanatory and we omit subscripts.

(1) $\operatorname{im} i_* \subset \ker p_*$.

$$p_* i_* = (pi)_* = 0_* = 0.$$

(2) $\ker p_* \subset \operatorname{im} i_*$.

If $p_*(z + B) = pz + B'' = B''$, then $pz = d''a''$. But p epic gives $a'' = pa$ for some a, so that $pz = d''pa = p\,da$, and $p(z - da) = 0$. By exactness, there is a' with $ia' = z - da$. Note that $a' \in Z'$, for $i\,d'a' = dia' = dz - d\,da = 0$ (z is a cycle). Since i is monic, $d'a' = 0$. Therefore

$$i_*(a' + B') = ia' + B = z - da + B = z + B.$$

(3) $\operatorname{im} p_* \subset \ker \partial$.

$$\partial p_*(z + B) = \partial(pz + B'') = x' + B',$$

where $ix' = dp^{-1} pz = dz = 0$ (this computation is correct modulo B'' since ∂ is independent of all choices). Since i is monic, $x' = 0$ and $\partial p_* = 0$.

(4) $\ker \partial \subset \operatorname{im} p_*$.

If $\partial(z'' + B'') = B'$, then $x' = i^{-1} dp^{-1} z'' \in B'$. Hence $x' = d'a'$. Now $ix' = i\,d'a' = dia' = dp^{-1} z''$, so that $d(p^{-1} z'' - ia') = 0$ and $p^{-1} z'' - ia' \in Z$. Therefore

$$p_*(p^{-1} z'' - ia' + B) = pp^{-1} z'' - pia' + B'' = z'' + B''.$$

(5)　$\operatorname{im} \partial \subset \ker i_*$.

$$i_* \partial(z'' + B'') = ix' + B, \qquad \text{where} \quad ix' = dp^{-1} z'' \in B.$$

(6)　$\ker i_* \subset \operatorname{im} \partial$.

Suppose $i_*(z' + B') = iz' + B = B$, so that $iz' = da$. Then $d''pa = p\,da = piz' = 0$ and $pa \in Z''$. But $\partial(pa + B'') = x' + B'$, where $ix' = dp^{-1} pa \equiv da = iz' \bmod B$. Since i is monic, $x' = z'$ and $\partial(pa + B'') = z' + B'$. $\blacksquare$

Theorem 6.3 is often called the **Exact Triangle** because of the mnemonic diagram

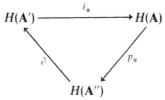

Theorem 6.4　(Naturality of ∂):　*Consider the commutative diagram of complexes with exact rows:*

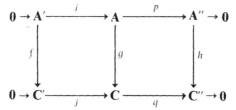

Then there is a commutative diagram of modules with exact rows:

$$\cdots \to H_n(\mathbf{A}') \xrightarrow{\ i_*\ } H_n(\mathbf{A}) \xrightarrow{\ p_*\ } H_n(\mathbf{A}'') \xrightarrow{\ \partial\ } H_{n-1}(\mathbf{A}') \to \cdots$$

$$\cdots \to H_n(\mathbf{C}') \xrightarrow{\ j_*\ } H_n(\mathbf{C}) \xrightarrow{\ q_*\ } H_n(\mathbf{C}'') \xrightarrow{\ \partial'\ } H_{n-1}(\mathbf{C}') \to \cdots$$

with vertical maps f_*, g_*, h_*, f_*.

Proof:　Exactness of the rows is Theorem 6.3. The first two squares commute because H_n is a functor. A routine but long chase gives commutativity of the square involving connecting homomorphisms. $\blacksquare$

If one were to choose several results to call Fundamental Lemmas of Homological Algebra, then he would include the last three theorems on his list. He should also include Theorems 6.5, 6.8, 6.9, and 6.20.

There is a valuable reformulation of the construction of the connecting homomorphisms, which yields another proof of the Long Exact Sequence.

Theorem 6.5 (Snake Lemma): *Consider the commutative diagram of modules with exact rows:*

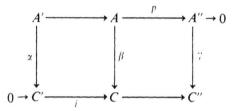

There is an exact sequence

$$\ker \alpha \to \ker \beta \to \ker \gamma \xrightarrow{\hat{c}} \operatorname{coker} \alpha \to \operatorname{coker} \beta \to \operatorname{coker} \gamma,$$

where $\partial : a'' \mapsto i^{-1}\beta p^{-1}a'' + \operatorname{im} \alpha$.

Moreover, if $A' \to A$ *is monic, then* $\ker \alpha \to \ker \beta$ *is monic, and if* $C \to C''$ *is epic, then* $\operatorname{coker} \beta \to \operatorname{coker} \gamma$ *is epic.*

Proof: The only nonobvious thing is that ∂ is well defined, and this has already been demonstrated. ∎

Exercises: 6.11.   Consider the commutative diagram of modules with exact rows (we are focusing on a portion of an exact sequence of complexes $0 \to \mathbf{A}' \to \mathbf{A} \to \mathbf{A}'' \to \mathbf{0}$):

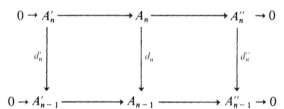

Show there is a commutative diagram with exact rows:

$$
\begin{array}{ccccccc}
A'_n/\operatorname{im} d'_{n+1} & \longrightarrow & A_n/\operatorname{im} d_{n+1} & \longrightarrow & A''_n/\operatorname{im} d''_{n+1} & \to 0 \\
\downarrow{\scriptstyle \Delta'} & & \downarrow{\scriptstyle \Delta} & & \downarrow{\scriptstyle \Delta''} & \\
0 \to Z_{n-1}(\mathbf{A}') & \longrightarrow & Z_{n-1}(\mathbf{A}) & \longrightarrow & Z_{n-1}(\mathbf{A}'') &
\end{array}
$$

where, e.g., $\Delta' : a'_n + \operatorname{im} d'_{n+1} \mapsto d'_n a'_n$.

6.12.   Give a second proof of the Long Exact Sequence by applying the Snake lemma to the diagram of Exercise 6.11.

6.13.   **(Mapping Cylinder)**   Let $f:(\mathbf{A}, d) \to (\mathbf{A}', d')$ be a chain map. For each n, define

$$M_n = A_{n-1} \oplus A'_n;$$

define $\Delta_n:M_n \to M_{n-1}$ by

$$\Delta_n:(a_{n-1}, a'_n) \mapsto (-d_{n-1}a_{n-1}, d'_n a'_n + f_{n-1}a_{n-1}).$$

Prove that $\mathbf{M}$ as just defined is a complex (it is called the **mapping cylinder** of f and is usually denoted $\mathbf{M}(f)$).

6.14.   If $\mathbf{A}$ is a complex, let $\mathbf{A}^+$ denote the complex obtained from $\mathbf{A}$ by increasing all indices by 1:

$$(\mathbf{A}^+)_n = A_{n-1}.$$

Show that $H_n(\mathbf{A}^+) = H_{n-1}(\mathbf{A})$.

6.15.   (i)   If $f:\mathbf{A} \to \mathbf{A}'$ is a chain map, there is an exact sequence

$$0 \to \mathbf{A}' \xrightarrow{i} \mathbf{M}(f) \xrightarrow{p} \mathbf{A}^+ \to 0,$$

where $i_n:A'_n \to A_{n-1} \oplus A'_n$ is given by $a'_n \mapsto (0, a'_n)$, and $p_n:A_{n-1} \oplus A'_n \to A_{n-1}$ is given by $(a_{n-1}, a'_n) \mapsto a_{n-1}$.

(ii)   The corresponding long exact sequence is

$$\cdots \to H_{n+1}(\mathbf{A}') \to H_{n+1}(\mathbf{M}(f)) \to H_n(\mathbf{A}) \xrightarrow{\partial_n} H_n(\mathbf{A}') \to H_n(\mathbf{M}(f)) \to \cdots.$$

(iii)   The connecting homomorphism $\partial_n:H_n(\mathbf{A}) \to H_n(\mathbf{A}')$ is f_*, the map induced by f.

(iv) f_* is an isomorphism for each n if and only if $\mathbf{M}(f)$ is acyclic.

6.16.   **(3 × 3 Lemma):**   Consider the commutative diagram of modules:

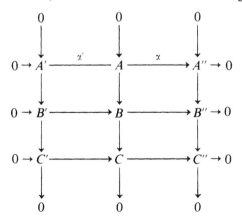

If the columns are exact and if the bottom two (or the top two) rows are exact, then the top row (or the bottom row) is exact. (Hint: Either use the Snake lemma or proceed as follows: first show that $\alpha\alpha' = 0$, then regard each row as a complex and the diagram as a short exact sequence of complexes, and apply Theorem 6.3.)

Remark: The 3×3 Lemma is often called the **9-Lemma**.

6.17. Let $0 \to \mathbf{A'} \to \mathbf{A} \to \mathbf{A''} \to 0$ be an exact sequence of complexes. If $\mathbf{A'}$ and $\mathbf{A''}$ are acyclic, then $\mathbf{A}$ is acyclic. (Hint: Theorem 6.3).

Let us pause before pursuing our main goal: the construction of "derived functors".

Lemma 6.6 (Mayer–Vietoris): *Consider the commutative diagram of modules with exact rows:*

$$\cdots \to A_n \xrightarrow{i_n} B_n \xrightarrow{p_n} C_n \xrightarrow{\hat{c}_n} A_{n-1} \longrightarrow B_{n-1} \longrightarrow C_{n-1} \to \cdots$$

$$\downarrow \alpha_n \qquad \downarrow \beta_n \qquad \downarrow \gamma_n \qquad \downarrow \alpha_{n-1} \qquad \downarrow \beta_{n-1} \qquad \downarrow \gamma_{n-1}$$

$$\cdots \to A'_n \xrightarrow{j_n} B'_n \xrightarrow{q_n} C'_n \longrightarrow A'_{n-1} \longrightarrow B'_{n-1} \longrightarrow C'_{n-1} \to \cdots$$

If every γ_n is an isomorphism, there is an exact sequence

$$\cdots \to A_n \xrightarrow{(\alpha_n, i_n)} A'_n \oplus B_n \xrightarrow{j_n - \beta_n} B'_n \xrightarrow{\hat{c}_n \gamma_n^{-1} q_n} A_{n-1} \to A'_{n-1} \oplus B_{n-1} \to B'_{n-1} \to \cdots.$$

Proof: This, too, is automatic; we merely define the maps in more detail. The map $A_n \to A'_n \oplus B_n$ sends $a_n \mapsto (\alpha_n a_n, i_n a_n)$ and the map $A'_n \oplus B_n \to B'_n$ sends $(a'_n, b_n) \mapsto j_n a'_n - \beta_n b_n$. ∎

This last lemma is useful in algebraic topology. If Y is a subspace of a space X, then every continuous map $\Delta_n \to Y$ (where Δ_n is the standard n-simplex) may be regarded as a continuous map $\Delta_n \to X$. In this way, we obtain an exact sequence of complexes (where $S(X)$ is the singular complex of X)

$$0 \to S(Y) \to S(X) \to S(X)/S(Y) \to 0.$$

By definition, the nth **relative homology group of X mod Y** is

$$H_n(X, Y) = H_n(S(X)/S(Y)).$$

If $Y = \varnothing$, then $H_n(X, \varnothing) = H_n(X)$ as defined in Chapter 1. Algebraic topology considers the category whose objects are all pairs (X, Y), where Y is a subspace of X (perhaps empty) and whose morphisms $f:(X, Y) \to (X', Y')$ are all continuous maps $f: X \to X'$ for which $f(Y) \subset Y'$.

Assume X is a topological space with subspaces X_1 and X_2. Using Theorems 6.3 and 6.4, we have a commutative diagram with exact rows

$$\cdots \to H_n(X_1 \cap X_2) \longrightarrow H_n(X_1) \longrightarrow H_n(X_1, X_1 \cap X_2) \longrightarrow H_{n-1}(X_1 \cap X_2) \to \cdots$$

$$\Big\downarrow \qquad\qquad \Big\downarrow \qquad\qquad\quad \Big\downarrow \qquad\qquad\qquad \Big\downarrow$$

$$\cdots \to H_n(X_2) \longrightarrow H_n(X_1 \cup X_2) \longrightarrow H_n(X_1 \cup X_2, X_2) \longrightarrow H_{n-1}(X_2) \to \cdots$$

where all arrows other than connecting homomorphisms are induced by inclusions. Now the **Excision Axiom** is the theorem stating that if $X = X_1 \cup X_2 = X_1^0 \cup X_2^0$ (where X_i^0 is the interior of X_i), then the maps $H_n(X_1, X_1 \cap X_2) \to H_n(X_1 \cup X_2, X_2)$ are isomorphisms for every n. Lemma 6.6 now applies, for every third vertical arrow is an isomorphism.

Theorem 6.7 (**Mayer–Vietoris**): *Let* X_1 *and* X_2 *be subspaces of* X *with* $X = X_1^0 \cup X_2^0$. *There is an exact sequence*

$$\cdots \to H_n(X_1 \cap X_2) \to H_n(X_1) \oplus H_n(X_2) \to H_n(X) \to H_{n-1}(X_1 \cap X_2) \to \cdots.$$

Proof: We have sketched the proof in the paragraph above. We remark that Lemma 6.6 allows one to be more precise; explicit formulas for the maps in this exact sequence are given. ∎

It is an important fact in topology that homotopic continuous maps induce the same homomorphisms in homology. We extract the (simple) algebraic part of this result.

Definition: Let $f : A \to A'$ be a chain map; f is **nullhomotopic** if there are maps $s_n : A_n \to A'_{n+1}$ such that

$$f_n = d'_{n+1} s_n + s_{n-1} d_n, \qquad \text{all } n.$$

If f and g are chain maps $A \to A'$, then f is **homotopic** to g if $f - g$ is null-homotopic. The maps $\{s_n : n \in \mathbf{Z}\}$ form a **homotopy**.

It is easy to see that homotopy is an equivalence relation on $\text{Hom}(A, A')$, the group of all chain maps $A \to A'$.

Theorem 6.8:   *If f and g are homotopic chain maps $\mathbf{A} \to \mathbf{A}'$, then*

$$f_* = g_* : H_n(\mathbf{A}) \to H_n(\mathbf{A}') \qquad \text{for all} \quad n \in \mathbf{Z}.$$

Proof: We omit subscripts. If z is an n-cycle, then

$$fz - gz = d'sz + s\,dz.$$

Since $dz = 0$, we have $fz - gz \in B_n(\mathbf{A}')$ and so $f_* = g_*$.   ∎

Exercises: 6.18.   Theorem 6.8 is true if the maps s_n are merely $\mathbf{Z}$-maps.

6.19.   A complex $\mathbf{A}$ has a **contracting homotopy** if its identity map $1_{\mathbf{A}}$ is nullhomotopic. (This is the algebraic analog of a contractible topological space.) Prove that if $\mathbf{A}$ has a contracting homotopy, then $\mathbf{A}$ is acyclic.

6.20.   If f and g are homotopic chain maps $\mathbf{A} \to \mathbf{A}'$ and if F is an additive functor, then Ff and $Fg : F\mathbf{A} \to F\mathbf{A}'$ are also homotopic.

DERIVED FUNCTORS

Given a functor T between categories of modules, we construct a sequence of new functors, called derived functors, as suggested by our study of group extensions. Here is a rough sketch of how to evaluate these functors on a module M: choose a resolution of M, apply the functor T, and take homology of the resulting complex. It is clear, even from this brief description, that we are obliged to compare different resolutions in order to prove independence of the choice.

Notation: Let $\mathbf{X}$ be a complex of the form

$$\mathbf{X} = \cdots \to X_1 \to X_0 \to M \to 0.$$

The complex obtained by suppressing M is

$$\mathbf{X}_M = \cdots \to X_1 \to X_0 \to 0$$

and is called the **deleted complex** of $\mathbf{X}$.

Similarly, we define the deleted complex $\mathbf{Y}_N$ of the complex

$$\mathbf{Y} = 0 \to N \to Y^0 \to Y^1 \to \cdots$$

by suppressing N.

Deleted complexes arise in practice from either a projective or injective resolution of a module. If we suppress M from a projective resolution

$$\cdots \to P_1 \to P_0 \to M \to 0,$$

we really have not lost any information, for $M = \operatorname{coker}(P_1 \to P_0)$. Furthermore, if we regard projective resolutions as generalizing the notion of generators and relations, then suppressing M is a rather natural thing to do, i.e., we retain the generators and relations (as is the usual procedure in group theory).

Theorem 6.9   (Comparison Theorem):   *Consider the diagram*

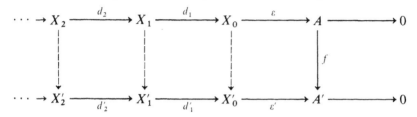

where the rows are complexes. If each X_n in the top row is projective and if the bottom row is exact, then there is a chain map $\bar{f}: \mathbf{X}_A \to \mathbf{X}'_{A'}$ (the dashed arrows) making the completed diagram commute.

Moreover, any two such chain maps are homotopic.

Remark: The dual theorem is true and the proof is similar; the sequences go to the right, the top row is assumed exact, and each term in the bottom row (save the first) is assumed injective.

Proof: (i)   The existence of $\bar{f}$.   We do an induction on n. If $n = 0$, we have the diagram

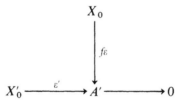

Since ε' is epic and X_0 is projective, there is a map $\bar{f}_0: X_0 \to X'_0$ with $\varepsilon' \bar{f}_0 = f\varepsilon$.

For the inductive step, consider the diagram

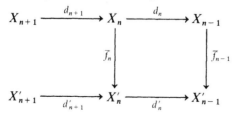

If we show that $\operatorname{im}\bar{f}_n d_{n+1} \subset \operatorname{im} d'_{n+1}$, then we will have the diagram

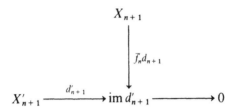

and projectivity of X_{n+1} will provide a map $\bar{f}_{n+1}: X_{n+1} \to X'_{n+1}$ with $d'_{n+1}\bar{f}_{n+1} = \bar{f}_n d_{n+1}$. To check the inclusion, note that exactness of the bottom row in the original diagram gives $\operatorname{im} d'_{n+1} = \ker d'_n$, so it suffices to prove $d'_n\bar{f}_n d_{n+1} = 0$. But $d'_n\bar{f}_n d_{n+1} = \bar{f}_{n-1}d_n d_{n+1} = 0$.

(ii) Uniqueness of $\bar{f}$ to homotopy. Assume $h: X_A \to X'_{A'}$ is a second chain map satisfying $\varepsilon'h_0 = f\varepsilon$. We construct a homotopy s by induction. Begin by defining $s_{-1}: X_{-1} \to X'_0$ as the zero map (there is no choice here since $X_{-1} = 0$). For the inductive step (and also for s_0), we shall show that

$$\operatorname{im}(h_{n+1} - \bar{f}_{n+1} - s_n d_{n+1}) \subset \operatorname{im} d'_{n+2};$$

this will give the picture

$$
\begin{array}{c}
X_{n+1} \\
\downarrow {\scriptstyle h_{n+1} - \bar{f}_{n+1} - s_n d_{n+1}} \\
X'_{n+2} \xrightarrow{\ d'_{n+2}\ } \operatorname{im} d'_{n+2} \longrightarrow 0
\end{array}
$$

and projectivity of X_{n+1} will give a map $s_{n+1}: X_{n+1} \to X'_{n+2}$ satisfying the desired equation. To verify the inclusion, the exactness of the bottom row in the original diagram gives $\operatorname{im} d'_{n+2} = \ker d'_{n+1}$, so it suffices to show

$$d'_{n+1}(h_{n+1} - \bar{f}_{n+1} - s_n d_{n+1}) = 0.$$

But the left side is

$$
\begin{aligned}
d'_{n+1}(h_{n+1} - \bar{f}_{n+1}) - d'_{n+1}s_n d_{n+1} &= d'_{n+1}(h_{n+1} - \bar{f}_{n+1}) - (h_n - \bar{f}_n - s_{n-1}d_n)d_{n+1} \\
&= d'_{n+1}(h_{n+1} - \bar{f}_{n+1}) - (h_n - \bar{f}_n)d_{n+1},
\end{aligned}
$$

and this is zero because h and $\bar{f}$ are chain maps. $\blacksquare$

Definition: If $\bar{f}: X_A \to X'_{A'}$ is a chain map for which

$$f\varepsilon = \varepsilon'\bar{f}_0$$

(notation as in the Comparison Theorem 6.9), then we say $\bar{f}$ is a chain map **over** f.

Given a functor T, we now describe its **left derived functors** L_nT. For each module A, choose, once for all, a projective resolution of A, and let $\mathbf{P}_A$ be the corresponding deleted complex; form $T(\mathbf{P}_A)$ as in Example 6 and take homology.

Definition: For each module A,

$$(L_nT)A = H_n(T\mathbf{P}_A) = \ker Td_n/\operatorname{im} Td_{n+1}.$$

To complete the definition of L_nT, we must describe its action on $f:A \to B$. By the Comparison Theorem 6.9, there is a chain map $\bar{f}:\mathbf{P}_A \to \mathbf{P}_B$ over f. Define

$$(L_nT)f:(L_nT)A \to (L_nT)B$$

by

$$(L_nT)f = H_n(T\bar{f}) = (T\bar{f})_*,$$

i.e., if $z_n \in \ker Td_n$, then

$$(L_nT)f:z_n + \operatorname{im} Td_{n+1} \mapsto (T\bar{f}_n)z_n + \operatorname{im} Td'_{n+1}.$$

Let us draw a picture:

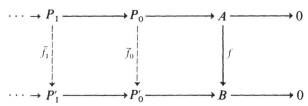

First, fill in the dashed arrows; second, apply T to this diagram; finally, take the maps induced by $T\bar{f}$ in homology.

Theorem 6.10: *Given a functor T, then L_nT is an additive functor for every n.*

Proof: We will only show $(L_nT)f$ is well defined, where $f:A \to B$, and leave the remaining verifications to the reader. Assume $h:\mathbf{P}_A \to \mathbf{P}_B$ is a second chain map over f. The Comparison Theorem says $\bar{f}$ and h are homotopic, so that $T\bar{f}$ and Th are homotopic (Exercise 6.19), and Theorem 6.8 says that $T\bar{f}$ and Th induce the same maps in homology. ∎

Definition: If $T = \otimes_R B$, then we define $L_nT = \operatorname{Tor}_n^R(\ ,B)$. In particular,

$$\operatorname{Tor}_n^R(A, B) = \ker(d_n \otimes 1)/\operatorname{im}(d_{n+1} \otimes 1),$$

where

$$\cdots \to P_2 \overset{d_2}{\to} P_1 \overset{d_1}{\to} P_0 \to A \to 0$$

is the projective resolution of A that we chose once for all.

Let us now deal with the dependence of L_nT on the fixed choices of projective resolutions. Had we begun with other choices $\cdots \to \hat{P}_2 \to \hat{P}_1 \to \hat{P}_0 \to A \to 0$ of projective resolutions (one for each module A), we would have obtained functors we temporarily denote $\hat{L}_nT$. Our next project is to show L_nT and $\hat{L}_nT$ are essentially the same.

Theorem 6.11: *For any functor T, the left derived functors L_nT and $\hat{L}_nT$ are naturally equivalent. In particular, for each A,*

$$(L_nT)A \cong (\hat{L}_nT)A,$$

i.e., these modules are independent of the choice of projective resolution of A.

Proof: Consider the diagram

$$\cdots \to P_2 \to P_1 \to P_0 \to A \longrightarrow 0$$
$$\downarrow 1_A$$
$$\cdots \to \hat{P}_2 \to \hat{P}_1 \to \hat{P}_0 \to A \longrightarrow 0$$

where the top row is the chosen projective resolution of A used to define L_nT, and the bottom row that used to define $\hat{L}_nT$. By the Comparison Theorem, there is a chain map $i: \mathbf{P}_A \to \hat{\mathbf{P}}_A$ over 1_A (unique to homotopy) and, applying T gives a chain map $Ti: T\mathbf{P}_A \to T\hat{\mathbf{P}}_A$ over 1_{TA}; this latter chain map induces maps (one for each n)

$$\tau_A = (Ti)_*: (L_nT)A \to (\hat{L}_nT)A.$$

We claim each τ_A is an isomorphism (thereby establishing the second sentence of the theorem). To obtain the inverse of τ_A, turn the above diagram upside down so that the hatted row is now on top. The Comparison Theorem gives a chain map $j: \hat{\mathbf{P}}_A \to \mathbf{P}_A$. The composite $ji: \mathbf{P}_A \to \mathbf{P}_A$ is thus a chain map over 1_A; since $1_{\mathbf{P}}: \mathbf{P}_A \to \mathbf{P}_A$ is also a chain map over 1_A, the Comparison Theorem says ji and $1_{\mathbf{P}}$ are homotopic; therefore $1 = (ji)_* = j_*i_*$. Similarly, $i_*j_* = 1$, so that i_* is an isomorphism and hence $\tau_A = (Ti)_*$ is an isomorphism.

Let us now prove the isomorphisms τ_A constitute a natural transormation: if $f: A \to B$, we must show commutativity of

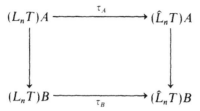

To evaluate clockwise, consider the picture

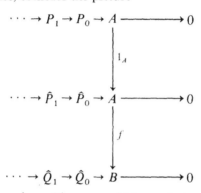

Applying the Comparison Theorem yields a chain map $\mathbf{P}_A \to \hat{\mathbf{Q}}_B$ over $f1_A = f$. Going counterclockwise also gives a chain map $\mathbf{P}_A \to \hat{\mathbf{Q}}_B$ over $1_B f = f$, so these two chain maps are homotopic. Now apply T to obtain homotopic chain maps $T\mathbf{P}_A \to T\hat{\mathbf{Q}}_B$ over Tf; the induced maps are equal, by Theorem 6.8. ∎

Corollary 6.12: *The definition of* $\operatorname{Tor}_n^R(A, B)$ *is independent of the choice of projective resolution of* A.

Corollary 6.13: *Let* $\cdots \to P_2 \xrightarrow{d_2} P_1 \xrightarrow{d_1} P_0 \xrightarrow{\varepsilon} A \to 0$ *be a projective resolution, and define* $K_0 = \ker \varepsilon$ *and* $K_n = \ker d_n$ *for* $n \geq 1$. *Then if* T *is covariant,*

$$(L_{n+1}T)A \cong (L_n T)K_0 \cong (L_{n-1}T)K_1 \cong \cdots \cong (L_1 T)K_{n-1}.$$

In particular,

$$\operatorname{Tor}_{n+1}(A, B) \cong \operatorname{Tor}_n(K_0, B) \cong \cdots \cong \operatorname{Tor}_1(K_{n-1}, B).$$

Proof: It is clear that $\cdots \to P_2 \to P_1 \to K_0 \to 0$ is a projective resolution of K_0. Since the indices are no longer correct, define

$$Q_{n-1} = P_n \quad \text{and} \quad \Delta_{n-1} = d_n, \quad n \geq 1;$$

the resolution now reads

$$\cdots \to Q_2 \xrightarrow{\Delta_2} Q_1 \xrightarrow{\Delta_1} Q_0 \to K_0 \to 0.$$

By definition,

$$(L_n T)K_0 \cong H_n(TQ_{K_0}) = \ker T\Delta_n / \operatorname{im} T\Delta_{n+1}$$
$$= \ker Td_{n+1} / \operatorname{im} Td_{n+2} = H_{n+1}(T\mathbf{P}_A) = (L_{n+1}T)A.$$

The remaining isomorphisms are obtained by iteration. ∎

We still have to grapple with the following question. Suppose we begin with the functor $A \otimes_R$ and construct its left derived functors; if we denote these functors $\operatorname{Tor}_n^R(A, \ )$, then is $\operatorname{Tor}_n^R(A, B) = \operatorname{Tor}_n^R(A, B)$? There is no misprint. We ask whether the value of $\operatorname{Tor}_n^R(A, \ )$ on B is the same as the value of $\operatorname{Tor}_n^R(\ , B)$ on A. Equivalently, is $H_n(\mathbf{P}_A \otimes B) \cong H_n(A \otimes \mathbf{P}_B)$? Before we worry about this (by the way, the answer is "yes", Theorem 7.9), let us construct right derived functors. The recipe is almost the same as for left derived functors, so that our sketch of the construction should serve as a review of what we have just done. There will be two definitions, depending on the variance of T (a similar dichotomy exists for left derived functors, but we shall not discuss $L_n T$ for contravariant T).

For each module A, choose, once for all, an injective resolution $0 \to A \to E^0 \xrightarrow{d^0} E^1 \xrightarrow{d^1} E^2 \to \cdots$. Note that we have already used the convention of raising indices (Example 5) to avoid negative indices. Let $\mathbf{E}_A$ denote the corresponding deleted resolution.

Definition: If T is covariant, its **right derived functors** $R^n T$ are defined on a module A by

$$(R^n T)A = H^n(T\mathbf{E}_A) = \ker(Td^n) / \operatorname{im}(Td^{n-1}).$$

The reader may lower all indices and observe that

$$(R^n T)A = H_{-n}(T\mathbf{E}_A) = \ker(Td_{-n}) / \operatorname{im}(Td_{-n+1}).$$

The definition of $(R^n T)f$, for $f: A \to B$, is just as for left derived functors. The dual of the Comparison Theorem asserts the existence of a chain map $\bar{f}: \mathbf{E}_A \to \mathbf{E}_B$, unique to homotopy, and so a unique map is induced in homology $H^n(T\mathbf{E}_A) \to H^n(T\mathbf{E}_B)$.

Theorem 6.14: *If T is covariant, each right derived functor $R^n T$ is an additive functor whose definition is independent of the choices of injective resolutions.*

Proof: Dual to the proof of Theorem 6.11. ∎

There is an easy explanation for the adjectives "left" and "right" modifying "derived functors": given T, the complexes $T\mathbf{P}_A$ go to the left, while the complexes $T\mathbf{E}_A$ go to the right.

Definition: If $T = \operatorname{Hom}_R(C, \ )$, then

$$R^n T = \operatorname{Ext}_R^n(C, \ ).$$

In particular,

$$\operatorname{Ext}_R^n(C, A) = \ker d_*^n / \operatorname{im} d_*^{n-1},$$

where $0 \to A \to E^0 \overset{d^0}{\to} E^1 \overset{d^1}{\to} E^2 \to \cdots$ is the chosen injective resolution of A.

Corollary 6.15:   *The definition of* $\operatorname{Ext}_R^n(C, A)$ *is independent of the choice of injective resolution of* A.

Corollary 6.16:   *Let* $0 \to A \overset{\varepsilon}{\to} E^0 \overset{d^0}{\to} E^1 \overset{d^1}{\to} E^2 \to \cdots$ *be an injective resolution, and define* $L^0 = \operatorname{im} \varepsilon$ *and* $L^n = \operatorname{im} d^{n-1}$ *for* $n \geq 1$. *Then if* T *is covariant,*

$$(R^{n+1} T)A \cong (R^n T)L^0 \cong (R^{n-1} T)L^1 \cong \cdots \cong (R^1 T)L^{n-1}.$$

In particular,

$$\operatorname{Ext}^{n+1}(C, A) \cong \operatorname{Ext}^n(C, L^0) \cong \cdots \cong \operatorname{Ext}^1(C, L^{n-1}).$$

Finally, let us define $R^n T$ when T is contravariant; since we want the functored complex to go to the right, the following definition is forced on us.

Definition: If T is contravariant, then

$$(R^n T)C = \ker T d_{n+1} / \operatorname{im} T d_n,$$

where $\cdots \to P_2 \overset{d_2}{\to} P_1 \overset{d_1}{\to} P_0 \to C \to 0$ is the projective resolution of C chosen once for all.

The reader should look again at Example 7 to convince himself that the indices are, indeed, correct. We also let the reader provide the definition $(R^n T)f$ for $f : C \to C'$.

Theorem 6.17   *If* T *is contravariant, then each* $R^n T$ *is an additive contravariant functor whose definition is independent of the choices of projective resolutions.*

Definition: If $T = \text{Hom}_R(\ , A)$, then $R^n T = \text{Ext}^n(\ , A)$. In particular,

$$\text{Ext}^n_R(C, A) = \ker d_{n+1*}/\text{im } d_{n*},$$

where $\cdots \to P_2 \xrightarrow{d_2} P_1 \xrightarrow{d_1} P_0 \to C \to 0$ is the chosen projective resolution of C.

Corollary 6.18:   *The definition of $\text{Ext}^n_R(C, A)$ is independent of the choice of projective resolution of C.*

Corollary 6.19:   *Let $\cdots \to P_2 \xrightarrow{d_2} P_1 \xrightarrow{d_1} P_0 \xrightarrow{\varepsilon} C \to 0$ be a projective resolution, and define $K_0 = \ker \varepsilon$ and $K_n = \ker d_n$ for $n \geq 1$. Then if T is contravariant,*

$$(R^{n+1}T)C \cong (R^n T)K_0 \cong (R^{n-1}T)K_1 \cong \cdots \cong (R^1 T)K_{n-1}.$$

In particular,

$$\text{Ext}^{n+1}(C, A) \cong \text{Ext}^n(K_0, A) \cong \cdots \cong \text{Ext}^1(K_{n-1}, A).$$

As for Tor, we ask whether the value of $\text{Ext}^n_R(C,\ )$ on A is the same as the value of $\text{Ext}^n_R(\ , A)$ on C; the answer is "yes" and is Theorem 7.8.

In summary, one chooses a resolution $\mathbf{X}$ of each module and the question of whether it is a projective or injective resolution is dictated by whether one wants $T\mathbf{X}$ to go left or to go right.

Résumé: $L_n T$; T covariant; use projective resolutions.
$R^n T$; T covariant; use injective resolutions.
$R^n T$; T contravariant; use projective resolutions.

The reader may wonder why we have not considered left derived functors of Hom or right derived functors of tensor. The reason is that we want to use derived functors of T to investigate T, and this investigation is most fruitful when either $L_0 T = T$ or $R^0 T = T$. For any covariant functor T, however, we shall see that $L_0 T$ is right exact and $R^0 T$ is left exact, so that $L_n \text{Hom}$, for example, conveys no obvious information about Hom.

Let us glance back at Chapter 5 to see that derived functors did appear there. Since some subtle points are needed, however, we defer a detailed discussion to Chapter 10. Let G be a group and let $R = \mathbf{Z}G$ be its group ring. As in Chapter 5, regard $\mathbf{Z}$ as a trivial G-module, i.e., $x \cdot n = n$ for all $x \in G$ and $n \in \mathbf{Z}$; furthermore, recall we constructed the first few terms of a projective (even free) resolution of $\mathbf{Z}$:

$$F_3 \to F_2 \xrightarrow{d_2} F_1 \xrightarrow{d_1} F_0 \to \mathbf{Z} \to 0.$$

If A is a G-module and $T = \text{Hom}_R(\ , A)$, then we calculated $(R^n T)\mathbf{Z}$ for $n = 0, 1, 2$, and observed fixed points, derivations, and factor sets. If $T = \ \otimes_R A$, then we gave a brief account of $(L_n T)\mathbf{Z}$ for $n = 0, 1$.

Lemma 6.20 **(Horseshoe Lemma):** *Consider the diagram*

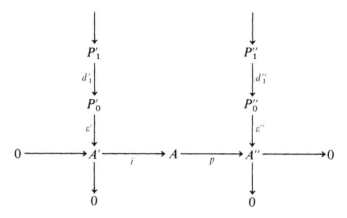

where the columns are projective resolutions and the row is exact. Then there exists a projective resolution of A and chain maps so that the columns form an exact sequence of complexes.

Remark: The dual theorem is also true. The Lemma is so named because one is given a horseshoe and is required to fill it in.

Proof: By induction, it suffices to complete the 3×3 diagram

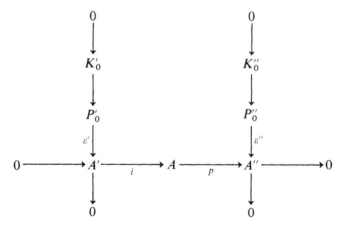

where the rows and columns are exact and P_0', P_0'' are projective. Define $P_0 = P_0' \oplus P_0''$, $i_0 : P_0' \to P_0$ by $x' \mapsto (x', 0)$, and $p_0 : P_0 \to P_0''$ by $(x', x'') \mapsto x''$. It is clear that P_0 is projective and that

$$0 \to P_0' \xrightarrow{i_0} P_0 \xrightarrow{p_0} P_0'' \to 0$$

is exact. Since P_0'' is projective, there is a map $\sigma: P_0'' \to A$ with $p\sigma = \varepsilon'$. Define $\varepsilon: P_0 \to A$ by

$$\varepsilon: (x', x'') \mapsto i\varepsilon'x' + \sigma x''.$$

It is an easy verification that, if $K_0 = \ker \varepsilon$, the resulting 3×3 diagram commutes. Exactness of the top row is the 3×3 Lemma, Exercise 6.16. ∎

Theorem 6.21: *Let $0 \to A' \to A \to A'' \to 0$ be an exact sequence of modules. If T is a covariant functor, there is an exact sequence*

$$\cdots \to L_n T A' \to L_n T A \to L_n T A'' \xrightarrow{\hat{c}} L_{n-1} T A' \to \cdots$$
$$\cdots \to L_0 T A' \to L_0 T A \to L_0 T A'' \to 0.$$

Proof: Let $\cdots \to P_1' \to P_0' \to A' \to 0$ and $\cdots \to P_1'' \to P_0'' \to A'' \to 0$ be the chosen projective resolutions of A' and of A''. We are now in the situation of Lemma 6.20, so there is a projective resolution of A, $\cdots \to \hat{P}_1 \to \hat{P}_0 \to A \to 0$, that we may fit in the middle. Upon deleting, there is an exact sequence of complexes

$$0 \to \mathbf{P}_{A'}' \to \hat{\mathbf{P}}_A \to \mathbf{P}_{A''}'' \to 0.$$

Applying T gives another exact sequence of complexes

$$0 \to T\mathbf{P}_{A'}' \to T\hat{\mathbf{P}}_A \to T\mathbf{P}_{A''}'' \to 0,$$

for each row $0 \to P_n' \to \hat{P}_n \to P_n'' \to 0$ is split and Exercise 6.10 applies. (In general, the exact sequence of complexes is not split, for there is no reason why the splittings $P_n'' \to \hat{P}_n$ should constitute a chain map $\mathbf{P}_{A''}'' \to \hat{\mathbf{P}}_A$.) There is a long exact sequence

$$\cdots \to H_n(T\mathbf{P}_{A'}') \to H_n(T\hat{\mathbf{P}}_A) \to H_n(T\mathbf{P}_{A''}'') \xrightarrow{\hat{c}} H_{n-1}(T\mathbf{P}_{A'}') \to \cdots,$$

i.e., an exact sequence

$$\cdots \to L_n T A' \to \hat{L}_n T A \to L_n T A'' \to L_{n-1} T A' \to \cdots.$$

Notice that we have $\hat{L}_n T A$ instead of $L_n T A$, for the projective resolution of A constructed with the Horseshoe Lemma need not be the projective resolution originally chosen. This is no cause for concern; Theorem 6.11 says that $L_n T$ and $\hat{L}_n T$ are naturally equivalent. We may thus replace $\hat{L}_n T$ by $L_n T$, adjust maps by isomorphisms, and obtain an exact sequence

$$\cdots \to L_n T A' \to L_n T A \to L_n T A'' \to L_{n-1} T A' \to \cdots.$$

Finally, the sequence does terminate at 0, for $L_n T = 0$ for negative n: indeed, every P_n, hence every TP_n, is 0 for negative n. ∎

Corollary 6.22: *For every covariant functor T, the derived functor $L_0 T$ is right exact.*

Proof: We have just seen that exactness of $0 \to A' \to A \to A'' \to 0$ yields exactness of $L_0 T A' \to L_0 T A \to L_0 T A'' \to 0$. ∎

We need an elementary (three-dimensional!) diagram lemma.

Lemma 6.23: *Consider the commutative cube (all faces commute)*

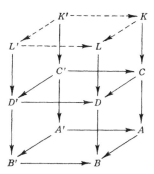

in which L', K', L, K are kernels of obvious arrows. Then the dashed arrows exist and every new square commutes.

Proof: Existence of the dashed arrows is easy: for example, Exercise 2.7 provides a unique dashed arrow making the diagram below commute:

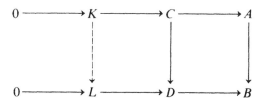

It only remains to prove commutativity of the (top) dashed square. We leave this to the reader, with the hint to compose $K' \to L' \to L$ and $K' \to K \to L$ with the map $L \to D$. After showing these coincide, one may cancel $L \to D$ because it is monic. ∎

The following lemma generalizes the Horseshoe Lemma 6.20; it asserts one may construct projective resolutions over a certain commutative diagram.

Lemma 6.24:   *Assume we are given a commutative diagram of modules with exact rows*

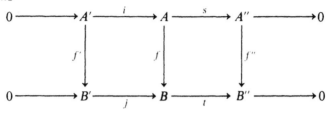

projective resolutions $\mathbf{P}'$, $\mathbf{P}''$, $\mathbf{Q}'$, $\mathbf{Q}''$ *of the corners* A', A'', B', B'', *respectively, and chain maps* $F':\mathbf{P}' \to \mathbf{Q}'$ *over* f' *and* $F'':\mathbf{P}'' \to \mathbf{Q}''$ *over* f''.

Then there exist projective resolutions $\mathbf{P}$ *of* A *and* $\mathbf{Q}$ *of* B *and a chain map* $F:\mathbf{P} \to \mathbf{Q}$ *over* f *giving a commutative diagram of complexes with exact rows*

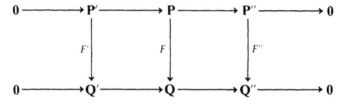

Remark: The dual theorem with injective resolutions is also true.

Proof: By induction on n, it suffices to complete the following three-dimensional diagram,

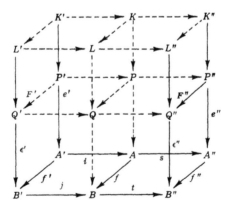

where P and Q are projectives to be constructed, all dashed arrows must be constructed, and each three-term row and column is a short exact sequence (i.e., certain zeros have been omitted). The modules L, K, ... are kernels as defined below.

Define $P = P' \oplus P''$ and $Q = Q' \oplus Q''$; define $P' \to P$ by $x' \mapsto (x', 0)$ and $P \to P''$ by $(x', x'') \mapsto x''$; similarly, define $Q' \to Q$ and $Q \to Q''$. For any choice of $\gamma : P'' \to Q'$ (to be constrained later), define $F : P \to Q$ by

$$F : (x', x'') \mapsto (F'x' + \gamma x'', F''x'').$$

With these definitions, the "P-Q level" is a commutative diagram with exact rows.

Define $e : P \to A$ by

$$e(x', x'') = ie'x' + \sigma x'',$$

where $\sigma : P'' \to A$ satisfies $s\sigma = e''$ (σ exists since P'' is projective). Similarly, define $\varepsilon : Q \to B$ by

$$\varepsilon(y', y'') = j\varepsilon'y' + \tau y'',$$

where $\tau : Q'' \to B$ satisfies $t\tau = \varepsilon''$. This gives commutativity of front and rear walls: the Five Lemma now shows e and ε are epic.

A short computation shows that $fe = \varepsilon F$ if γ satisfies

$$j\varepsilon'\gamma = -\tau F'' + f\sigma.$$

To see that γ can be so chosen, consider the diagram with exact row

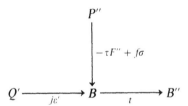

Now $t(-\tau F'' + f\sigma) = -\varepsilon''F'' + f''s\sigma = -\varepsilon''F'' + f''e'' = 0$, so that

$$\operatorname{im}(-\tau F'' + f\sigma) \subset \ker t = \operatorname{im} j\varepsilon'.$$

Projectivity of P'' guarantees the existence of a desired map $\gamma : P'' \to Q'$. We have proved commutativity of the bottom cubes.

The modules L, K, etc., are, by definition, kernels of the obvious arrows. Applying Lemma 6.23 to each cube supplies the needed dashed arrows and guarantees commutativity. Finally, the row of K's and the row of L's are exact, as two applications of the 3×3 Lemma show (one in each plane). ∎

There is a good reason why this last proof resembles that of the Horseshoe Lemma 6.20: that lemma holds in greater generality than for complexes of modules; it holds for complexes whose terms are objects in suitable "abelian" categories. In particular, the category $\mathfrak{C}$ with objects all module maps and with morphisms all ordered pairs of maps giving commutative squares, is such a category, and the 2 × 3 diagram of the hypothesis is just a short exact sequence in $\mathfrak{C}$.

Theorem 6.25:   *The connecting homomorphisms are natural, i.e., given a commutative diagram with exact rows*

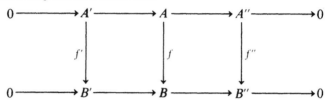

then the following diagram commutes for all $n \geq 1$:

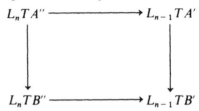

Proof: Erect the chosen projective resolutions on the four corners; call them $\mathbf{P}', \mathbf{P}'', \mathbf{Q}', \mathbf{Q}''$, and use the Comparison Theorem 6.9 to construct chain maps $F':\mathbf{P}'_{A'} \to \mathbf{Q}'_{B'}$ over f' and $F'':\mathbf{P}''_{A''} \to \mathbf{Q}''_{B''}$ over f''. By Lemma 6.24, there are projective resolutions $\hat{\mathbf{P}}$ of A and $\hat{\mathbf{Q}}$ of B and a chain map $F:\hat{\mathbf{P}}_A \to \hat{\mathbf{Q}}_B$ over f giving a commutative diagram of complexes with exact rows

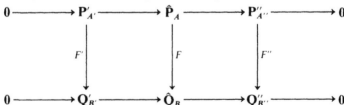

This diagram remains commutative and exact after applying T, for each nth row of modules is split. The result follows now from Theorem 6.4 (note that the hatted complexes do not enter into the conclusion). ∎

Theorem 6.26:   *If $0 \to A' \to A \to A'' \to 0$ is an exact sequence of modules and T is covariant, there is an exact sequence*

$$0 \to R^0 T A' \to R^0 T A \to R^0 T A'' \to \cdots$$

$$\cdots \to R^n T A' \to R^n T A \to R^n T A'' \overset{\partial}{\to} R^{n+1} T A' \to \cdots$$

with natural connecting homomorphisms ∂.

Proof: Only one point needs comment: why does ∂ raise the index by 1? If $\mathbf{E}''_{A''}$ and $\mathbf{E}'_{A'}$ are deleted injective resolutions, then $\partial: H_{-n}(T\mathbf{E}''_{A''}) \to H_{-n-1}(T\mathbf{E}'_{A'})$. The index raising convention involving change of sign thus gives $\partial: R^n T A'' \to R^{n+1} T A'$. ∎

By looking at the beginning of the exact sequence, one sees that if T is covariant, then $R^0 T$ is left exact.

Theorem 6.27: *If $0 \to A' \to A \to A'' \to 0$ is an exact sequence of modules and T is contravariant, there is an exact sequence*

$$0 \to R^0 T A'' \to R^0 T A \to R^0 T A' \to \cdots$$

$$\cdots \to R^n T A'' \to R^n T A \to R^n T A' \xrightarrow{\partial} R^{n+1} T A'' \to \cdots$$

with natural connecting homomorphisms.

If T is contravariant, we see that $R^0 T$ is left exact.

We end this chapter with some words about "cohomology", which usually means a contravariant homology theory: the derived functors of a contravariant functor are thus called cohomology functors. In this case, all the usual terminology is equipped with the prefix "co": cochain, cocycle, coboundary, cohomology; also, all indices are raised. For example, the material in Chapter 5 is called "cohomology of groups", for it arises as derived functors of the contravariant functors $\mathrm{Hom}_G(\ , A)$. This clear distinction—homology covariant, cohomology contravariant—is often blurred, however, for we shall see the derived functors of the contravariant $\mathrm{Hom}(\ , B)$ have the same value on A that the derived functors of the covariant $\mathrm{Hom}(A, \ )$ have on B. As a result, one often calls the derived functors of Hom, either variance, cohomology functors.

7 Ext

Let us examine Ext more closely. At the moment, we have two definitions for $\text{Ext}^n(A, B)$:

$$\text{Ext}^n(A, B) = H_{-n}(\text{Hom}(A, \mathbf{E}_B)),$$

where $\mathbf{E}_B$ is a deleted injective resolution of B;

$$\text{Ext}^n(A, B) = H_{-n}(\text{Hom}(\mathbf{P}_A, B)),$$

where $\mathbf{P}_A$ is a deleted projective resolution of A.

Until we prove these two definitions coincide, let us use a temporary notation for the second:

$$\text{ext}^n(A, B) = H_{-n}(\text{Hom}(\mathbf{P}_A, B)).$$

ELEMENTARY PROPERTIES

Theorem 7.1: *If n is negative, then $\text{Ext}^n(A, B) = 0 = \text{ext}^n(A, B)$ for all A, B.*

Proof: If $\mathbf{E}_B = \cdots \to 0 \xrightarrow{d_2} 0 \xrightarrow{d_1} E_0 \xrightarrow{d_0} E_{-1} \to \cdots$, then $\text{Hom}(A, \mathbf{E}_B)$ has only 0's to the left of $\text{Hom}(A, E_0)$, and so all homology groups are 0 there as well. A similar argument works in the other variable. ∎

Theorem 7.2: $\text{Ext}^0(A, \ )$ *is naturally equivalent to* $\text{Hom}(A, \ )$.

Proof: If $\mathbf{E}_B = \cdots \to 0 \overset{d_1}{\to} E_0 \overset{d_0}{\to} E_{-1} \to \cdots$, then

$$\text{Ext}^0(A, B) = \ker d_{0*}/\text{im } d_{1*} = \ker d_{0*}.$$

If the (nondeleted) injective resolution is

$$0 \to B \overset{\varepsilon}{\to} E_0 \overset{d_0}{\to} E_{-1} \to \cdots,$$

then left exactness of $\text{Hom}(A,)$ gives an exact sequence

$$0 \to \text{Hom}(A, B) \overset{\varepsilon_*}{\to} \text{Hom}(A, E_0) \xrightarrow{d_{0*}} \text{Hom}(A, E_{-1}),$$

so that $\ker d_{0*} = \text{im } \varepsilon_*$, i.e., $\varepsilon_* : \text{Hom}(A, B) \to \text{Ext}^0(A, B)$ is an isomorphism. We let the reader check that the maps ε_*, one for each module B, define a natural equivalence. ∎

Remark: The only property of $\text{Hom}(A,)$ used in the proof of Theorem 7.2 is that it is left exact. Thus, if T is any covariant left exact functor, then $R^0 T \cong T$.

Theorem 7.3: *If $0 \to B' \to B \to B'' \to 0$ is an exact sequence, there is a long exact sequence with natural connecting homomorphisms*

$$0 \to \text{Hom}(A, B') \to \text{Hom}(A, B) \to \text{Hom}(A, B'') \overset{\partial}{\to} \text{Ext}^1(A, B') \to \cdots.$$

Proof: Theorems 6.26 and 7.2. ∎

Observe that Ext thus repairs the exactness we may have lost when we applied Hom.

Theorem 7.4: $\text{ext}^0(, B)$ *is naturally equivalent to* $\text{Hom}(, B)$. *More generally, if T is any contravariant left exact functor, then $R^0 T \cong T$.*

Theorem 7.5: *If $0 \to A' \to A \to A'' \to 0$ is an exact sequence, then there is a long exact sequence with natural connecting homomorphisms*

$$0 \to \text{Hom}(A'', B) \to \text{Hom}(A, B) \to \text{Hom}(A', B) \overset{\partial}{\to} \text{ext}^1(A'', B) \to \cdots.$$

Exercise: 7.1. Suppose A' is a submodule of A and $f : A' \to B$. Call $\partial f \in \text{ext}^1(A/A', B)$ the **obstruction** of f. Prove that f can be extended to A if and only if its obstruction is 0. (Of course, one may conclude $\partial f = 0$ when $\text{ext}^1(A/A', B) = 0$.)

Theorem 7.6: *If B is injective, then $\text{Ext}^n(A, B) = 0$ for all modules A and all $n \geq 1$.*

Proof: If B is injective, then $0 \to B \overset{\varepsilon}{\to} E^0 \to 0$ is an injective resolution, where $E^0 = B$ and $\varepsilon = 1_B$. With respect to this choice of injective resolution, it is clear that $\text{Ext}^n(A, B) = H_{-n}(\text{Hom}(A, \mathbf{E}_B)) = 0$ whenever $n \geq 1$. ∎

Theorem 7.7:   *If A is projective, then $\text{ext}^n(A, B) = 0$ for all modules B and all $n \geq 1$.*

We now prove that $\text{Ext} = \text{ext}$.

Definition: Let $\mathfrak{A}$, $\mathfrak{B}$, and $\mathfrak{C}$ be categories. A function $T : \mathfrak{A} \times \mathfrak{B} \to \mathfrak{C}$ is a **bifunctor** if $T(A, \ ) : \mathfrak{B} \to \mathfrak{C}$ is a functor for each $A \in \text{obj} \, \mathfrak{A}$, $T(\ , B) : \mathfrak{A} \to \mathfrak{C}$ is a functor for each $B \in \text{obj} \, \mathfrak{B}$, and, for each pair of morphisms $A' \to A$ in $\mathfrak{A}$ and $B' \to B$ in $\mathfrak{B}$, there is a commutative diagram

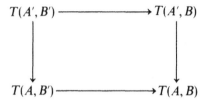

Tensor product is a bifunctor $\mathfrak{M}_R \times {}_R\mathfrak{M} \to \textbf{Ab}$. Of course, one may allow either or both variables to be contravariant; once appropriate changes are made in the definition, one sees Hom is a bifunctor.

Theorem 7.8:   *Let $0 \to A \to E^0 \to E^1 \to \cdots$ be an injective resolution and $\cdots \to P_1 \to P_0 \to C \to 0$ be a projective resolution. Then for all $n \geq 0$*

$$H^n(\text{Hom}(\mathbf{P}_C, A)) \cong H^n(\text{Hom}(C, \mathbf{E}_A)).$$

Thus the two definitions of Ext_R^n have the same value on (C, A).

Remark: This proof is due to A. Zaks.

Proof: Let us display the kernels and images of the resolutions:

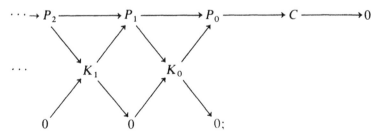

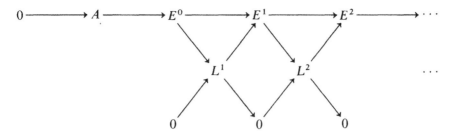

Since Hom is a bifunctor, the exact sequences $0 \to K_0 \to P_0 \to C \to 0$ and $0 \to A \to E^0 \to L^1 \to 0$ give a commutative diagram

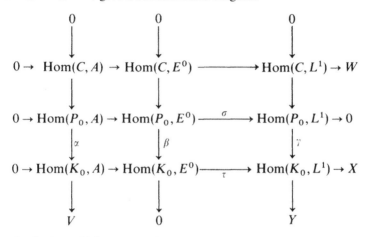

Zeros flank the middle row and column because P_0 is projective and E^0 is injective; the other zeros result from left exactness of Hom. The modules W, V, X, Y are, by definition, cokernels of obvious arrows. Applying the Snake lemma (for the maps α, β, γ) yields an exact sequence

$$\ker \beta \to \ker \gamma \to \operatorname{coker} \alpha \to \operatorname{coker} \beta,$$

i.e.,

$$\operatorname{Hom}(C, E^0) \to \operatorname{Hom}(C, L^1) \to V \to 0.$$

Since $W = \operatorname{coker}(\operatorname{Hom}(C, E^0) \to \operatorname{Hom}(C, L^1))$, we conclude that

$$W \cong V.$$

We can identify W as $\operatorname{Ext}^1(C, A)$, for we are looking at the beginning of the long exact sequence; after three Homs follow three Ext^1's, the second of which is $\operatorname{Ext}^1(C, E^0) = 0$ because E^0 is injective. Similarly, looking at the first column identifies V as $\operatorname{ext}^1(C, A)$. The theorem is thus proved for $n = 1$, i.e., $\operatorname{Ext}^1 \cong \operatorname{ext}^1$.

The diagram contains more information. Since β and σ are epic, it is easy to see $X \cong Y$, i.e.,

$$\operatorname{Ext}^1(C, L^1) \cong \operatorname{Ext}^1(K_0, A).$$

The master diagram arose from two short exact sequences, one with middle term projective, the other with middle term injective. Thus, for any i, j, we may begin with $0 \to K_j \to P_j \to K_{j-1} \to 0$ and $0 \to L^i \to E^i \to L^{i+1} \to 0$ and conclude

$$\operatorname{Ext}^1(K_j, L^i) \cong \operatorname{Ext}^1(K_{j-1}, L^{i+1}).$$

Note that this remains valid if we interpret $K_{-1} = C$ and $L^0 = A$. Combine these isomorphisms with Corollaries 6.16 and 6.19:

$$\operatorname{Ext}^{n+1}(C, A) \cong \operatorname{Ext}^1(C, L^{n-1}) \cong \operatorname{Ext}^1(K_0, L^{n-2}) \cong \operatorname{Ext}^1(K_1, L^{n-3})$$
$$\cdots \cong \operatorname{Ext}^1(K_{n-2}, L^0) \cong \operatorname{ext}^1(K_{n-1}, A) \cong \operatorname{ext}^{n+1}(C, A).$$

We conclude that $\operatorname{Ext}^n(C, A) \cong \operatorname{ext}^n(C, A)$, as desired.   ∎

The reader should observe that all was proved from the following data: Hom is a left exact bifunctor whose first right derived functors vanish on appropriate modules. It is now a simple matter to give the dual treatment to the right exact bifunctor tensor. That Tor_1 vanishes on projectives is proved as in Theorem 7.6 (details are given in Theorem 8.4).

Theorem 7.9:   *If* $\cdots \to P_1 \to P_0 \to A \to 0$ *and* $\cdots \to Q_1 \to Q_0 \to B \to 0$ *are projective resolutions, then for all* $n \geq 0$,

$$H_n(\mathbf{P}_A \otimes B) \cong H_n(A \otimes \mathbf{Q}_B).$$

Thus the two definitions of Tor_n^R *have the same value on* (A, B).

Corollary 7.10:   *Assume* $\operatorname{Tor}_1(A, B) = 0$ *if either* A *or* B *is flat. If* $\cdots \to P_1 \to P_0 \to A \to 0$ *and* $\cdots \to Q_1 \to Q_0 \to B \to 0$ *are flat resolutions, then for all* $n \geq 0$,

$$H_n(\mathbf{P}_A \otimes B) \cong H_n(A \otimes \mathbf{Q}_B).$$

Proof: In the proof of Theorem 7.9, projectivity is used only in ensuring that Tor_1 vanish.   ∎

Remarks: 1.   In the next chapter, we shall prove that $\operatorname{Tor}_1(A, B)$ does vanish if either A or B is flat (Theorem 8.7).
2.   The force of Corollary 7.10 is that one may use flat resolutions, not merely projective resolutions, to compute Tor.

Let us return to Ext, having disposed of the possible distinction between Ext and ext.

Definition: An **extension** of A by C is an exact sequence

$$0 \to A \to B \to C \to 0.$$

Theorem 7.11:   *If* $\mathrm{Ext}^1(C, A) = 0$, *then every extension of A by C is split.*

Proof: Let $0 \to A \overset{i}{\to} B \to C \to 0$ be an extension. Applying $\mathrm{Hom}(\ , A)$ gives exactness of

$$\mathrm{Hom}(B, A) \overset{i^*}{\to} \mathrm{Hom}(A, A) \to \mathrm{Ext}^1(C, A) = 0,$$

so that i^* is epic. In particular, there is $g \in \mathrm{Hom}(B, A)$ with $1_A = i^*(g) = gi$, and this says the original sequence splits.   ∎

Later, we shall prove the converse of Theorem 7.11; now we prove the converses of Theorems 7.6 and 7.7.

Corollary 7.12:   *If* $\mathrm{Ext}^1(C, A) = 0$ *for all* A, *then* C *is projective; if* $\mathrm{Ext}^1(C, A) = 0$ *for all* C, *then* A *is injective.*

Proof: Theorems 3.13 and 3.19.   ∎

The next two results show that Ext behaves as Hom does on sums and products.

Theorem 7.13:   *For all n,* $\mathrm{Ext}^n(\coprod A_k, B) \cong \prod \mathrm{Ext}^n(A_k, B)$.

Proof: We do an induction on n, the case $n = 0$ being Theorem 2.4. For each k, construct an exact sequence

$$0 \to L_k \to P_k \to A_k \to 0,$$

where P_k is projective. There results an exact sequence

$$0 \to \coprod L_k \to \coprod P_k \to \coprod A_k \to 0,$$

and $\coprod P_k$ is projective. There is a commutative diagram with exact rows

$$
\begin{array}{ccccccc}
\mathrm{Hom}(\coprod P_k, B) & \longrightarrow & \mathrm{Hom}(\coprod L_k, B) & \longrightarrow & \mathrm{Ext}^1(\coprod A_k, B) & \longrightarrow & \mathrm{Ext}^1(\coprod P_k, B) = 0 \\
\downarrow & & \downarrow & & \downarrow & & \\
\prod \mathrm{Hom}(P_k, B) & \longrightarrow & \prod \mathrm{Hom}(L_k, B) & \longrightarrow & \prod \mathrm{Ext}^1(A_k, B) & \longrightarrow & \prod \mathrm{Ext}^1(P_k, B) = 0
\end{array}
$$

(the vertical arrows are the isomorphisms of Theorem 2.4; the maps in the bottom row are the maps of Theorem 7.5 at each coordinate). Diagram-chasing provides the dashed arrow

$$\mathrm{Ext}^1(\coprod A_k, B) \to \prod \mathrm{Ext}^1(A_k, B)$$

which must be an isomorphism. The theorem is thus true for $n = 1$ (many inductions involving Ext^n start slowly). If $n > 1$, there is a diagram

$$\text{Ext}^{n-1}(\coprod P_k, B) \longrightarrow \text{Ext}^{n-1}(\coprod L_k, B) \overset{\partial}{\longrightarrow} \text{Ext}^n(\coprod A_k, B) \longrightarrow \text{Ext}^n(\coprod P_k, B)$$

$$\downarrow \psi$$

$$\prod \text{Ext}^{n-1}(P_k, B) \longrightarrow \prod \text{Ext}^{n-1}(L_k, B) \underset{\partial'}{\longrightarrow} \prod \text{Ext}^n(A_k, B) \longrightarrow \prod \text{Ext}^n(P_k, B)$$

The first and last terms in each row are 0 (Theorem 7.6), so exactness gives ∂ and ∂' isomorphisms. By induction, there is an isomorphism ψ, and $\partial' \psi \partial^{-1} : \text{Ext}^n(\coprod A_k, B) \to \prod \text{Ext}^n(A_k, B)$ is an isomorphism. $\blacksquare$

Theorem 7.14: *For all n, $\text{Ext}^n(A, \prod B_k) \cong \prod \text{Ext}^n(A, B_k)$.*

Proof: Dual to the proof of Theorem 7.13. In particular, begin with exact sequences $0 \to B_k \to E_k \to Q_k \to 0$, where E_k is injective, and then apply the long exact sequence for $\text{Ext}(A, \ )$ to $0 \to \prod B_k \to \prod E_k \to \prod Q_k \to 0$. $\blacksquare$

Given a complex of R-modules and R-maps, the complex obtained after applying Hom_R is a complex of abelian groups and $\mathbb{Z}$-maps. If R is commutative, however, we may consider R-modules as $(R - R)$-bimodules and equip each Hom_R with an R-module structure (Theorem 1.15). Also, the maps between the Hom's are then R-maps.

Theorem 7.15: *If R is commutative, then $\text{Ext}^n_R(A, B)$ is an R-module.*

Proof: Ext^n is, by definition, a quotient of two R-submodules of $\text{Hom}_R(P_n, B)$, where P_n is the nth term of a projective resolution of A. $\blacksquare$

We generalize this result. Let $r \in Z(R)$, the center of R, and let A be an R-module. Then $\mu_r : A \to A$ defined by $a \mapsto ra$ is an R-map, called **multiplication** by r (or **homothety**). If $r \notin Z(R)$, then μ_r may not be an R-map (e.g., take $A = R$).

Theorem 7.16: *If $\mu : A \to A$ is multiplication by r, where $r \in Z(R)$, then $\mu^* : \text{Ext}^n_R(A, B) \to \text{Ext}^n_R(A, B)$ is also multiplication by r. The same is true in the second variable.*

Proof: Consider the diagram

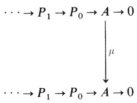

where the rows are identical projective resolutions of A. Recall the definition of μ^*: first fill in a chain map g over μ, say $g = \{g_n : P_n \to P_n\}$, then apply $\text{Hom}_R(\ , B)$, and μ^* is the induced map

$$\mu^*(z_n + \text{boundaries}) = g_n^* z_n + \text{boundaries}.$$

We also know that any choice of chain map over μ gives the same μ^*. In particular, define g by letting $g_n : P_n \to P_n$ be multiplication by r. This is a chain map over μ, and

$$\mu^*(z_n + \text{boundaries}) = r z_n + \text{boundaries} = r(z_n + \text{boundaries}). \quad \blacksquare$$

Exercises: 7.2.   Use Theorem 7.16 to show that $\text{Ext}_R^n(A, B)$ is a $Z(R)$-module.

7.3.   Let R be a domain and A an R-module. Prove that A is torsion-free if and only if every multiplication $\mu_r : A \to A$ (with $r \neq 0$) is monic; A is divisible if and only if every such multiplication is epic.

7.4.   Let R be a domain and A a divisible R-module. If $rA = 0$ for some $r \in R$, $r \neq 0$, then $A = 0$.

7.5.   Let T be a functor that **preserves multiplications**: whenever $\mu_r : A \to A$ is multiplication by $r \in Z(R)$, then $T\mu_r : TA \to TA$ is also multiplication by r. If R is a domain and A is a torsion-free, divisible R-module, then TA is also torsion-free divisible.

7.6.   Let R be commutative and A an R-module with $rA = 0$ for some $r \in R$. If T preserves multiplications, then $rTA = 0$.

7.7.   Every additive functor $T : \mathbf{Ab} \to \mathbf{Ab}$ preserves multiplications. More generally, if R is a subring of $\mathbf{Q}$, then every additive functor $T : {}_R\mathfrak{M} \to {}_R\mathfrak{M}$ preserves multiplications.

The theorems we have just proved are useful in computing Ext. We give the next result for abelian groups, although it obviously generalizes to modules over any domain.

Theorem 7.17:   *If B is an abelian group, then*

$$\text{Ext}_{\mathbf{Z}}^1(\mathbf{Z}/m\mathbf{Z}, B) \cong B/mB.$$

Proof: Apply $\text{Hom}_{\mathbf{Z}}(\ , B)$ to $0 \to \mathbf{Z} \overset{m}{\to} \mathbf{Z} \to \mathbf{Z}/m\mathbf{Z} \to 0$, where the first map is multiplication by m, to obtain exactness of

$$\text{Hom}_{\mathbf{Z}}(\mathbf{Z}, B) \overset{m}{\to} \text{Hom}_{\mathbf{Z}}(\mathbf{Z}, B) \to \text{Ext}_{\mathbf{Z}}^1(\mathbf{Z}/m\mathbf{Z}, B) \to \text{Ext}_{\mathbf{Z}}^1(\mathbf{Z}, B).$$

Since $\mathbf{Z}$ is projective, even free, the last Ext $= 0$; also, the first map is still multiplication by m, while $\text{Hom}_{\mathbf{Z}}(\mathbf{Z}, B)$ may be identified with B (Theorem 1.16). It follows that $\text{Ext}_{\mathbf{Z}}^1(\mathbf{Z}/m\mathbf{Z}, B) \cong B/mB$. $\blacksquare$

If A and B are f.g. abelian groups, recall the Fundamental Theorem asserting each of them is a sum of cyclic groups; the results above permit explicit computation of $\operatorname{Ext}^1_{\mathbf{Z}}(A, B)$.

Exercises: 7.8. If A and B are arbitrary abelian groups, then $\operatorname{Ext}^n_{\mathbf{Z}}(A, B) = 0$ for all $n \geq 2$. (Hint: Use Corollary 4.19: every subgroup of a free abelian group is free abelian.)

   7.9. If A is an abelian group with $mA = A$ for some $m \in \mathbf{Z}$, then every exact sequence $0 \to A \to E \to \mathbf{Z}/m\mathbf{Z} \to 0$ splits.

   7.10. If A and C are abelian groups with $mA = 0 = nC$, where $(m, n) = 1$, then every extension of A by C is split.

   7.11. If A is a torsion abelian group, then $\operatorname{Ext}^1_{\mathbf{Z}}(A, \mathbf{Z}) \cong \operatorname{Hom}(A, \mathbf{R}/\mathbf{Z})$, where $\mathbf{R}/\mathbf{Z}$ is the circle group.

   7.12. (i)   If G is a nonzero divisible group, then $\operatorname{Hom}(\mathbf{Q}, G) \neq 0$.

           (ii) If p_m is the mth prime, then $(\prod_m \mathbf{Z}/p_m\mathbf{Z})/(\coprod_m \mathbf{Z}_m/p_m\mathbf{Z})$ is divisible.

   7.13. Using Exercise 7.12, prove there is a nonsplit extension

$$0 \to \coprod_m \mathbf{Z}/p_m\mathbf{Z} \to E \to \mathbf{Q} \to 0.$$

(We shall see later (Theorem 9.3) that this result is false if the index set of the sum is finite.)

EXT¹ AND EXTENSIONS

We seek an interpretation of $\operatorname{Ext}^1_R(C, A)$, and we begin with a definition motivated by Theorem 5.9.

Definition: Two extensions ξ and ξ' of A by C are **equivalent** if there is a commutative diagram

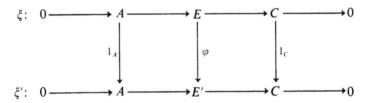

It is easy to check that equivalence is, in fact, an equivalence relation (to prove symmetry, use the Five Lemma to show that φ must be an isomorphism). The equivalence class of an extension $\xi:0 \to A \to B \to C \to 0$ is denoted $[\xi]$, and the set of all equivalence classes is denoted $e(C, A)$. Our aim is to identify $e(C, A)$ with $\text{Ext}^1(C, A)$ (thus providing the etymology of Ext). We remind the reader that equivalent extensions have isomorphic "middle modules", but, unfortunately, the converse may fail. Recall that an example was given, after Theorem 5.9, of inequivalent extensions having isomorphic middle modules.

Exercise: 7.14. Any two split extensions of A by C are equivalent.

Suppose $\xi:0 \to A \to B \to C \to 0$ is an extension. Given a projective resolution of C, we may form the diagram

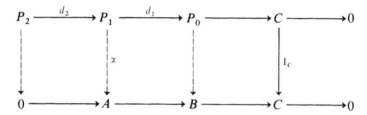

By the Comparison Theorem 6.9, we may fill in the dashed arrows and obtain a commutative diagram. In particular, there is a map $\alpha:P_1 \to A$ with $\alpha d_2 = 0$. By definition, $\text{Ext}^1(C, A) = \ker d_2^* / \text{im } d_1^*$, where $d_2^* = \text{Hom}(d_2, A)$. Since $\alpha d_2 = d_2^*(\alpha)$, we see that $\alpha \in \ker d_2^*$, i.e. α is a cocycle. Second, we know that α is unique to homotopy: if $\alpha':P_1 \to A$ is a second such map, then there are maps s_0, s_1 with $\alpha - \alpha' = 0 \cdot s_1 + s_0 d_1$ (for there is only the zero map $0 \to A$).

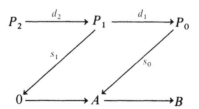

Thus, $\alpha - \alpha' \in \text{im } d_1^*$, and ξ determines a unique element of $\text{Ext}^1(C, A)$. It is easy to check that if ξ and ξ' are equivalent, they determine the same element. In sum, we have defined a function $\psi:e(C, A) \to \text{Ext}^1(C, A)$, namely, $[\xi] \mapsto \alpha + \text{im } d_1^*$.

Exercise: 7.15. If ξ is a split extension, then $\psi[\xi] = 0$.

The next lemma arises from analyzing the diagram occurring in the definition of the function ψ.

Lemma 7.18:   *Consider the diagram*

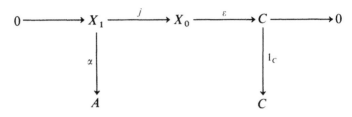

where the row is an extension ξ.

(i)   *There exists an extension of A by C making the following diagram commute:*

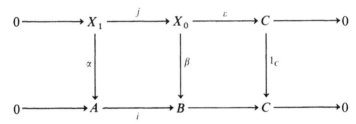

(ii)   *In any such diagram, the first square is a pushout.*
(iii)   *Any two extensions completing the diagram are equivalent.*

Proof: (i)   This is Lemma 3.21.
(ii)   First show the map $\sigma: A \oplus X_0 \to B$ defined by $(a, x_0) \mapsto ia + \beta x_0$ is epic; then show $\ker \sigma = \{(\alpha x_1, -jx_1): x_1 \in X_1\}$ and use Exercise 2.29.
(iii)   Any two extensions completing the diagram begin with A and end with C. Since we have pushouts, the map between the middle modules is provided by the corresponding universal property. ∎

Notation: The extension just constructed is denoted $\alpha\xi$.

Observe that the dual of Lemma 7.18 is also true. Now begin with

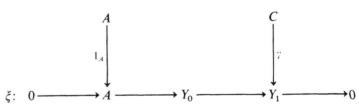

There does exist a completion, unique to equivalence, and the second square is a pullback.

Notation: The extension just described is denoted $\xi\gamma$.

Theorem 7.19: *The function* $\psi: e(C, A) \to \text{Ext}^1(C, A)$ *is a one–one correspondence.*

Proof: We exhibit a function $\theta: \text{Ext}^1(C, A) \to e(C, A)$ inverse to ψ. Let $\cdots \to P_2 \xrightarrow{d_2} P_1 \xrightarrow{d_1} P_0 \to C \to 0$ be the projective resolution of C chosen in the definition of Ext. If $\alpha: P_1 \to A$ is a cocycle, then $\alpha d_2 = 0$ and so α induces a map $\bar{\alpha}: P_1/\text{im } d_2 \to A$, namely, $\bar{\alpha}(x + \text{im } d_2) = \alpha x$. If Ξ denotes the extension

$$\Xi: 0 \to P_1/\text{im } d_2 \to P_0 \to C \to 0,$$

define θ by $\alpha + \text{im } d_1^* \mapsto [\bar{\alpha}\Xi]$. By definition, $\theta(\alpha + \text{im } d_1^*)$ is the bottom row of the following commutative diagram:

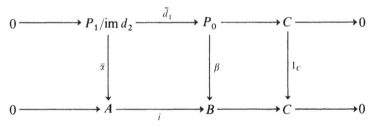

Let us show θ is independent of the choice of representative α. Suppose $\alpha' = \alpha + sd_1$, where $s: P_0 \to A$. The following diagram commutes:

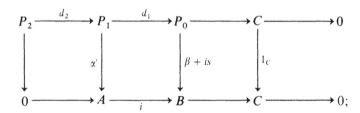

it follows that there is another commutative diagram

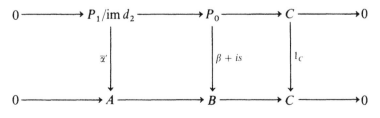

The uniqueness result of Lemma 7.18 shows that $\bar{\alpha}'\Xi$ is equivalent to the bottom row $\alpha\xi$, i.e., $\theta(\alpha + \operatorname{im} d_1^*) = \theta(\alpha' + \operatorname{im} d_1^*)$.

We claim $\psi\theta = 1$. Starting with $\alpha + \operatorname{im} d_1^*$, construct the diagram

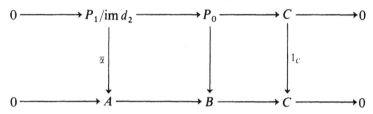

Visibly, ψ of the bottom row is $\alpha + \operatorname{im} d_1^*$.

Finally, we claim $\theta\psi = 1$. Starting with $[\xi]$, observe that $\psi[\xi] = \alpha + \operatorname{im} d_1^*$, as in the diagram just above. Now $\theta(\alpha + \operatorname{im} d_1^*) = [\bar{\alpha}\Xi] = [\xi]$, by the uniqueness in Lemma 7.18. ∎

Corollary 7.20: $\operatorname{Ext}^1(C, A) = 0$ *if and only if every extension of A by C is split.*

Proof: Necessity is Theorem 7.11. Conversely, if every extension of A by C is split, then Exercise 7.14 shows $e(C, A)$ is a set with one element, and the Theorem gives $\operatorname{Ext}^1(C, A) = 0$. ∎

We may now tie a loose end from Chapter 5. After proving Theorem 5.9, we claimed that $e(\mathbf{Z}/p\mathbf{Z}, \mathbf{Z}/p\mathbf{Z}) \cong \mathbf{Z}/p\mathbf{Z}$. In Chapter 5, $e(C, A)$ meant all equivalence classes of extensions (not necessarily abelian) of the abelian group A by the group C. In the special case $A = C = \mathbf{Z}/p\mathbf{Z}$, however, we pointed out there that every middle group of an extension is necessarily abelian; thus, the two meanings of $e(\mathbf{Z}/p\mathbf{Z}, \mathbf{Z}/p\mathbf{Z})$ coincide, and may be identified with $\operatorname{Ext}^1_{\mathbf{Z}}(\mathbf{Z}/p\mathbf{Z}, \mathbf{Z}/p\mathbf{Z})$, which has order p (Theorem 7.17).

$\operatorname{Ext}^1(C, A)$ has been treated via projective resolutions of C; we know that it could also be treated via injective resolutions of A. This approach yields a function $\psi': e(C, A) \to \operatorname{Ext}^1(C, A)$ defined by $[\xi] \mapsto \gamma + \operatorname{im} d_*^0$, where

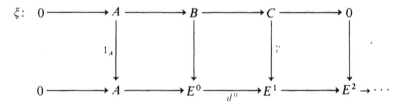

Just as above, one may show ψ' is a one–one correspondence; its inverse is θ', where $\theta'(\gamma + \operatorname{im} d_*^0) = [\Omega\bar{\gamma}]$,

$$\Omega = 0 \to A \to E^0 \to E^1/\operatorname{im} d^0 \to 0,$$

and

$$\bar{\gamma}: C \to E^1/\mathrm{im}\, d^0$$

is the map induced by the cocycle γ.

If G is a group, X a set, and $\psi: X \to G$ a one–one correspondence, then there is a unique addition on X making X a group and ψ an isomorphism. What is this addition on $e(C, A)$? We prepare by presenting three book-keeping formulas. We abuse notation so that $\psi[\xi]$ denotes any coset representative of $\alpha + \mathrm{im}\, d_1^*$.

I. If $[\xi] \in e(C, A)$ and $h: A \to A'$, then

$$\psi[h\xi] = h\psi[\xi].$$

This formula follows from the diagram

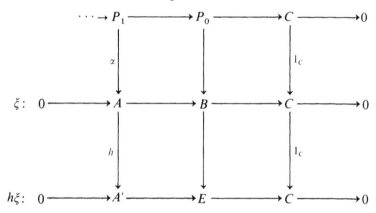

II. If $[\xi] \in e(C, A)$ and $k: C' \to C$, then

$$\psi[\xi k] = \psi[\xi]k_1,$$

where k_1 is part of a chain map over k (see diagram below). Consider the diagram

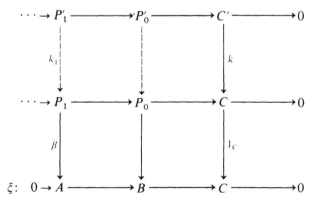

where k_1 is as advertised (existing by the Comparison Theorem), and β is a coset representative of $\psi[\xi]$.

Having pictured $\psi[\xi]k_1 = \beta k_1$, let us now picture $\psi[\xi k]$. Consider the diagram

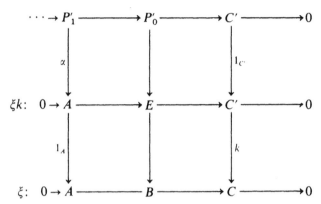

Here α is a coset representative of $\psi[\xi k]$. Since βk_1 and α are each the first constituent of a chain map over k, uniqueness to homotopy puts them in the same coset mod im d_1^*.

The most obvious way to put two extensions ξ and ξ' together is to form their **sum** $\xi \oplus \xi'$:

$$0 \to A \oplus A' \xrightarrow{i \oplus i'} B \oplus B' \xrightarrow{p \oplus p'} C \oplus C' \to 0.$$

III. $\psi[\xi \oplus \xi'] = \psi[\xi] \oplus \psi[\xi']$. (The right hand side is the sum of maps as pictured below.) This formula follows from the diagram

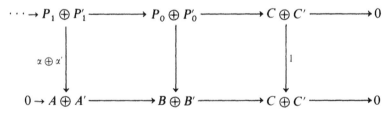

The clue indicating how to assemble these formulas is the realization that Ext^1 is a generalized Hom (after all, Hom $= \text{Ext}^0$), and addition of homomorphisms can be described in terms of composition and sum. Explicitly, let $f, g \in \text{Hom}(C, A)$. If we define $\Delta_C : C \to C \oplus C$ by $c \mapsto (c, c)$ and $\nabla_A : A \oplus A \to A$ by $(a, a') \mapsto a + a'$, then

$$f + g = \nabla(f \oplus g)\Delta,$$

for the right side sends $c \mapsto (c, c) \mapsto (fc, gc) \mapsto fc + gc$.

Exercise: 7.16. If $\cdots \to P_1 \to P_0 \to C \to 0$ is a projective resolution of C and $\cdots \to P_1 \oplus P_1 \to P_0 \oplus P_0 \to C \oplus C \to 0$ is a projective resolution of $C \oplus C$, then $\{f_n : P_n \to P_n \oplus P_n\}$ is a chain map over Δ, where $f_n = \Delta_{P_n}$ for all n. (A similar result holds for ∇.) Conclude that we may take $\Delta_1 = \Delta$ in bookkeeping formula II.

Theorem 7.21: $e(C, A)$ *is an abelian group under* **Baer sum**,

$$[\xi] + [\xi'] = [(\nabla(\xi \oplus \xi'))\Delta],$$

and $\psi : e(C, A) \to \text{Ext}^1(C, A)$ *is an isomorphism.*

Remark: It is true that associativity $[(\alpha\xi)\gamma] = [\alpha(\xi\gamma)]$ holds; however, we do not use this fact in the proof.

Proof: The formula for Baer sum defines a relation

$$\rho : e(C, A) \times e(C, A) \to e(C, A)$$

which we do not yet know is a function. The bookkeeping formulas give

$$
\begin{aligned}
\psi[(\nabla(\xi \oplus \xi'))\Delta] &= \psi[\nabla(\xi \oplus \xi')]\Delta_1 = \nabla\psi[\xi \oplus \xi']\Delta_1 \\
&= \nabla(\psi[\xi] \oplus \psi[\xi'])\Delta_1 \\
&= \nabla(\psi[\xi] \oplus \psi[\xi'])\Delta \qquad \text{(Exercise 7.16)} \\
&= \psi[\xi] + \psi[\xi'].
\end{aligned}
$$

There are two conclusions from this computation. First, $\psi\rho$ is a function, so that, since ψ is a one–one correspondence, $\rho = \psi^{-1}(\psi\rho)$ is also a function. Second, $\psi([\xi] + [\xi']) = \psi[\xi] + \psi[\xi']$, so that ψ is a homomorphism and ρ is the good addition on $e(C, A)$. ∎

One could define $\text{Ext}^1(C, A)$ as $e(C, A)$, and prove directly that it is an abelian group under Baer sum that repairs the loss of exactness after applying Hom. This approach has the advantage that it avoids choosing resolutions of either variable; indeed, no projectives or injectives need be mentioned. This is MacLane's viewpoint; paraphrasing him, this definition of Ext^1 relegates resolutions to their proper place, namely, as aids to computation. Yoneda has shown how to generalize $e(C, A)$ and Baer sum to recapture $\text{Ext}^n(C, A)$. Briefly, an element of $\text{Ext}^n(C, A)$ corresponds to an equivalence class of exact sequences of the form

$$0 \to A \to B_n \to \cdots \to B_1 \to C \to 0.$$

Details may be found in [MacLane, 1963, pp. 82–87]. We have chosen a standard presentation since it follows the historical development (even though Baer's work introducing Baer sum is quite early), hence is more natural once one has seen computations as in Chapter 5.

Exercises: 7.17.   Consider the diagram with exact rows

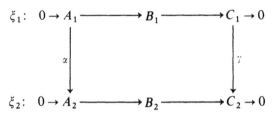

Prove there is a map $\beta: B_1 \to B_2$ making the diagram commute if and only if $[\alpha\xi_1] = [\xi_2\gamma]$.

7.18.   Show that $-[\xi] = [(-1_A)\xi] = [\xi(-1_C)]$. Generalize by replacing -1_A and -1_C with multiplication by an element in the center of the ring of scalars.

7.19.   Show that $[\xi] = [0 \to A \xrightarrow{i} B \to C \to 0]$ has finite order in $e(C, A)$ if and only if there is $m \in \mathbf{Z}$, $m \neq 0$, and a map $s: B \to A$ with $si = m \cdot 1_A$.

7.20. $e(C, \ )$ is a functor: if $h: A \to A'$, define

$$h_{\#}: e(C, A) \to e(C, A')$$

by $[\xi] \mapsto [h\xi]$. Show that $e(C, \ )$ and $\mathrm{Ext}^1(C, \ )$ are naturally equivalent.

7.21.   If χ is the extension $0 \to A' \xrightarrow{i} A \xrightarrow{\pi} A'' \to 0$, define a map $D: \mathrm{Hom}(C, A'') \to e(C, A')$ by $g \mapsto [\chi g]$. Prove exactness of

$$\mathrm{Hom}(C, A) \xrightarrow{\pi_*} \mathrm{Hom}(C, A'') \xrightarrow{D} e(C, A') \xrightarrow{i_*} e(C, A) \xrightarrow{\pi_*} e(C, A'').$$

7.22.   Prove commutativity of

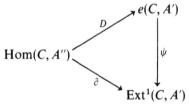

7.23. $e(\ , A)$ is a contravariant functor: if $k: C' \to C$, then $k^{\#}: e(C, A) \to e(C', A)$ is given by $[\xi] \mapsto [\xi k]$. Prove that $e(\ , A)$ and $\mathrm{Ext}^1(\ , A)$ are naturally equivalent.

7.24.   If Γ is the extension $0 \to C' \to C \to C'' \to 0$, define $D': \mathrm{Hom}(C', A) \to e(C'', A)$ by $g \mapsto [g\Gamma]$. Prove exactness of

$$\mathrm{Hom}(C, A) \to \mathrm{Hom}(C', A) \xrightarrow{D'} e(C'', A) \to e(C, A) \to e(C', A).$$

7.25.   Prove commutativity of

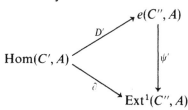

The naive approach to extensions that we followed in Chapter 5 can also be followed here, and, as there, it leads to a special free resolution of the first variable. Thus, given an extension of left R-modules $0 \to A \to B \to C \to 0$, let us choose a lifting $\lambda : C \to B$ with $\lambda 0 = 0$. Each element of B has a unique expression of the form $a + \lambda c$, and, when we try to add two elements, a factor set emerges:

$$(a + \lambda c) + (a' + \lambda c') = a + a' + f(c, c') + \lambda(c + c');$$

here $f : C \times C \to A$ satisfies identities:

  (i) $f(c, 0) = 0 = f(0, c')$;
  (ii) $f(c', c'') - f(c + c', c'') + f(c, c' + c'') - f(c, c') = 0$;
  (iii) $f(c, c') = f(c', c)$.

The second identity arises from associativity; the third arises from commutativity. This is not enough, for such a function f only carries the information that B is an abelian group. To ensure that B is a module, observe there is a second function $g : R \times C \to A$ defined by $g(r, c) = r\lambda c - \lambda rc$. There are more identities, arising from the module axioms.

  (iv) $g(1, c) = 0 = g(r, 0)$;
  (v) $rg(s, c) = g(rs, c) - g(r, sc)$;
  (vi) $g(r + s, c) + f(rc, sc) = g(r, c) + g(s, c)$;
  (vii) $g(r, c + c') + f(rc, rc') = g(r, c) + g(r, c')$.

The ordered pair (f, g) conveys all the data. We only remark that the set of all such (f, g) forms an abelian group (each coordinate acts via pointwise addition), and if one chooses a second lifting $\lambda' : C \to B$, then he determines a subgroup. Obviously, the resolution and boundary formula are more complicated than before (although they are simple for $\mathbf{Z}$-modules when g can be forgotten).

AXIOMS

In this section, we characterize the sequence of functors $\operatorname{Ext}^n(C, \ )$, $n \geq 0$, by some properties we already know them to have. Note that $\operatorname{Ext}^0(C, \ ) =$

Hom(C,) is characterized in Theorem 3.38. The dual theorem, characterizing the sequence Ext^n(, A), $n \geq 0$, is also true and is left as an exercise; note that Ext^0(, A) = Hom(, A) is characterized in Theorem 3.36.

Definition: A sequence of functors $\{T_n : n \in \mathbf{Z}\}$ is **connected** if, for every exact sequence of modules $0 \to A' \to A \to A'' \to 0$, there exist maps

$$\Delta_n : T_n(A'') \to T_{n-1}(A')$$

such that

   (i) $\cdots \to T_n(A') \to T_n(A) \to T_n(A'') \overset{\Delta_n}{\to} T_{n-1}(A') \to T_{n-1}(A) \to T_{n-1}(A'') \to \cdots$

is a complex;

   (ii)   the "connecting maps" Δ_n are natural: given a commutative diagram with exact rows

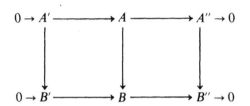

there are commutative squares for all n:

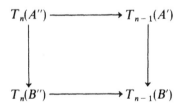

Remarks: 1.   The sequence is **positive** if $T_n = 0$ for all $n < 0$; the sequence is **negative** if $T_n = 0$ for $n > 0$. In the latter case, we employ the usual convention of raising indices (and $\Delta^n : T^n(A'') \to T^{n+1}(A')$).

2.   The definition can be easily generalized so that the functors are evaluated on complexes instead of modules.

Definition: A connected sequence of functors is **strongly connected** if the complex in (i) is acyclic, i.e., it is an exact sequence.

Theorem 7.22   **(Axioms):**   *Let $\{T^n\}$ be a negative sequence of functors such that*

   (i)   *$\{T^n\}$ is strongly connected;*

(ii) $T^0 \cong \mathrm{Hom}(C, \ )$ *for some module* C;
(iii) $T^n(E) = 0$ *for all injective modules* E *and all* $n \geq 1$.

Then $T^n \cong \mathrm{Ext}^n(C, \ )$ *for all* $n \geq 0$.

Proof: We do an induction on n. The hypothesis gives the result for $n = 0$ and, as usual with inductions here, we must treat the case $n = 1$ before doing the inductive step.

Given a module A, choose an exact sequence

$$0 \to A \to E \to D \to 0$$

with E injective. There is a commutative diagram with exact rows

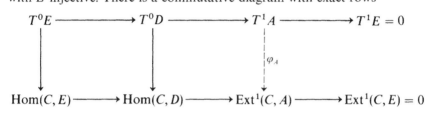

where the downward arrows arise from the given natural equivalence $T^0 \cong \mathrm{Hom}(C, \ )$. Diagram-chasing provides a map $\varphi_A : T^1 A \to \mathrm{Ext}^1(C, A)$ that is necessarily an isomorphism.

It appears that φ_A depends on the choice of imbedding into an injective. Assume B is a module with

$$0 \to B \to G \to H \to 0$$

an exact sequence with G injective; thus, φ_B is defined. If $f : A \to B$, injectivity provides dashed arrows,

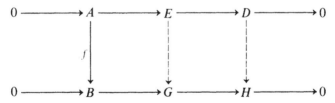

There is a three-dimensional diagram:

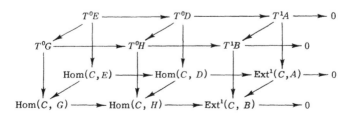

Now every face of the left cube commutes. In the right cube, the top and bottom faces commute by naturality of the connecting homomorphisms; the front and back faces commute, by the construction of φ_A and φ_B. It follows that the right face commutes, i.e., φ is natural. Further, if we set $A = B$ and $f = 1_A$, then it follows that each φ_A is independent of the choice of imbedding in an injective.

The inductive step is similar (and easier, for the connecting maps are now isomorphisms). ∎

Exercises: 7.26.   Let $0 \to A' \to A \to A'' \to 0$ be exact. Show commutativity of

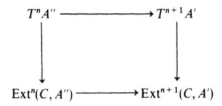

where the vertical maps are the natural isomorphisms constructed in the above proof.

7.27.   State and prove the (dual) characterization of the sequence $\{\text{Ext}^n(\, , A)\}$. Condition (iii) may be relaxed so that it reads "$T^n(B) = 0$ for all free modules B and all $n \geq 1$" (thus, "projective" may be replaced by "free").

7.28.   Let T be a left exact functor, and let $\{T^n\}$ be a negative sequence of functors such that

(i) $\{T^n\}$ is strongly connected;
(ii) $T^0 \cong T$;
(iii) $T^n(E) = 0$ for every injective module E and all $n \geq 1$.

Prove that $T^n \cong R^n T$ for all $n \geq 0$.

Characterizations of $\text{Ext}_R^n(C, \,)$ for fixed $n > 0$ are rare. Here is an instance of such a theorem [Griffith, 1974]. Let R be Dedekind with quotient field Q. If $T : {}_R\mathfrak{M} \to {}_R\mathfrak{M}$ is a half exact functor that preserves products, annihilates injectives, and $\text{Hom}_R(Q, TP) = 0$ for every projective P, then $T \cong \text{Ext}_R^1(C, \,)$ for some C.

The next construction will not be used for a while, but this is an appropriate place for it. Assume an exact sequence of modules

$$0 \to A_p \to \cdots \to A_1 \to A_0 \to 0$$

and a negative connected sequence of functors $\{T^n\}$. "Factor" the exact sequence into short exact sequences

$$0 \to K_i \to A_i \to K_{i-1} \to 0, \qquad 1 \le i \le p-1,$$

where $K_i = \ker(A_i \to A_{i-1})$; note that $K_{p-1} = A_p$ and $K_0 = A_0$. For each i and n, there are connecting homomorphisms $T^n(K_{i-1}) \to T^{n+1}(K_i)$. There is thus a composite

$$T^n(A_0) \to T^{n+1}(K_1) \to T^{n+2}(K_2) \to \cdots \to T^{n+p-1}(K_{p-1}) = T^{n+p-1}(A_p).$$

Definition: The composite above, $T^n(A_0) \to T^{n+p-1}(A_p)$, is called an **iterated connecting homomorphism**.

Of course, there is a dual construction for a positive connected sequence of functors.

Lemma 7.23: *Consider the commutative diagram with exact rows*

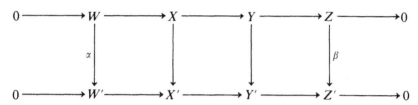

If $\{T^n\}$ is a negative connected sequence of functors, then

$$D'T^{n-1}(\beta) = T^{n+1}(\alpha)D : T^{n-1}(Z) \to T^{n+1}(W'),$$

where $D : T^{n-1}(Z) \to T^{n+1}(W)$ and $D' : T^{n-1}(Z') \to T^{n+1}(W')$ are iterated connecting homomorphisms.

Proof: A simple calculation using naturality of connecting homomorphisms. ∎

Definition: A diagram

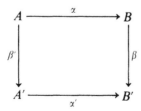

is **anticommutative** if $\beta\alpha = -\alpha'\beta'$.

Theorem 7.24:   *Let $\{T^n\}$ be a negative connected sequence of functors, and assume all rows and columns of the following commutative diagram are exact:*

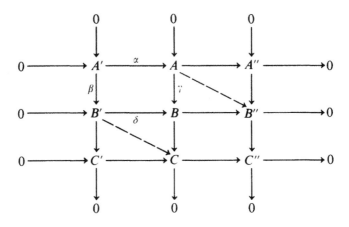

Then there is an anticommutative diagram

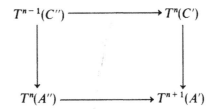

where the maps are connecting homomorphisms.

Proof: If one follows the dashed arrows, there are two exact sequences of length 4:

$$0 \to A' \to A \to B'' \to C'' \to 0; \qquad 0 \to A' \to B' \to C \to C'' \to 0.$$

These sequences induce iterated connecting homomorphisms D and D', respectively, $T^{n-1}(C'') \to T^{n+1}(A')$; D is the clockwise composite we are examining; D' is the counterclockwise one.

We now construct a new four-term exact sequence which will be used to compare the ones above. Define

$$A' \xrightarrow{\sigma} A \oplus B' \xrightarrow{\tau} B$$

by $\sigma a' = (\alpha a', \beta a')$ and $\tau(a, b') = \gamma a - \delta b'$. It is easy to check that

$$0 \to A' \to A \oplus B' \to B \to C'' \to 0$$

is exact, where $B \to C''$ is the map obtained from the 3×3 diagram. If we define $\varepsilon: A' \to A'$ by $\varepsilon: a' \mapsto -a'$, then there is a commutative diagram

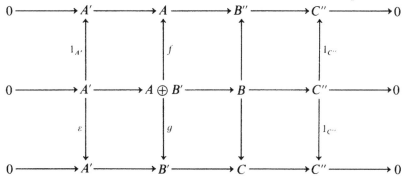

where $f:(a, b') \mapsto a$ and $g:(a, b') \mapsto -b'$ (other maps are from the 3×3 diagram). Applying Lemma 7.23 twice yields $T^{n+1}(\varepsilon)D = D'$. As $\varepsilon = -1_{A'}$, we have $D' = T^{n+1}(\varepsilon)D = -D$, as claimed. ∎

Corollary 7.25: *Let $\{T^n\}$ be a negative connected sequence of functors, and consider a 3×3 commutative diagram of complexes with exact rows and columns. Then there results an anticommutative diagram*

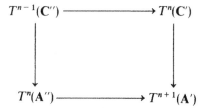

Proof: The proof just given can painlessly be adapted to complexes. ∎

Theorem 7.26: *Consider the commutative diagram of complexes with exact rows and columns*

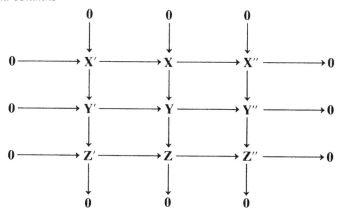

For each n, there is an anticommutative diagram

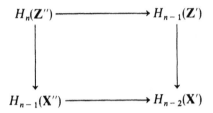

$$H_n(\mathbf{Z}'') \longrightarrow H_{n-1}(\mathbf{Z}')$$
$$\downarrow \qquad\qquad\qquad \downarrow$$
$$H_{n-1}(\mathbf{X}'') \longrightarrow H_{n-2}(\mathbf{X}')$$

where the maps are connecting homomorphisms.

Proof: We may use Corollary 7.25, for the homology functors $\{H_n\}$ form a connected sequence of functors defined on complexes. ∎

This discussion will be continued later to prove certain anticommutativity results for Tor and Ext (Theorems 11.23 and 11.24). In order to prove these theorems, it is necessary to define new complexes whose homology yields Tor or Ext (namely, resolve both variables simultaneously). Although one could give an elementary proof here, it would be out of context, and so we wait until a discussion of bicomplexes and spectral sequences.

Exercises: 7.29. Given two short exact sequences with the same middle term,

$$0 \to A \to B \to C \to 0 \quad \text{and} \quad 0 \to D \to B \to E \to 0,$$

then there is a commutative 3×3 diagram with exact rows and columns

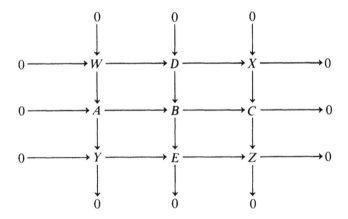

(Hint: Let the upper left corner be the pullback.)

7.30.   Give an example of a commutative 3×3 diagram bordered by zeros, and with exact rows and columns, in which the upper left corner is not a pullback.

8 Tor

Now we turn to Tor. First, recall the definition on modules:

$$\text{Tor}_n^R(A, B) = H_n(\mathbf{P}_A \otimes_R B) = H_n(A \otimes_R \mathbf{Q}_B),$$

where $\mathbf{P}_A$ and $\mathbf{Q}_B$ are deleted projective resolutions; the second equation is Theorem 7.9.

ELEMENTARY PROPERTIES

Theorem 8.1: *If n is negative, then* $\text{Tor}_n(A, B) = 0$ *for all A, B.*

Proof: If $\mathbf{Q}_B = \cdots \to Q_1 \to Q_0 \to 0$, then $A \otimes \mathbf{Q}_B$ has only 0's to the right of $A \otimes Q_0$. ∎

Theorem 8.2: $\text{Tor}_0^R(A,)$ *is naturally equivalent to* $A \otimes_R$; $\text{Tor}_0^R(, B)$ *is naturally equivalent to* $\otimes_R B$.

Proof: If $\mathbf{Q}_B = \cdots \to Q_1 \xrightarrow{d_1} Q_0 \xrightarrow{d_0} 0$, then

$$\text{Tor}_0(A, B) = \ker(1 \otimes d_0)/\text{im}(1 \otimes d_1) = \text{coker}(1 \otimes d_1).$$

If the (nondeleted) resolution is $\cdots \to Q_1 \xrightarrow{d_1} Q_0 \xrightarrow{\varepsilon} B \to 0$, then right exactness of $A \otimes$ gives an exact sequence

$$A \otimes Q_1 \xrightarrow{\ 1 \otimes d_1\ } A \otimes Q_0 \xrightarrow{\ 1 \otimes \varepsilon\ } A \otimes B \to 0,$$

so that $1 \otimes \varepsilon$ induces an isomorphism coker $1 \otimes d_1 \to A \otimes B$. We let the reader check that these maps, one for each B, define a natural equivalence. ∎

Remark: We have really proved that if T is right exact, then $L_0 T \cong T$.

Theorem 8.3: *If $0 \to B' \to B \to B'' \to 0$ is an exact sequence, then there is a long exact sequence*

$$\cdots \to \mathrm{Tor}_1(A, B'') \to A \otimes B' \to A \otimes B \to A \otimes B'' \to 0$$

with natural connecting homomorphisms; similarly in the other variable.

Proof: Theorems 6.21 and 8.2. ∎

Observe that Tor thus repairs the exactness we may have lost by tensoring.

Theorem 8.4: *If P is projective, then $\mathrm{Tor}_n(P, B) = 0$ for all B and all $n \geq 1$; similarly in the other variable.*

Proof: Just as the corresponding result for Ext: the definition of Tor is independent of the choice of projective resolution of P. ∎

Observe that we have already used Theorem 8.4 in our proof of Theorem 7.9.

The following theorem enables us to stop saying "similarly in the other variable"; there is no analogue of this result for Ext. Let R^{op} be the opposite ring of R.

Theorem 8.5: *For all $n \geq 0$ and all A, B,*

$$\mathrm{Tor}_n^R(A, B) \cong \mathrm{Tor}_n^{R^{\mathrm{op}}}(B, A).$$

Proof: If $\mathbf{P}_A = \cdots \to P_1 \to P_0 \to 0$, then it is easily checked that

$$t_n : P_n \otimes_R B \to B \otimes_{R^{\mathrm{op}}} P_n$$

given by

$$x_n \otimes b \mapsto b \otimes x_n$$

are isomorphisms that constitute a chain map $t : \mathbf{P}_A \otimes B \to B \otimes \mathbf{P}_A$. By Exercise 6.6, t is an isomorphism of complexes, so that

$$H_n(\mathbf{P}_A \otimes B) \cong H_n(B \otimes \mathbf{P}_A)$$

for all $n \geq 0$. Since $\mathbf{P}_A$ may be construed as a (deleted) projective resolution of the left R^{op}-module A, the proof is completed by applying Theorem 7.9. ∎

Corollary 8.6:   *If R is commutative, then*

$$\text{Tor}_n^R(A, B) \cong \text{Tor}_n^R(B, A)$$

for all $n \geq 0$ and all A, B.

We now improve Theorem 8.4.

Theorem 8.7:   *If F is flat, then $\text{Tor}_n(F, B) = 0$ for all $n \geq 1$ and all B.*

Proof: Let $\mathbf{Q}_B = \cdots \to Q_1 \to Q_0 \to 0$. Since F is flat, the functor $F \otimes$ is exact. It follows from Exercise 6.4 that the complex $F \otimes \mathbf{Q}_B$ has all homology groups 0 except the zeroth.   ∎

Corollary 8.8:   *The definition of* Tor *is independent of the choice of flat resolution of either variable.*

Proof: We have just proved that the hypothesis of Corollary 7.10 always holds.   ∎

Let us give a second proof of Theorem 8.7 by induction on n. As we have already seen in the last chapter, such proofs involve a short exact sequence whose middle module is special (projective, injective—or flat) and the corresponding long exact sequence.

Definition: An inductive proof as just described is called **dimension shifting**.

Here is a proof by dimension shifting that $\text{Tor}_n(F, B) = 0$ for $n \geq 1$, all B, and flat F. Choose an exact sequence

$$0 \to K \overset{i}{\to} P \to B \to 0$$

with P projective. There is an exact sequence

$$\text{Tor}_1(F, P) \to \text{Tor}_1(F, B) \to F \otimes K \overset{1 \otimes i}{\longrightarrow} F \otimes P.$$

Since P is projective, the first Tor $= 0$. Since F is flat, $1 \otimes i$ is monic, and so $\text{Tor}_1(F, B) = 0$.

For the inductive step, look further out in the long exact sequence:

$$\text{Tor}_{n+1}(F, P) \to \text{Tor}_{n+1}(F, B) \to \text{Tor}_n(F, K).$$

The two outside terms are 0, one by Theorem 8.4, the other by induction, and so exactness forces $\text{Tor}_{n+1}(F, B) = 0$.

There is a strong converse.

Theorem 8.9:   *If $\text{Tor}_1(F, B) = 0$ for all B, then F is flat.*

Proof: If $0 \to B' \xrightarrow{i} B \to B'' \to 0$ is exact, then so is

$$\text{Tor}_1(F, B'') \to F \otimes B' \xrightarrow{1 \otimes i} F \otimes B.$$

Since $\text{Tor}_1(F, B'') = 0$, it follows that $1 \otimes i$ is monic, and hence F is flat.   ∎

Theorem 8.10: $\text{Tor}_n(\coprod A_k, B) \cong \coprod \text{Tor}_n(A_k, B)$, all $n \geq 0$.

Proof: Dimension shifting, using Theorem 2.8.   ∎

Theorem 8.11:   *If the index set is directed, then*

$$\text{Tor}_n(\varinjlim A_k, B) \cong \varinjlim \text{Tor}_n(A_k, B), \qquad all \quad n \geq 0.$$

Proof: Dimension shifting, using Theorem 2.18.   ∎

One must assume the index set is directed in Theorem 8.11, otherwise it may be false, e.g., $\text{Tor}_1(\ , B)$ need not be right exact.

Theorem 8.12:   *If R is commutative, then $\text{Tor}_n^R(A, B)$ is an R-module.*

Proof: In this case, the complex defining Tor is comprised of R-modules and R-maps.   ∎

As with Ext, this last result may be generalized.

Theorem 8.13:   *If $r \in Z(R)$, the center of R, and $\mu : A \to A$ is multiplication by r, then so is $\mu_* : \text{Tor}_n(A, B) \to \text{Tor}_n(A, B)$.*

Proof: Similar to the proof of Theorem 7.16.   ∎

Exercises: 8.1.   If $0 \to A' \to A \to A'' \to 0$ is exact and A' and A'' are flat, then A is flat.

8.2.   If A'' is flat, then the exact sequence

$$0 \to A' \to A \to A'' \to 0$$

is pure exact.

8.3.   Use Theorem 8.13 to show $\text{Tor}_n^R(A, B)$ is a $Z(R)$-module.

8.4.   Prove that $\text{Tor}_1^Z(\mathbf{Z}/n\mathbf{Z}, B) \cong B[n]$, where $B[n] = \{b \in B : nb = 0\}$.

8.5.   If A and B are f.g. abelian groups, compute $\text{Tor}_1^Z(A, B)$.

8.6.   For any abelian groups A and B, prove that

$$\text{Tor}_2^Z(A, B) = 0.$$

8.7.   If A and B are abelian groups with $mA = 0 = nB$, where $(m, n) = 1$, then $\text{Tor}_1^Z(A, B) = 0$. Conclude that, in this case, exactness of $0 \to D \to C \to B \to 0$ implies exactness of $0 \to A \otimes D \to A \otimes C \to A \otimes B \to 0$.

8.8.   If A and B are finite abelian groups, then $\text{Tor}_1^Z(A, B) \cong A \otimes_Z B$.

TOR AND TORSION

In this section, R denotes a domain, Q denotes its quotient field, and K denotes the module

$$K = Q/R.$$

Recall that the **torsion submodule** tA of a module A is defined by

$$tA = \{a \in A : ra = 0 \text{ for some nonzero } r \in R\}.$$

Were R not a domain, then tA might not be a submodule: if $ra = 0 = r'a'$, then $rr'(a + a') = 0$, but it is possible that $rr' = 0$.

Definition: A module A is **torsion** if $tA = A$; a module A is **torsion-free** if $tA = 0$.

We have remarked, in Chapter 4, that the module A/tA is always torsion-free; thus every module (over a domain) is an extension of a torsion module by a torsion-free module. Exercise 7.13 shows that such extensions need not split, even when $R = \mathbf{Z}$.

The torsion submodule actually defines a functor: if $f : A \to B$, define $tf = f \mid tA$. One reason for the name Tor is that $t \cong \operatorname{Tor}_1^R(K, \ )$, as we shall now prove.

Lemma 8.14: *There is a natural isomorphism* $\operatorname{Tor}_1(K, A) \cong A$ *for all torsion modules A.*

Proof: Exactness of $0 \to R \to Q \to K \to 0$ gives exactness of

$$\operatorname{Tor}_1(Q, A) \to \operatorname{Tor}_1(K, A) \xrightarrow{\partial} R \otimes A \to Q \otimes A.$$

The first term is 0 since Q is flat (Corollary 3.48) and $Q \otimes A = 0$ because A is torsion (Exercise 2.27). It follows that $\partial = \partial_A$ is an isomorphism. If B is torsion and $f : A \to B$, naturality of the connecting homomorphism gives commutativity of

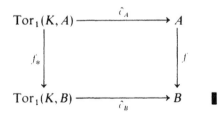

Lemma 8.15: $\operatorname{Tor}_n(K, A) = 0$ *for every A and all $n \geq 2$.*

Proof: We have exactness of

$$\operatorname{Tor}_n(Q, A) \to \operatorname{Tor}_n(K, A) \to \operatorname{Tor}_{n-1}(R, A),$$

and the outside terms are 0 since $n - 1 \geq 1$ and both Q and R are flat. ∎

Lemma 8.16: *If A is torsion-free, then $\operatorname{Tor}_1(K, A) = 0$.*

Proof: We saw in Lemma 4.31 that A can be imbedded in a vector space E over Q; as E is a sum of copies of Q, it is flat. Exactness of $0 \to A \to E \to E/A \to 0$ gives exactness of

$$\operatorname{Tor}_2(K, E/A) \to \operatorname{Tor}_1(K, A) \to \operatorname{Tor}_1(K, E).$$

The first term is 0 by Lemma 8.15; the last term is 0 since E is flat. ∎

Theorem 8.17: *The functors $\operatorname{Tor}_1(K, \ )$ and t are naturally equivalent.*

Proof: Exactness of $0 \to tA \xrightarrow{i} A \to A/tA \to 0$ gives exactness of

$$\operatorname{Tor}_2(K, A/tA) \to \operatorname{Tor}_1(K, tA) \xrightarrow{i_*} \operatorname{Tor}_1(K, A) \to \operatorname{Tor}_1(K, A/tA).$$

The first term is 0 by Lemma 8.15; the last term is 0 by Lemma 8.16. It follows that i_* is an isomorphism, hence that $\sigma = i_* \partial^{-1} : tA \to \operatorname{Tor}_1(K, A)$ is an isomorphism. It is a simple matter, using Lemma 8.14, to show that the maps $\sigma = \sigma_A$ constitute a natural equivalence. ∎

Corollary 8.18: *For every module A, there is an exact sequence*

$$0 \to tA \to A \to Q \otimes A \to K \otimes A \to 0.$$

Corollary 8.19: *A module A is torsion if and only if $Q \otimes A = 0$.*

Proof: We have known necessity for a long time. For the converse, use Corollary 8.18. (One could also use Theorem 3.71.) ∎

Another reason for the name Tor is that Tor is always a torsion module; let us prove this.

Lemma 8.20: *If B is torsion, then $\operatorname{Tor}_n(A, B)$ is torsion for all A and for all $n \geq 0$.*

Proof: We do a dimension shift. If $n = 0$, each generator $a \otimes b$ is torsion, whence $\operatorname{Tor}_0(A, B) = A \otimes B$ is torsion.

If $n = 1$, there is an exact sequence $0 \to N \to P \to A \to 0$ with P projective, and this gives exactness of

$$0 = \operatorname{Tor}_1(P, B) \to \operatorname{Tor}_1(A, B) \to N \otimes B.$$

Since $N \otimes B$ is torsion, by the case $n = 0$, so is its submodule $\text{Tor}_1(A, B)$.

For the inductive step, look further out in the sequence. We have exactness of

$$0 = \text{Tor}_{n+1}(P, B) \to \text{Tor}_{n+1}(A, B) \to \text{Tor}_n(N, B) \to \text{Tor}_n(P, B) = 0.$$

Since $\text{Tor}_n(N, B)$ is torsion, by induction, so is $\text{Tor}_{n+1}(A, B)$. ∎

Theorem 8.21: $\text{Tor}_n(A, B)$ *is torsion for all* A, B *and all* $n \geq 1$.

Proof: Let $n = 1$, and consider the special case when B is torsion-free. By Corollary 8.18, there is an exact sequence

$$0 \to B \to E \to X \to 0,$$

where E is a vector space over Q and $X = E/B$ is torsion. There results an exact sequence

$$\text{Tor}_2(A, X) \to \text{Tor}_1(A, B) \to \text{Tor}_1(A, E).$$

Now $\text{Tor}_2(A, X)$ is torsion, by Lemma 8.20, while $\text{Tor}_1(A, E) = 0$ because E is flat (being a sum of copies of Q). Thus, $\text{Tor}_1(A, B)$ is a quotient of a torsion module, hence is itself torsion.

Now let B be arbitrary. Exactness of $0 \to tB \to B \to B/tB \to 0$ gives exactness of

$$\text{Tor}_1(A, tB) \to \text{Tor}_1(A, B) \to \text{Tor}_1(A, B/tB).$$

The outside terms are torsion, for tB is torsion and B/tB is torsion-free. It follows easily that $\text{Tor}_1(A, B)$ is torsion. The proof is completed by dimension shifting. ∎

Of course, Theorem 8.21 need not hold for $n = 0$, i.e., $A \otimes B$ need not be torsion.

Exercises: 8.9. Let $R = k[x, y]$, where k is a field, and let I be the ideal (x, y). Show that $x \otimes y - y \otimes x \in I \otimes_R I$ is nonzero and is annihilated by x. Conclude that the tensor product of torsion-free modules over a domain need not be torsion-free. (Hint: Look in $(I/I^2) \otimes (I/I^2)$.) Compare Exercise 3.38.

8.10. **(Axioms)** Let $\{T_n : n \geq 0\}$ be a positive sequence of functors such that

  (i) $\{T_n\}$ is strongly connected;
  (ii) $T_0 \cong A \otimes$ for some module A;
  (iii) $T_n(P) = 0$ for all free modules P and all $n \geq 1$.

Then $T_n \cong \text{Tor}_n(A, \ )$ for all $n \geq 0$.

8.11.   Let R be a domain with quotient field Q, and let $K = Q/R$. If t is the torsion submodule functor, prove that its right derived functors $R^i t$ are given by

$$R^0 t = t, \qquad R^1 t = K \otimes_R, \qquad R^i t = 0 \qquad \text{for} \quad i > 1.$$

(Hint: Use Exercise 7.28.)

8.12.   **(MacLane)** If A and B are abelian groups, show that $\operatorname{Tor}_1^Z(A, B)$ may be generated by those $(a, n, b) \in A \times \mathbf{Z} \times B$ such that $na = 0$ in A and $nb = 0$ in B, subject to the relations (whenever both sides are defined):

(i) $(a_1 + a_2, n, b) = (a_1, n, b) + (a_2, n, b)$;
(ii) $(a, n, b_1 + b_2) = (a, n, b_1) + (a, n, b_2)$;
(iii) $(ma, n, b) = (a, mn, b) = (a, m, nb)$.

(Hint: Choose an exact sequence $0 \to K \xrightarrow{i} F \to B \to 0$ with F free and show the group described is isomorphic to $\ker(1_A \otimes i)$.) MacLane [1963, pp. 150–159] gives generators and relations for $\operatorname{Tor}_n^R(M, N)$.

If I is a right ideal in a ring R and J a left ideal, then $IJ = \{\sum a_k b_k : a_k \in I, b_k \in J\}$. It is easy to see that $IJ \subset I \cap J$; also, there is a map $I \otimes_R J \to IJ$ with $a \otimes b \mapsto ab$. In a later chapter, we shall prove (Corollary 11.27)

$$\operatorname{Tor}_1^R(R/I, R/J) \cong I \cap J/IJ$$

and

$$\operatorname{Tor}_2^R(R/I, R/J) \cong \ker(I \otimes J \to IJ).$$

UNIVERSAL COEFFICIENT THEOREMS

In Chapter 1, we defined the homology groups $H_n(X)$ of a topological space X as

$$H_n(X) = H_n(\mathbf{S}(X)),$$

where $\mathbf{S}(X)$ is the singular complex of X. It is often convenient to modify this construction by allowing "coefficients" in an abelian group G:

$$H_n(X; G) = H_n(\mathbf{S}(X) \otimes G)$$

(one says "coefficients" because a typical element of degree n in $\mathbf{S}(X) \otimes G$ has the form $\sum s_i \otimes g_i$, where $s_i \in S_n(X)$ and $g_i \in G$). For example, "obstruction theory" deals with the problem of extending a continuous map defined on a "nice" subspace of X and involves cohomology with coefficients in a certain homotopy group. One might hope that $H_n(X; G) \cong H_n(X) \otimes G$, but this is usually not the case. The next theorem allows one to compute homology with coefficients from unadorned homology, and hence is called a universal coefficient theorem. Afterwards, we shall give the dual result for cohomology.

Theorem 8.22   (Universal Coefficient Theorem for Homology): *Let R be right hereditary, A an R-module, and $(\mathbf{K}, d)$ a complex of projective R-modules. There is a split exact sequence*

$$0 \to H_n(\mathbf{K}) \otimes_R A \xrightarrow{\lambda} H_n(\mathbf{K} \otimes_R A) \xrightarrow{\mu} \mathrm{Tor}_1^R(H_{n-1}(\mathbf{K}), A) \to 0$$

in which λ and μ are natural. Thus,

$$H_n(\mathbf{K} \otimes_R A) \cong H_n(\mathbf{K}) \otimes_R A \oplus \mathrm{Tor}_1^R(H_{n-1}(\mathbf{K}), A).$$

Remark: The theorem is true if $\mathbf{K}$ is a complex of flat modules. This, and much more, is proved in the Kunneth formula, Theorem 11.31.

Proof: For each n, there are exact sequences

(∗) $$0 \to Z_n(\mathbf{K}) \overset{i_n}{\hookrightarrow} K_n \xrightarrow{d_n} B_{n-1}(\mathbf{K}) \to 0$$

and

$$0 \to B_{n-1}(\mathbf{K}) \hookrightarrow Z_{n-1}(\mathbf{K}) \to H_{n-1}(\mathbf{K}) \to 0$$

(the first is just the definition of cycles and boundaries; the second is just the definition of homology). Splice these two sequences together to obtain an exact sequence

(∗∗)
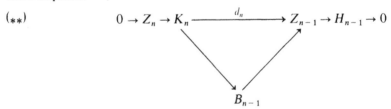

Since every K_n is projective and R is hereditary, Corollary 4.18 shows the submodules Z_n (of K_n) and B_{n-1} (of K_{n-1}) are also projective. There are two consequences: the exact sequence (∗) is split (Theorem 3.12); the exact sequence (∗∗) is a projective resolution of H_{n-1}.

Let $\mathbf{L}$ denote the complex obtained from (∗∗) by suppressing H_{n-1}; $\mathbf{L}$ is thus a deleted resolution and

$$\mathbf{L} \otimes A = 0 \to Z_n \otimes A \xrightarrow{i_n \otimes 1} K_n \otimes A \xrightarrow{d_n \otimes 1} Z_{n-1} \otimes A \to 0$$

is a complex with homology

$$H_j(\mathbf{L} \otimes A) = \mathrm{Tor}_j^R(H_{n-1}, A).$$

Now $\text{Tor}_2^R(H_{n-1}, A) = 0$ (Exercise 8.6), whence $i \otimes 1$ is monic (alternatively, that (∗) is split implies $Z_n \otimes A$ is even a summand of $K_n \otimes A$ with injection $i \otimes 1$). We can thus identify $Z_n \otimes A$ (via $i \otimes 1$) with a submodule of $K_n \otimes A$. The remaining computations are:

$$(\ast\ast\ast) \quad \begin{aligned} \text{Tor}_1^R(H_{n-1}, A) &= H_1(\mathbf{L} \otimes A) = (\ker d_n \otimes 1)/Z_n \otimes A\,; \\ H_{n-1} \otimes A &= \text{Tor}_0^R(H_{n-1}, A) = H_0(\mathbf{L} \otimes A) = Z_{n-1} \otimes A/\text{im}(d_n \otimes 1). \end{aligned}$$

Consider now $K_{n+1} \xrightarrow{\;d_{n+1}\;} K_n \xrightarrow{\;d_n\;} K_{n-1}$. Examining elements, one verifies the inclusions

$$\text{im } d_{n+1} \otimes 1 \subset Z_n \otimes A \subset \ker d_n \otimes 1 \subset K_n \otimes A.$$

The Third Isomorphism Theorem gives

$$(\ker d_n \otimes 1/\text{im } d_{n+1} \otimes 1)/[(Z_n \otimes A)/\text{im } d_{n+1} \otimes 1] \cong \ker d_n \otimes 1/Z_n \otimes A,$$

which may be rewritten as an exact sequence

$$0 \to Z_n \otimes A/\text{im } d_{n+1} \otimes 1 \xrightarrow{\;\lambda\;} \ker d_n \otimes 1/\text{im } d_{n+1} \otimes 1 \xrightarrow{\;\mu\;} \ker d_n \otimes 1/Z_n \otimes A \to 0.$$

The middle term is just $H_n(\mathbf{K} \otimes A)$, while we have already computed that the first term is $H_n(\mathbf{K}) \otimes A$ (item (∗∗∗) with $n - 1$ replaced by n) and the last term is $\text{Tor}_1^R(H_{n-1}(\mathbf{K}), A)$.

To see that this sequence splits, observe that Z_n is a summand of K_n (for (∗) splits), so that $Z_n \otimes A$ is a summand of $K_n \otimes A$ and hence of $\ker d_n \otimes 1$ (Exercise 2.23); it follows that $Z_n \otimes A/\text{im } d_{n+1} \otimes 1$ is a summand of $\ker d_n \otimes 1/\text{im } d_{n+1} \otimes 1$. ∎

Remark: The splitting may not be natural.

Exercise: 8.13. Show that the map $\lambda: H_n(\mathbf{K}) \otimes A \to H_n(\mathbf{K} \otimes A)$ is given by

$$[z_n] \otimes a \mapsto [z_n \otimes a],$$

where [ ] denotes homology class.

Corollary 8.23: *If X is a topological space and G an abelian group, then for all n,*

$$H_n(X; G) \cong H_n(X) \otimes_{\mathbf{Z}} G \oplus \text{Tor}_1^{\mathbf{Z}}(H_{n-1}(X), G).$$

Proof: By definition, $H_n(X) = H_n(\mathbf{S}(X))$ and $H_n(X; G) = H_n(\mathbf{S}(X) \otimes G)$. The Universal Coefficient Theorem applies at once, for $\mathbf{S}(X)$ is a complex of free abelian groups. ∎

Corollary 8.24:   *If* **K** *is a complex of vector spaces over a field R, and V is a vector space over R, then for all n,*

$$H_n(\mathbf{K} \otimes_R V) \cong H_n(\mathbf{K}) \otimes_R V.$$

Proof: Every field R is hereditary and every vector space over R is a free R-module. ∎

Corollary 8.25:   *Let* **K** *be a complex of free abelian groups. If either* $H_{n-1}(\mathbf{K})$ *or A is torsion-free, then*

$$H_n(\mathbf{K} \otimes_{\mathbf{Z}} A) \cong H_n(\mathbf{K}) \otimes_{\mathbf{Z}} A.$$

Proof: Either assumption forces $\operatorname{Tor}_1^{\mathbf{Z}}(H_{n-1}, A) = 0$. ∎

Of course, any hypothesis forcing $\operatorname{Tor}_1(H_{n-1}, A)$ to vanish (e.g., Exercise 8.7) will give a similar result.

Let us now turn to cohomology; recall that the nth **cohomology group** of a topological space X with coefficients in an abelian group A is defined by

$$H^n(X; A) = H_{-n}(\operatorname{Hom}(\mathbf{S}(X), A)).$$

Theorem 8.26: (Universal Coefficient Theorem for Cohomology):   *Let R be hereditary, A an R-module, and* $(\mathbf{K}, d)$ *a complex of projective R-modules. There is a split exact sequence*

$$0 \to \operatorname{Ext}_R^1(H_{n-1}(\mathbf{K}), A) \xrightarrow{\lambda} H^n(\operatorname{Hom}_R(\mathbf{K}, A)) \xrightarrow{\mu} \operatorname{Hom}_R(H_n(\mathbf{K}), A) \to 0$$

in which λ *and* μ *are natural. Thus*

$$H^n(\operatorname{Hom}_R(\mathbf{K}, A)) \cong \operatorname{Hom}_R(H_n(\mathbf{K}), A) \oplus \operatorname{Ext}_R^1(H_{n-1}(\mathbf{K}), A).$$

Proof: The proof of Theorem 8.22 applies here; the only change is that one now uses the contravariant functor $\operatorname{Hom}_R(\ , A)$ instead of the covariant functor $\otimes_R A$. ∎

The next result shows that the homology groups $H_n(X)$ of a space X determine its cohomology groups.

Corollary 8.27:   *If X is a topological space and G an abelian group, then for all n,*

$$H^n(X; G) \cong \operatorname{Hom}_{\mathbf{Z}}(H_n(X), G) \oplus \operatorname{Ext}_{\mathbf{Z}}^1(H_{n-1}(X), G).$$

It is known that for any sequence of abelian groups $A_0, A_1, A_2, \dots$, there exists a topological space X with $H_n(X) \cong A_n$ for all n. In contrast, if one defines $H^n(X) = H_{-n}(\operatorname{Hom}(\mathbf{S}(X), \mathbf{Z})) = H^n(X; \mathbf{Z})$, Nunke–Rotman [1962] prove that if $H''(X)$ is countable, then it is a sum of a finite group and a free abelian group.

Corollary 8.28: *Let* **K** *be a complex of free abelian groups. If either* $H_{n-1}(\mathbf{K})$ *is free or A is divisible, then*

$$H''(\operatorname{Hom}_{\mathbf{Z}}(\mathbf{K}, A)) \cong \operatorname{Hom}_{\mathbf{Z}}(H_n(\mathbf{K}), A).$$

Proof: Either hypothesis forces $\operatorname{Ext}^1_{\mathbf{Z}}(H_{n-1}, A) = 0.$ ∎

Of course, variations on this theme are played by assuming other hypotheses guaranteeing that Ext^1 vanish.

Corollary 8.29: *If* **K** *is a complex of vector spaces over a field R, and if V is a vector space over R, then for all n*

$$H''(\operatorname{Hom}_R(\mathbf{K}, V)) \cong \operatorname{Hom}_R(H_n(\mathbf{K}), V).$$

In particular,

$$H''(\operatorname{Hom}_R(\mathbf{K}, R)) \cong H_n(\mathbf{K})^*,$$

*where * denotes dual space.*

9 Son of Specific Rings

This chapter is intended to indicate that Ext and Tor are valuable tools in studying rings; the next chapter will indicate their value in studying groups. We begin with a minor illustration that these functors are good for something, and so we use Ext to prove a result (Theorem 9.3) that does not mention Ext in either its hypothesis or conclusion.

Theorem 9.1: *If R is left hereditary, then $\text{Ext}_R^n(A, B) = 0$ for all left R-modules A, B and all $n \geq 2$.*

Proof: There is an exact sequence $0 \to P_1 \to P_0 \to A \to 0$ with P_0 projective. By Corollary 4.18, P_1 is also projective, and the short exact sequence above is a projective resolution of A. The result follows from the definition of Ext. ∎

Theorem 9.2: *If R is a Dedekind ring and A is a torsion-free R-module, then $\text{Ext}_R^1(A, B)$ is divisible for every R-module B.*

Proof: By Lemma 4.31, there is an exact sequence

$$0 \to A \to E \to X \to 0$$

with E torsion-free divisible; this gives rise to an exact sequence

$$\text{Ext}^1(E, B) \to \text{Ext}^1(A, B) \to \text{Ext}^2(X, B).$$

The last term is 0, by Theorem 9.1, so that the first map is epic. Since $\text{Ext}^1(E, B)$ is divisible (Exercise 7.5), its image $\text{Ext}^1(A, B)$ is also divisible. ∎

In Exercise 7.13 we saw that the torsion subgroup of an abelian group may not be a summand.

Theorem 9.3: *Let R be Dedekind, and let B be an R-module with torsion submodule tB. If there is some nonzero $r \in R$ with $r(tB) = 0$, then tB is a summand of B.*

Proof: We must show that $0 \to tB \to B \to B/tB \to 0$ splits. Since B/tB is torsion-free, it suffices to prove $\text{Ext}^1(A, T) = 0$ whenever A is torsion-free and $T \cong tB$. Now $\text{Ext}^1(A, T)$ is divisible, by Theorem 9.2; on the other hand, $rT = 0$ implies $r \, \text{Ext}^1(A, T) = 0$ (Theorem 7.16). It follows that $\text{Ext}^1(A, T) = 0$, for divisibility implies multiplication by r ($r \neq 0$) is epic. ∎

Note that the hypothesis $r(tB) = 0$ holds if tB is f.g.

DIMENSIONS

We begin by measuring how far a module is from being projective.

Definition: If A is a left R-module, then $\text{pd}(A) \leq n$ (pd abbreviates **projective dimension**) if there is a projective resolution

$$0 \to P_n \to \cdots \to P_1 \to P_0 \to A \to 0.$$

If no such finite resolution exists, define $\text{pd}(A) = \infty$; otherwise, if n is the least such integer, define $\text{pd}(A) = n$.

Examples: 1. $\text{pd}(A) = 0$ if and only if A is projective.
2. If R is left hereditary, then $\text{pd}(A) \leq 1$ for every R-module A (this is contained in the proof of Theorem 9.1).

Let us now give a name to submodules that arise in analyzing a projective resolution.

Definition: Let $\cdots \to P_1 \xrightarrow{d_1} P_0 \xrightarrow{\varepsilon} A \to 0$ be a projective resolution; denote $\ker \varepsilon$ by K_0 and, for $n \geq 1$, denote $\ker d_n$ by K_n. For $n \geq 0$, K_n is the nth **syzygy** of A.

Obviously, the syzygies of A depend on the choice of projective resolution.

Definition: Two modules A and B are **projectively equivalent** if there are projectives P and Q with $A \oplus P \cong B \oplus Q$.

It is obvious that this is an equivalence relation.

Theorem 9.4: *Let $\{K_n\}$ and $\{K'_n\}$ be syzygies of A defined by two projective resolutions of A.*

   (i)   *For each $n \geq 0$, K_n and K'_n are projectively equivalent;*
   (ii) *for every module B, $\operatorname{Ext}^1(K_n, B) \cong \operatorname{Ext}^1(K'_n, B)$;*
   (iii) *for every module B and every $n \geq 1$, $\operatorname{Ext}^{n+1}(A, B) \cong \operatorname{Ext}^1(K_{n-1}, B)$.*

Proof: (i) This follows at once from the generalized Schanuel lemma, Exercise 3.37.
   (ii) Theorems 7.7 and 7.13.
   (iii) Corollary 6.19. ∎

As a result of Theorem 9.4, one often abuses language and speaks of *the* nth syzygy of A, even though such a module is only defined to projective equivalence.

A short digression. The word "syzygy" is used in astronomy, in the context of several bodies b_0, b_1, $b_2, \ldots$, with b_1 rotating about b_0, b_2 rotating around b_1 (hence around b_0), and so forth. For example, sun, earth, moon. "Syzygy", from the Greek "syzygos" meaning "yoked together", describes certain positions of these bodies relative to each other. In describing a module A by generators and relations, say, $P_0/K_0 \cong A$, let us (poetically) regard the relations K_0 as a yoke between P_0 and A; further regard K_1, relations on relations, as yoking P_1 to K_0 (hence to A), and so forth.

Theorem 9.5: *The following are equivalent for a left R-module A:*

   (i) $\operatorname{pd}(A) \leq n$;
   (ii) $\operatorname{Ext}^k(A, B) = 0$ *for all modules B and all $k \geq n + 1$;*
   (iii) $\operatorname{Ext}^{n+1}(A, B) = 0$ *for all modules B;*
   (iv) *there is a projective resolution of A with projective $(n-1)$st syzygy;*
   (v)   *every projective resolution of A has a projective $(n-1)$st syzygy.*

Proof: (i) $\Rightarrow$ (ii) There is a projective resolution of A with $P_k = 0$ for all $k \geq n + 1$. Therefore, $\operatorname{Hom}(P_k, B) = 0$ for all $k \geq n + 1$, and so $\operatorname{Ext}^k(A, B) = 0$ for all $k \geq n + 1$.
   (ii) $\Rightarrow$ (iii) Trivial.
   (iii) $\Rightarrow$ (iv) Take a projective resolution of A with $(n-1)$st syzygy K_{n-1}. By hypothesis, $0 = \operatorname{Ext}^{n+1}(A, B) = \operatorname{Ext}^1(K_{n-1}, B)$ for all B; by Corollary 7.12, K_{n-1} is projective.
   (iv) $\Rightarrow$ (v) Assume K_{n-1} and K'_{n-1} are $(n-1)$st syzygies of A given by two projective resolutions. By Theorem 9.4(i), there are projectives P and

Q with $K_{n-1} \oplus P \cong K'_{n-1} \oplus Q$. If we further assume K_{n-1} is projective, then $K_{n-1} \oplus P$ is projective; hence K'_{n-1} is projective, being a summand of a projective.

(v) $\Rightarrow$ (i)   Let $\cdots \to P_1 \to P_0 \to A \to 0$ be a projective resolution. By hypothesis, K_{n-1} is projective, so that

$$0 \to K_{n-1} \to P_{n-1} \to \cdots \to P_1 \to P_0 \to A \to 0$$

is a projective resolution exhibiting $\mathrm{pd}(A) \leq n$. ∎

We temporarily adopt an overcomplicated notation.

Definition: If R is a ring, its **left projective global dimension** $l\mathrm{pD}(R)$ is defined by

$$l\mathrm{pD}(R) = \sup\{\mathrm{pd}(A) : A \in {}_R\mathfrak{M}\}.$$

Corollary 9.6: $l\mathrm{pD}(R) \leq n$ *if and only if* $\mathrm{Ext}_R^{n+1}(A, B) = 0$ *for all left R-modules A and B.*

Proof: Immediate from Theorem 9.5(iii). ∎

Exercises: 9.1. $l\mathrm{pD}(R) = 0$ if and only if R is semisimple.

9.2.   If R is quasi-Frobenius, prove that $l\mathrm{pD}(R) = 0$ or ∞. Conclude that $l\mathrm{pD}(\mathbf{Z}/n\mathbf{Z}) = \infty$ if n is not square-free.

9.3. $l\mathrm{pD}(R) \leq 1$ if and only if R is left hereditary.

9.4.   If I is a left ideal of R, then either R/I is projective or $\mathrm{pd}(R/I) = \mathrm{pd}(I) + 1$ (where $\infty = \infty + 1$).

9.5.   Let R_n be the ring of all $n \times n$ matrices over a ring R. Prove that $l\mathrm{pD}(R) = l\mathrm{pD}(R_n)$. (Hint: Corollaries 3.41 and 3.42.)

9.6.   If M is an R-module with $\mathrm{pd}(M) = n < \infty$, then prove that $\mathrm{Ext}_R^n(M, F) \neq 0$ for some free R-module F.

All may be repeated using injective modules instead of projectives; we need only give the definitions and results.

Definition: If B is a left R-module, then $\mathrm{id}(B) \leq n$ (id abbreviates **injective dimension**) if there is an injective resolution

$$0 \to B \to E^0 \to E^1 \to \cdots \to E^n \to 0.$$

If no such finite resolution exists, define $\mathrm{id}(B) = \infty$; otherwise, if n is the least such integer, define $\mathrm{id}(B) = n$.

Of course, B is injective if and only if $\mathrm{id}(B) = 0$.

Definition: Let $0 \to B \xrightarrow{\varepsilon} E^0 \xrightarrow{d^0} E^1 \to \cdots$ be an injective resolution; denote im ε by L^0 and, for $n \geq 1$, denote im d^{n-1} by L^n. For $n \geq 0$, L^n is the nth **cosyzygy** of B.

Definition: Two modules B and C are **injectively equivalent** if there are injectives E and E' with $B \oplus E \cong C \oplus E'$.

Since the dual of the generalized Schanuel lemma holds, we may assert the following result.

Theorem 9.7: *Let $\{L^n\}$ and $\{M^n\}$ be cosyzygies of B defined by two injective resolutions of B.*

(i) *For each $n \geq 0$, L^n and M^n are injectively equivalent;*

(ii) *for every module A, $\operatorname{Ext}^1(A, L^n) \cong \operatorname{Ext}^1(A, M^n)$;*

(iii) *for every module A and every $n \geq 1$, $\operatorname{Ext}^{n+1}(A, B) \cong \operatorname{Ext}^1(A, L^{n-1})$.*

Thoerem 9.8: *The following are equivalent for a left R-module B:*

(i) *$\operatorname{id}(B) \leq n$;*

(ii) *$\operatorname{Ext}^k(A, B) = 0$ for all modules A and all $k \geq n + 1$;*

(iii) *$\operatorname{Ext}^{n+1}(A, B) = 0$ for all modules A;*

(iv) *every injective resolution of B has an injective $(n - 1)$st cosyzygy.*

Definition: If R is a ring, its **left injective global dimension** $l\mathrm{iD}(R)$ is defined by

$$l\mathrm{iD}(R) = \sup\{\operatorname{id}(B) : B \in {}_R\mathfrak{M}\}.$$

Corollary 9.9: *$l\mathrm{iD}(R) \leq n$ if and only if $\operatorname{Ext}_R^{n+1}(A, B) = 0$ for all left R-modules A and B.*

We now combine Corollaries 9.6 and 9.9.

Theorem 9.10: *For any ring R, $l\mathrm{pD}(R) = l\mathrm{iD}(R)$.*

Proof: The two corollaries yield the same criterion. ∎

In view of Theorem 9.10, one defines the **left global dimension** $l\mathrm{D}(R)$ as the common value of $l\mathrm{pD}(R)$ and $l\mathrm{iD}(R)$. If one considers right R-modules, he may define the **right global dimension** $r\mathrm{D}(R)$. The Wedderburn theorem implies left semisimple is the same as right semisimple; thus, $l\mathrm{D}(R) = 0$ if and only if $r\mathrm{D}(R) = 0$. The first example of a ring for which the left and right global dimensions differ was given by Kaplansky, who exhibited a ring R with $l\mathrm{D}(R) = 1$ and $r\mathrm{D}(R) = 2$. Jategaonkar [1969] proved that if $1 \leq m < n \leq \infty$, then there exists a ring R with $l\mathrm{D}(R) = m$ and $r\mathrm{D}(R) = n$. The same phenomenon (but with n finite) is also exhibited by Fossum–Griffith–Reiten [1975, pp. 74–75].

On the positive side, we shall prove (Corollary 9.23) that $l\mathrm{D}(R)$ coincides with $r\mathrm{D}(R)$ when R is left and right noetherian. Jensen [1966] showed that if all one-sided ideals of R are countably generated, then $|l\mathrm{D}(R) - r\mathrm{D}(R)| \leq 1$; this is generalized by Osofsky [1968] who showed that if every one-sided ideal of R can be generated by at most $\aleph_n$ elements, then $|l\mathrm{D}(R) - r\mathrm{D}(R)| \leq n + 1$. In this last result, set theory makes its presence known. Indeed, if R is a product of countably many fields, then $\mathrm{D}(R) = 2$ if and only if one accepts the Continuum Hypothesis [Osofsky, 1973, p. 60] (when R is commutative, one drops the unneeded letters l and r).

The next problem is to develop ways to compute global dimension.

Lemma 9.11: A left R-module B is injective if and only if $\mathrm{Ext}^1(R/I, B) = 0$ for all left ideals I.

Proof: Apply $\mathrm{Hom}(\ , B)$ to $0 \to I \to R \to R/I \to 0$ to obtain exactness of $\mathrm{Hom}(R, B) \to \mathrm{Hom}(I, B) \to \mathrm{Ext}^1(R/I, B) = 0$. The result now follows from Baer's criterion, Theorem 3.20. ∎

Theorem 9.12 (Auslander): *For any ring R,*

$$l\mathrm{D}(R) = \sup\{\mathrm{pd}(R/I) : I \text{ is a left ideal}\}.$$

Remark: This proof is due to Matlis.

Proof: If $\sup\{\mathrm{pd}(R/I)\} = \infty$, we are done. Assume that $\mathrm{pd}(R/I) \leq n$ for all ideals I, so that $\mathrm{Ext}^{n+1}(R/I, B) = 0$ for every module B. It suffices to prove $\mathrm{id}(B) \leq n$ for every B. Take an injective resolution of B with $(n-1)$st cosyzygy L^{n-1}. By Theorem 9.7(iii), $0 = \mathrm{Ext}^{n+1}(R/I, B) \cong \mathrm{Ext}^1(R/I, L^{n-1})$. But Lemma 9.11 gives L^{n-1} injective, and therefore $\mathrm{id}(B) \leq n$. ∎

Thus, to compute $l\mathrm{D}(R)$, it suffices to know projective dimensions of the cyclic modules. Note also that global dimension measures how far a ring is from being semisimple. Using Exercise 9.4, we see that if R is not semisimple, then $l\mathrm{D}(R) - 1 = \sup\{\mathrm{pd}(I) : I = \text{left ideal}\}$. It is now clear why the definition of left hereditary ring is natural; moreover, one expects some good theorems because these rings are only one step removed from semisimple rings.

Exercises: 9.7. If $0 \to A' \to A \to A'' \to 0$ is exact with A projective, then either all three modules are projective or

$$\mathrm{pd}(A'') = \mathrm{pd}(A') + 1.$$

(Of course, this generalizes Exercise 9.4.)

9.8. Given a family of modules $\{A_k : k \in K\}$, then

$$\mathrm{pd}(\coprod A_k) = \sup\{\mathrm{pd}(A_k) : k \in K\}.$$

Conclude that if $l\mathrm{D}(R) = \infty$, then there exists a module A with $\mathrm{pd}(A) = \infty$.

9.9.   If $0 \to A' \to A \to A'' \to 0$ is exact and two of the modules have finite projective dimension, then so does the third. Moreover, if $n < \infty$, $\mathrm{pd}(A') = n$, and $\mathrm{pd}(A'') \le n$, then $\mathrm{pd}(A) = n$.

The question arises whether there is an analogue of Lemma 9.11 to test for projectivity. The obvious candidates do not work. If we assume $\mathrm{Ext}_R^1(P, R/I) = 0$ for all I, then Theorem 9.3 shows that we may only conclude that P is torsion-free (when R is Dedekind). If we assume $\mathrm{Ext}_R^1(P, I) = 0$ for all ideals I, this, too, is not enough. Indeed, when $R = \mathbf{Z}$, then $I \ne 0$ implies $I \cong \mathbf{Z}$ and one is asking whether $\mathrm{Ext}_{\mathbf{Z}}^1(P, \mathbf{Z}) = 0$ implies P is free. This question is called **Whitehead's problem**. If P is countable, K. Stein proved P is free. If P is uncountable, then Shelah [1974] proved Whitehead's problem is undecideable: the statement "$\mathrm{Ext}^1(P, \mathbf{Z}) = 0$ and card $P = \aleph_1$ implies P is free" and its negation are each consistent with the (ZFC) axioms of set theory!

Exercises: 9.10.   Let R be a Dedekind ring with quotient field Q, and let $K = Q/R$. Prove that an R-module A is injective if and only if $\mathrm{Ext}_R^1(K, A) = 0$.
    9.11.   If R is Dedekind, then an R-module A is projective if and only if $\mathrm{Ext}^1(A, F) = 0$ for every free module F.

Let us now define another dimension, using flat modules and Tor.

Definition: If A is a right R-module, then $\mathrm{fd}(A) \le n$ (fd abbreviates **flat dimension**) if there is a flat resolution

$$0 \to F_n \to \cdots \to F_0 \to A \to 0.$$

If no such finite resolution exists, define $\mathrm{fd}(A) = \infty$; otherwise, if n is the least such integer, define $\mathrm{fd}(A) = n$.

Definition: Let $\cdots \to F_1 \overset{d_1}{\to} F_0 \overset{\varepsilon}{\to} A \to 0$ be a flat resolution; denote $\ker \varepsilon$ by Y_0 and, for $n \ge 1$, denote $\ker d_n$ by Y_n. For $n \ge 0$, Y_n is the nth **yoke** of A.

If each F_n is projective, then the nth yoke is the nth syzygy.
Of course, the nth yoke of A depends on the choice of flat resolution. Unfortunately, Schanuel's lemma does not hold for flat resolutions. For example, if $K = \mathbf{Q}/\mathbf{Z}$, then two flat resolutions of K are

$$0 \to \mathbf{Z} \to \mathbf{Q} \to K \to 0 \qquad \text{and} \qquad 0 \to S \to F \to K \to 0,$$

where F is free (of infinite rank). Clearly $\mathbf{Z} \oplus F$ is free, and hence cannot be isomorphic to $\mathbf{Q} \oplus S$ because $\mathbf{Q}$ is not $\mathbf{Z}$-projective. However, we can link flat dimension to Tor.

Theorem 9.13:   *The following are equivalent for a right R-module A:*

(i) $\text{fd}(A) \le n$;
(ii) $\text{Tor}_k(A, B) = 0$ for all $k \ge n + 1$ and all left R-modules B;
(iii) $\text{Tor}_{n+1}(A, B) = 0$ for all left R-modules B;
(iv)   *every flat resolution of A has a flat $(n - 1)$st yoke.*

Proof: (i) $\Rightarrow$ (ii)   Assume $0 \to F_n \to \cdots \to F_0 \to A \to 0$ is a flat resolution. By Corollary 8.8, $\text{Tor}(A, B)$ can be computed via flat resolutions, and the result follows.

(ii) $\Rightarrow$ (iii)   Trivial.

(iii) $\Rightarrow$ (iv)   Let Y_{n-1} be the $(n - 1)$st yoke of a flat resolution of A. Using Corollary 8.8, we see that $\text{Tor}_{n+1}(A, B) \cong \text{Tor}_1(Y_{n-1}, B)$. But $\text{Tor}_1(Y_{n-1}, B) = 0$ for all B gives Y_{n-1} flat, by Theorem 8:9.

(iv) $\Rightarrow$ (i)   As in the proof of Theorem 9.5.   ∎

Remark: The proof of (iii) $\Rightarrow$ (iv) just given could have been used in the proof of Theorem 9.5, thus avoiding projective equivalence and Schanuel's lemma.

Definition: The **right weak dimension** of a ring R is defined by

$$rwD(R) = \sup\{\text{fd}(A) : A \in \mathfrak{M}_R\}.$$

Theorem 9.14: $rwD(R) \le n$ *if and only if* $\text{Tor}_{n+1}(A, B) = 0$ *for all right R-modules A and all left R-modules B.*

Proof: Immediate from Theorem 9.13.   ∎

Definition: The **left weak dimension** of a ring R is defined by

$$lwD(R) = \sup\{\text{fd}(B) : B \in {}_R\mathfrak{M}\}.$$

Theorem 9.15:   *For any ring R,*

$$rwD(R) = lwD(R).$$

Proof: One may prove the left versions of Theorems 9.13 and 9.14, obtaining the same formula for $rwD(R)$ and for $lwD(R)$.   ∎

Definition: The **weak dimension** of R, $wD(R)$, is the common value of $rwD(R)$ and $lwD(R)$.

In the beginning, it seems a nuisance that $A \otimes_R B$, more generally, $\text{Tor}_n^R(A, B)$, requires A to be a right R-module and B to be a left R-module.

However, we now see that weak dimension requires no left–right distinction as does global dimension. This fact will soon be exploited.

Examples: 3. $wD(R) = 0$ if and only if R is von Neumann regular (Theorem 4.16).

4. $wD(R) \leq 1$ if and only if every submodule of a flat module is flat. We shall have more to say about this class of rings shortly.

The next result explains why this dimension is called "weak".

Theorem 9.16: *For any ring R,*

$$wD(R) \leq \min\{lD(R), rD(R)\}.$$

Proof: Given a right R-module A, we claim $fd(A) \leq pd(A)$. If $pd(A) \leq n$, there is a projective resolution

$$0 \to P_n \to \cdots \to P_0 \to A \to 0.$$

Since projective modules are flat, this is a flat resolution of A exhibiting $fd(A) \leq n$. A similar argument works for left R-modules. ∎

Since there are von Neumann regular rings that are not semisimple, the inequality may be strict.

Corollary 9.17: *If $Ext_R^{n+1}(A, B) = 0$ for all left R-modules A, B (or for all right R-modules A, B), then $Tor_{n+1}^R(C, D) = 0$ for all right R-modules C and all left R-modules D.*

In particular, we now see why $Tor_2^R \equiv 0$ for hereditary rings R.
We wish to compute $wD(R)$.

Lemma 9.18: *A left R-module B is flat if and only if $Tor_1^R(R/I, B) = 0$ for every right ideal I.*

Proof: Theorem 3.53 says that B is flat if and only if $\otimes_R B$ preserves exactness of $0 \to I \to R \to R/I \to 0$ for every right ideal. This easily translates into the vanishing of $Tor_1^R(R/I, B)$. ∎

Theorem 9.19: *For any ring R,*

$$wD(R) = \sup\{fd(R/I) : I \text{ a right ideal of } R\}$$
$$= \sup\{fd(R/J) : J \text{ a left ideal of } R\}.$$

Proof: The proof is just as that of Theorem 9.12, using Lemma 9.18 instead of Lemma 9.11. ∎

As global dimension, weak dimension is thus determined by the cyclic modules.

Lemma 9.20: *Let R be left noetherian and let A be a f.g. left R-module. There exists a projective resolution* $\cdots \to P_1 \to P_0 \to A \to 0$ *with each P_n, hence each syzygy, f.g.*

Proof: This is just a restatement of Corollary 4.4. ∎

This lemma has an interesting consequence.

Theorem 9.21: *If R is commutative noetherian and A and B are f.g. R-modules, then $\mathrm{Ext}_R^n(A, B)$ and $\mathrm{Tor}_n^R(A, B)$ are f.g. R-modules for all $n \geq 0$.*

Proof: First of all, the theorem is true when $n = 0$, i.e., $\mathrm{Hom}_R(A, B)$ and $A \otimes_R B$ are f.g. R-modules (commutativity is used to guarantee they are R-modules). For the inductive step, choose a projective resolution $\cdots \to P_1 \to P_0 \to A \to 0$ as in Lemma 9.20. Since $\mathrm{Hom}_R(P_n, B)$ is f.g. and $\mathrm{Ext}_R^n(A, B)$ is a quotient of a submodule of $\mathrm{Hom}_R(P_n, B)$, it follows from Theorem 4.1 that $\mathrm{Ext}_R^n(A, B)$ is f.g. The same argument works for $\mathrm{Tor}_n^R(A, B)$ if we replace $\mathrm{Hom}_R(P_n, B)$ by $P_n \otimes_R B$. ∎

Let us return to dimensions; for noetherian rings, weak dimension is nothing new.

Theorem 9.22: *If R is right noetherian, then $\mathrm{wD}(R) = \mathrm{rD}(R)$; if R is left noetherian, then $\mathrm{wD}(R) = l\mathrm{D}(R)$.*

Proof: We shall show that $\mathrm{fd}(A) = \mathrm{pd}(A)$ for every f.g. right R-module; this will suffice, for we may even compute $\mathrm{wD}(R)$ and $\mathrm{rD}(R)$ with cyclic modules. Now $\mathrm{fd}(A) \leq \mathrm{pd}(A)$, as we saw in the proof of Theorem 9.16. For the reverse inequality, assume $\mathrm{fd}(A) \leq n$; we must show $\mathrm{pd}(A) \leq n$. Take a projective resolution of A having f.g. terms (hence f.g. syzygies):

$$\cdots \to P_n \to \cdots \to P_0 \to A \to 0.$$

This is also a flat resolution, so that Theorem 9.13(iv) says that the $(n-1)$st yoke Y_{n-1} (which is the $(n-1)$st syzygy) is flat. Thus,

$$0 \to Y_{n-1} \to P_{n-1} \to \cdots \to P_0 \to A \to 0$$

is exact, the P's are projective, and Y_{n-1} is f.g. flat. Since R is right noetherian, Corollary 4.3 gives Y_{n-1} projective. We have exhibited a projective resolution of A showing that $\mathrm{pd}(A) \leq n$. ∎

Corollary 9.23 (Auslander): *If R is left and right noetherian, then $l\mathrm{D}(R) = \mathrm{rD}(R)$.*

Proof: In this case, both global dimensions equal $wD(R)$.   ∎

Osofsky [1973, p. 57] proves that if every right ideal of a ring R can be generated by $\aleph_n$ elements, then $rD(R) \leq wD(R) + n + 1$ (if $\aleph_{-1}$ means "finite", this generalizes Theorem 9.22).

Let us return to a class of rings mentioned in Chapter 4.

Theorem 9.24:   *The following are equivalent for a ring R:*

(i)   *every left ideal is flat;*
(ii)   *every right ideal is flat;*
(iii) $wD(R) \leq 1$.
(iv) $\mathrm{Tor}_2^R(A, B) = 0$ *for all* A_R *and* $_R B$.

Proof: This follows immediately from Theorems 9.19 and 9.14.   ∎

If R is left noetherian, we have seen in Theorem 4.34 that $wD(R) \leq 1$ implies R is left hereditary. This also follows from Theorem 9.22.

Theorem 9.25:   *The class of rings R with* $wD(R) \leq 1$ *contains all left or right semihereditary rings. In particular, if R is left or right semihereditary, every submodule of a flat module is flat.*

Proof: The last statement is just a restatement of the definition of weak dimension ≤ 1 (Example 4).

Since every ideal is a direct limit of f.g. ideals, R semihereditary implies every ideal is a direct limit of projectives, hence flat. By Theorem 9.24, $wD(R) \leq 1$.   ∎

It is shown in Exercise 9.26 that commutative domains R with $wD(R) \leq 1$ are semihereditary. Berstein [1958] computes the global dimension of a direct limit of rings over a countable directed index set.

HILBERT'S SYZYGY THEOREM

Global dimensions of polynomial rings are computed in this section. We begin with an easy result; there is a short proof avoiding Ext, using Schanuel's lemma, but I feel the coming proof explains the occurrence of 1 in the formula.

Lemma 9.26:   *If* $0 \to A' \to A \to A'' \to 0$ *is exact, then*

$$\mathrm{pd}(A'') \leq 1 + \max\{\mathrm{pd}(A), \mathrm{pd}(A')\}.$$

Proof: Clearly we may assume the right side is finite, say, $\mathrm{pd}(A) \leq n$ and $\mathrm{pd}(A') \leq n$. For every module B, there is an exact sequence

$$\cdots \to \mathrm{Ext}^{n+1}(A', B) \to \mathrm{Ext}^{n+2}(A'', B) \to \mathrm{Ext}^{n+2}(A, B) \to \cdots.$$

Using Theorem 9.5, the outer terms are 0. Therefore $\text{Ext}^{n+2}(A'', B) = 0$ for all B, and hence $\text{pd}(A'') \leq n + 1$. ∎

Exercise: 9.12. If $0 \to A' \to A \to A'' \to 0$ is exact, then

$$\text{pd}(A) \leq \max\{\text{pd}(A'), \text{pd}(A'')\}.$$

Moreover, if this sequence is not split and $\text{pd}(A') = \text{pd}(A'') + 1$, then the inequality is equality.

If R is a ring, then $R[t]$ is the polynomial ring: we allow R to be non-commutative, but we assume the variable t commutes with the coefficients in R (thus, t lies in the center of $R[t]$). If M is a left R-module, write

$$M[t] = R[t] \otimes_R M.$$

Since $R[t]$ is a free R-module (with basis $\{1, t, t^2, \ldots\}$) and since tensor product commutes with sums, we may regard the elements of $M[t]$ as "vectors" $(t^i \otimes m_i)$, $i \geq 0$, $m_i \in M$, with almost all $m_i = 0$.

Lemma 9.27: *For every left R-module M,*

$$\text{pd}_R(M) = \text{pd}_{R[t]}(M[t]).$$

Proof: It suffices to prove that if one dimension is finite and $\leq n$, then the other dimension is also $\leq n$.

Assume $\text{pd}_R(M) \leq n$; there is an R-projective resolution

$$0 \to P_n \to \cdots \to P_0 \to M \to 0.$$

Since $R[t]$ is a flat R-module (it is R-free), there is an exact sequence of $R[t]$-modules

$$0 \to R[t] \otimes P_n \to \cdots \to R[t] \otimes P_0 \to M[t] \to 0.$$

But $R[t] \otimes P_i$ is $R[t]$-projective, all i, so that $\text{pd}_{R[t]}(M[t]) \leq n$.

Assume $\text{pd}_{R[t]}(M[t]) \leq n$; there is an $R[t]$-projective resolution

$$0 \to Q_n \to \cdots \to Q_0 \to M[t] \to 0.$$

If we consider the terms only as R-modules, then we have an R-projective resolution of $\coprod M$ ($\aleph_0$ copies of M) of length n. By Exercise 9.8, $\text{pd}_R(M) \leq n$. ∎

Corollary 9.28: *If $l\text{D}(R) = \infty$, then $l\text{D}(R[t]) = \infty$.*

Proof: By Exercise 9.8, there exists an R-module M with $\text{pd}_R(M) = \infty$. But Lemma 9.27 now applies to give $\text{pd}_{R[t]}(M[t]) = \infty$, and so $l\text{D}(R[t]) = \infty$. ∎

We have seen that we may create an $R[t]$-module $M[t]$ from an R-module M; this construction is available, in particular, for an $R[t]$-module M if we forget the action of t.

Lemma 9.29: *If M is an $R[t]$-module, there is an $R[t]$-exact sequence*

$$0 \to M[t] \to M[t] \to M \to 0.$$

Proof: Define $e: M[t] \to M$ by $t^i \otimes m \mapsto t^i m$. There is an $R[t]$-exact sequence

$$0 \to K \to M[t] \overset{e}{\to} M \to 0,$$

and it suffices to prove $\ker e = K \cong M[t]$ as $R[t]$-modules. Define $\beta: M[t] \to K$ by $\sum t^i \otimes m_i \mapsto \sum t^i(1 \otimes t - t \otimes 1)m_i$; clearly β is an $R[t]$-map with $\operatorname{im} \beta \subset K$. To show β is an isomorphism, we write the formula in more detail:

$$\sum_{i=0}^{k} t^i \otimes m_i \mapsto 1 \otimes tm_0 + \sum_{i=1}^{k} t^i \otimes (tm_i - m_{i-1}) - t^{k+1} \otimes m_k.$$

If $\sum t^i \otimes m_i \in \ker \beta$, then

$$0 = -m_k = tm_k - m_{k-1} = \cdots = tm_1 - m_0,$$

so that each $m_i = 0$, and β is thus monic. If $\sum_{i=0}^{k} t^i \otimes v_i \in K$, then $\sum_{i=0}^{k} t^i v_i = 0$ in M; the equations

$$-v_0 = tm_0, \quad v_1 = tm_1 - m_0, \ldots, v_k = -m_{k-1}$$

can be solved recursively to show that β is epic. ∎

Corollary 9.30: *For every ring R,*

$$lD(R[t]) \leq lD(R) + 1.$$

Proof: If we agree that $\infty + 1 = \infty$, then we may assume that $lD(R) = n < \infty$. Let M be an $R[t]$-module. If we consider M as an R-module, then Lemma 9.27 gives

$$\operatorname{pd}_{R[t]}(M[t]) = \operatorname{pd}_R(M) \leq n$$

(we are assuming $lD(R) \leq n$). Coupling the $R[t]$-exact sequence

$$0 \to M[t] \to M[t] \to M \to 0$$

with Lemma 9.26, however, gives

$$\operatorname{pd}_{R[t]}(M) \leq 1 + \operatorname{pd}_{R[t]}(M[t]) \leq 1 + n. \quad ∎$$

Our aim is to replace the inequality of Corollary 9.30 by equality (Corollary 9.28 does this in the infinite-dimensional case). The proper context in which to proceed is that of "change of rings". Assume $\varphi: R \to R^*$ is a ring map. Every left R^*-module M^* acquires a left R-module structure via the formula

$$r \cdot m^* = \varphi(r)m^*, \qquad r \in R, \quad m^* \in M^*.$$

Every R^*-map $f: M^* \to N^*$ is also an R-map:

$$f(rm^*) = f(\varphi(r)m^*) = \varphi(r)f(m^*) = rf(m^*).$$

Theorem 9.31: *Every ring map $\varphi: R \to R^*$ defines an exact functor*

$$U: {}_{R^*}\mathfrak{M} \to {}_R\mathfrak{M}.$$

Remark: U is called a **change of rings functor**.

Proof: That U is a functor is automatic; that U is exact follows from observing that, as functions, $Uf = f$, and so Uf has the same kernel and image as does f. ∎

This topic will be discussed further in Chapter 11.

Exercise: 9.13. If $\varphi: R \to R^*$ is onto, then for every pair of R^*-modules M^*, N^*, there is equality

$$\mathrm{Hom}_{R^*}(M^*, N^*) = \mathrm{Hom}_R(M^*, N^*).$$

The following exposition is due to Kaplansky; afterwards, we shall present a more "homological" proof.

Theorem 9.32: *Let $\varphi: R \to R^*$ be a ring map and A^* a left R^*-module. Then*

$$\mathrm{pd}_R(A^*) \le \mathrm{pd}_{R^*}(A^*) + \mathrm{pd}_R(R^*).$$

Moreover, if equality holds for every nonzero A^* with $\mathrm{pd}_{R^*}(A^*) \le 1$, then equality holds for every nonzero A^* with $\mathrm{pd}_{R^*}(A^*)$ finite.

Proof: If $\mathrm{pd}_{R^*}(A^*) = \infty$, there is nothing to prove; we therefore assume $\mathrm{pd}_{R^*}(A^*) = n < \infty$ and proceed by induction. If $n = 0$, then A^* is R^*-projective; thus there exists a module B^* with $A^* \oplus B^* = \coprod R^*$. Exercise 9.8 applies to give

$$\mathrm{pd}_R(A^*) \le \mathrm{pd}_R(\coprod R^*) = \mathrm{pd}_R(R^*).$$

Suppose $n > 0$. There is an exact sequence of R^*-modules

$$0 \to K^* \to F^* \to A^* \to 0,$$

where F^* is R^*-free. By Lemma 9.26,

$$\mathrm{pd}_R(A^*) \le 1 + \max\{\mathrm{pd}_R(K^*), \mathrm{pd}_R(F^*)\} \le 1 + \max\{\mathrm{pd}_R(K^*), \mathrm{pd}_R(R^*)\}.$$

Since A^* is not R^*-projective, Exercise 9.7 gives $\mathrm{pd}_{R^*}(K^*) = n - 1$, so that induction gives

$$\mathrm{pd}_R(K^*) \le n - 1 + \mathrm{pd}_R(R^*).$$

Combining these inequalities:

$$\mathrm{pd}_R(A^*) \le 1 + \max\{n - 1 + \mathrm{pd}_R(R^*), \mathrm{pd}_R(R^*)\} \le n + \mathrm{pd}_R(R^*).$$

Finally, suppose we have equality whenever $A^* \ne 0$ and $\mathrm{pd}_{R^*}(A^*) \le 1$. Assume $\mathrm{pd}_{R^*}(A^*) = n \ge 2$, and let $\mathrm{pd}_R(R^*) = r$. By Exercise 9.7, the exact sequence of R^*-modules

$$0 \to K^* \to F^* \to A^* \to 0,$$

with F^* an R^*-free module, has $\mathrm{pd}_{R^*}(K^*) = n - 1 \ge 1$. In particular, $K^* \ne 0$ and so induction gives

$$\mathrm{pd}_R(K^*) = n - 1 + r.$$

Denoting $\mathrm{pd}_R(A^*)$ by a, Exercise 9.12 gives

$$r = \mathrm{pd}_R(F^*) \le \max\{n - 1 + r, a\}.$$

This inequality is strict (hence the sequence does not split), for

$$\max\{n - 1 + r, a\} \ge n - 1 + r > r.$$

The remainder of Exercise 9.12 gives

$$\mathrm{pd}_R(A^*) = \mathrm{pd}_R(K^*) + 1 = (n - 1 + r) + 1 = n + r. \quad \blacksquare$$

Theorem 9.33: *Let $x \in R$ be a central element that is neither a unit nor a zero divisor. If $R^* = R/Rx$, and if $lD(R^*) < \infty$, then*

$$lD(R) \ge lD(R^*) + 1.$$

Before proving this result, let us draw an instant consequence.

Theorem 9.34: *For any ring R,*

$$lD(R[t]) = lD(R) + 1.$$

Proof: Corollary 9.30 gives the inequality $\le$. Corollary 9.28 allows us to assume $lD(R) < \infty$; since $R[t]/tR[t] \cong R$ as rings, Theorem 9.33 applies to give the reverse inequality. $\blacksquare$

Corollary 9.35   (Hilbert's Theorem on Syzygies):   *If k is a field,*

$$D(k[t_1, \ldots, t_n]) = n.$$

Proof: Induction on n, after observing $D(k) = 0$.   ∎

Corollary 9.36:   *If k is a field and $R = k[t_1, \ldots, t_n]$, then every R-module A has a free resolution*

$$0 \to F_n \to \cdots \to F_0 \to A \to 0.$$

Proof: Corollary 9.35 and the Quillen–Suslin theorem.   ∎

Note that the hypothesis $lD(R^*) < \infty$ in Theorem 9.33 is essential. An example is provided by $R = \mathbf{Z}$ and $x = 4$: $D(\mathbf{Z}) = 1$ and $D(\mathbf{Z}/4\mathbf{Z}) = \infty$.

Proof of Theorem 9.33:   It suffices to show that if A^* is a nonzero R^*-module with $\mathrm{pd}_{R^*}(A^*) < \infty$, then

$$\mathrm{pd}_R(A^*) = \mathrm{pd}_{R^*}(A^*) + 1$$

(for then take sup of both sides, realizing that not every R-module need be of the form A^*).

Since x is central, multiplication by x is an R-map $R \to R$; moreover, x not a zero divisor gives an exact sequence of R-modules

$$0 \to R \xrightarrow{x} R \to R^* \to 0.$$

It follows that $\mathrm{pd}_R(R^*) \leq 1$; in fact, $\mathrm{pd}_R(R^*) = 1$, otherwise this sequence would split, forcing Rx to be a summand of R, and this would imply x is a zero divisor. By Theorem 9.32,

$$\mathrm{pd}_R(A^*) \leq \mathrm{pd}_{R^*}(A^*) + 1;$$

moreover, if we can prove equality whenever $A^* \neq 0$ and $\mathrm{pd}_{R^*}(A^*) \leq 1$, we are done. It suffices to eliminate strict inequality.

If $\mathrm{pd}_{R^*}(A^*) = 0$, then $\mathrm{pd}_R(A^*) \leq 1$ and strict inequality means $\mathrm{pd}_R(A^*) = 0$, i.e., A^* is R-projective. But $xA^* = 0$, while $xP \neq 0$ if P is any nonzero projective (P is a submodule of a free R-module, hence each element of P has coordinates in R; since x is not a zero divisor, x cannot annihilate any nonzero element of P). Since $A^* \neq 0$, $\mathrm{pd}_{R^*}(A^*) \neq 0$.

If $\mathrm{pd}_{R^*}(A^*) = 1$, then $\mathrm{pd}_R(A^*) \leq 2$ and strict inequality means $\mathrm{pd}_R(A^*) \leq 1$. We have already shown $\mathrm{pd}_R(A^*) \neq 0$; we claim $\mathrm{pd}_R(A^*) \neq 1$. Otherwise there is an exact sequence of R-modules

$$0 \to K \to F \to A^* \to 0,$$

where F is R-free and K is R-projective. Since $xA^* = 0$, there is an exact sequence of R^*-modules

$$0 \to K/xF \to F/xF \to A^* \to 0.$$

Now F/xF is R^*-free, so that $\mathrm{pd}_{R^*}(A^*) = 1$ shows that K/xF is R^*-projective. The exact sequence of R^*-modules

$$0 \to xF/xK \to K/xK \to K/xF \to 0$$

must split, whence $xF/xK \cong F/K \cong A^*$ is a summand of K/xK. As K is R-projective, K/xK is R^*-projective, and thus A^* is R^*-projective, contradicting $\mathrm{pd}_{R^*}(A^*) = 1$. $\blacksquare$

We now offer a second proof of Theorem 9.33. After all, the "moreover" clause of the crucial Theorem 9.32 does remind one of the slow starting inductions so characteristic of dimension shifting. An obvious way to prove $lD(R[t]) \geq m$ for some (finite) m is to exhibit a pair of $R[t]$-modules A and B with $\mathrm{Ext}^m(A, B) \neq 0$; we shall be able to do this once we have proved the next result.

Theorem 9.37   (Rees):   *Assume $x \in R$ is a central element that is neither a unit nor a zero divisor, and let $R^* = R/xR$; assume B is an R-module for which multiplication by x, $\mu = \mu_x : B \to B$, is monic. Then, for every R^*-module A, there are isomorphisms*

$$\mathrm{Ext}^n_{R^*}(A, B/xB) \cong \mathrm{Ext}^{n+1}_R(A, B).$$

Proof: First of all, the natural map $R \to R^*$ enables us, by change of rings, to regard A as an R-module; since an R^*-module is merely an R-module annihilated by x, the quotient B/xB is an R^*-module; thus, both sides make sense.

Next, we claim $\mathrm{Hom}_R(A, B) = 0$. If $a \in A$, then $xa = 0$, for A is an R^*-module. Consequently, if $f : A \to B$, then

$$xf(a) = f(xa) = f(0) = 0,$$

and the hypothesis on B forces $f = 0$.

To prove the theorem, we use the axioms characterizing the contravariant Ext functors (Exercise 7.27, proved by dimension shifting). Define contravariant functors $G^n : {}_{R^*}\mathfrak{M} \to \mathbf{Ab}$, $n \geq 0$, by

$$G^n(A) = \mathrm{Ext}^{n+1}_R(A, B).$$

Clearly $\{G^n\}$ is a strongly connected sequence of contravariant functors.

Exactness of $0 \to B \xrightarrow{\mu} B \to B/xB \to 0$ gives exactness of

$$\mathrm{Hom}_R(A, B) \to \mathrm{Hom}_R(A, B/xB) \to \mathrm{Ext}^1_R(A, B) \xrightarrow{\mu^*} \mathrm{Ext}^1_R(A, B).$$

We have already seen that $\text{Hom}_R(A, B) = 0$. On the other hand, μ^* is also multiplication by x (Theorem 7.16); since A is annihilated by x, we have $\text{im}\,\mu^* = 0$. We conclude that the connecting homomorphism is an isomorphism

$$\text{Hom}_R(A, B/xB) \xrightarrow{\sim} \text{Ext}^1_R(A, B).$$

However, $R \to R^*$ onto allows us to use Exercise 9.13 to conclude

$$\text{Hom}_{R^*}(A, B/xB) = \text{Hom}_R(A, B/xB).$$

There is thus a natural equivalence $G^0 \cong \text{Hom}_R(\,, B/xB)$.

It remains to prove that $G^n(P^*) = 0$ for all free R^*-modules P^* and all $n \geq 1$. Choose a basis of P^*, and let Q be the free R-module with the same basis. There is an R-exact sequence

$$0 \to Q \xrightarrow{x} Q \to P^* \to 0$$

(again we have used x a central non-zero-divisor). Exactness of

$$\cdots \to \text{Ext}^n_R(Q, B) \to \text{Ext}^{n+1}_R(P^*, B) \to \text{Ext}^{n+1}_R(Q, B) \to \cdots$$

shows that

$$G^n(P^*) = \text{Ext}^{n+1}_R(P^*, B) = 0, \qquad n \geq 1.$$

The axioms have been verified and the theorem is proved. ∎

Corollary 9.38 (= Theorem 9.33): *Assume $x \in R$ is a central element that is neither a unit not a zero divisor, and let $R^* = R/xR$. If $lD(R^*) = n < \infty$, then $lD(R) \geq lD(R^*) + 1$.*

Proof: Assume M is an R^*-module with $\text{pd}_{R^*}(M) = n$. By Exercise 9.6, there is a free R^*-module F^* with $\text{Ext}^n_R(M, F^*) \neq 0$. If we define B as the free R-module on the same basis as F^* (precisely as in the previous proof), then $B/xB \cong F^*$. By the theorem of Rees,

$$\text{Ext}^{n+1}_R(M, B) \cong \text{Ext}^n_{R^*}(M, F^*) \neq 0,$$

and this says $\text{pd}_R(M) \geq n + 1$. Therefore, $lD(R) \geq n + 1 = lD(R^*) + 1$. ∎

There is another comparison theorem that is often useful.

Definition: Let R be a (not necessarily commutative) ring. A right R-module M is **faithfully flat** if M is flat and whenever $M \otimes_R N = 0$, then $N = 0$.

Examples: 5. $R[t]$ is faithfully flat, for it is a free R-module.

6. If R is the ring of integers in an algebraic number field, then R is a faithfully flat **Z**-module (for R is **Z**-free).

7. **Q** is a flat **Z**-module, but it is not faithfully flat.

Exercise: 9.14.   Assume R is a subring of A with A_R flat. Prove that A_R is faithfully flat if and only if $A \otimes_R (R/I) \neq 0$ for every proper left ideal I of R. Conclude that A is faithfully flat if and only if $AI \neq A$ for every proper left ideal I of R.

Theorem 9.39   (McConnell–Roos):   *Assume R is a ring with $\mathrm{lD}(R) = n < \infty$, and that R is a subring of A, where A_R is faithfully flat. Assume either*

   (i) $_R A$ *is projective, or*
   (ii) $_R A$ *is flat and R is left noetherian.*

Then $\mathrm{lD}(R) \leq \mathrm{lD}(A)$.

Remark: Goodearl [1974] gives an example showing one must assume R has finite global dimension.

Proof: For any left R-module M, there is a map $\varphi : M \to A \otimes M$ given by $m \mapsto 1 \otimes m$; we claim φ is monic. If $K = \ker \varphi$, then $K = \{m \in M : 1 \otimes m = 0\}$. Since A_R is flat, exactness of $0 \to K \overset{i}{\to} M$ gives exactness of

$$0 \to A \otimes K \xrightarrow{\ 1 \otimes i\ } A \otimes M.$$

But the definition of K gives

$$\mathrm{im}(1 \otimes i) = \{1 \otimes m \in A \otimes M : m \in K\} = 0;$$

there is thus an exact sequence $0 \to A \otimes K \to 0$. Therefore $A \otimes K = 0$, so that A faithfully flat implies $K = 0$.

   Assume (i): $_R A$ is projective. Choose $_R M$ with $\mathrm{pd}_R(M) = n$. There is an R-exact sequence

$$0 \to M \to A \otimes M \to C \to 0.$$

Now $\mathrm{pd}_R(C) \leq n$, since $\mathrm{lD}(R) \leq n$, so Exercise 9.9 gives

$$\mathrm{pd}_R(A \otimes M) = n.$$

It suffices to prove $\mathrm{pd}_A(A \otimes M) = n$, for then $\mathrm{lD}(A) \geq n$, as claimed.
   Take an R-projective resolution of M:

$$0 \to P_n \to \cdots \to P_0 \to M \to 0.$$

Since A_R is flat, there is an A-exact sequence

$$0 \to A \otimes P_n \to \cdots \to A \otimes P_0 \to A \otimes M \to 0.$$

By Exercise 3.10, each $A \otimes P_i$ is A-projective. Hence $\mathrm{pd}_A(A \otimes M) \leq n$.
   For the reverse inequality, choose an A-projective resolution of $A \otimes M$:

$$\cdots \to Q_{n+1} \to Q_n \to \cdots \to Q_0 \to A \otimes M \to 0.$$

Since $_R A$ is projective, it is easy to see that each Q_i is also R-projective. Therefore, $\text{pd}_R(A \otimes M) = n$ implies (Theorem 9.5(v)) that we may assume

$$0 \to K_n \to Q_{n-1} \to \cdots \to Q_0 \to A \otimes M \to 0$$

is an R-projective resolution of $A \otimes M$ (where $K_n = \ker(Q_{n-1} \to Q_{n-2})$ considered as an R-module). Hence, $n \le \text{pd}_A(A \otimes M)$.

Assume (ii): $_R A$ is flat and R is left noetherian. The above argument may be repeated, yielding $\text{pd}_R(A \otimes M) = n$ and $\text{pd}_A(A \otimes M) \le n$. For the reverse inequality, we use R left noetherian to enable us to switch to flat resolutions (Theorem 9.22). Assume an A-flat resolution of $A \otimes M$:

$$\cdots \to F_{n+1} \to F_n \to \cdots \to F_0 \to A \otimes M \to 0.$$

Since $_R A$ is flat, each F_i is R-flat (for $\otimes_R F_i = \otimes_R (A \otimes_A F_i)$), and use Exercise 3.38). But now Theorem 9.13(iv) applies, and the proof is completed as in case (i). ∎

SERRE'S THEOREM

We now give an elementary proof of the theorem of Serre mentioned in Chapter 4; the basic idea of this proof is due to Borel, Serre, and Swan and the exposition provides details of the sketch given by Kaplansky [1970, p. 134].

Definition: A module M **has FFR (finite free resolution) of length** $\le n$ if there is an exact sequence

$$0 \to F_n \to F_{n-1} \to \cdots \to F_0 \to M \to 0$$

in which each F_i is f.g. free.

If M has FFR, then M is f.g., even finitely related. The reader may easily provide examples of such modules.

Exercises: 9.15. A projective module has FFR if and only if it is stably free. (Hint: Induction on length.)

9.16. Let R be left noetherian and let M be a left R-module with FFR of length $\le n$. Every free resolution of M, each of whose terms is f.g., has a stably free nth syzygy. (Hint: Generalized Schanuel lemma, Exercise 3.37.)

Lemma 9.40: *If M has a projective resolution*

$$0 \to P_n \to \cdots \to P_2 \xrightarrow{d_2} P_1 \xrightarrow{d_1} P_0 \xrightarrow{\varepsilon} M \to 0$$

in which each P_i is stably free, then M has FFR of length $\le n + 1$.

Proof: We do an induction on n. If $n = 0$, then ε is an isomorphism $P_0 \overset{\sim}{\to} M$. Since P_0 is stably free, there are f.g. free modules F_0 and F_0' with $F_0 \cong P_0 \oplus F_0'$, i.e., there is an exact sequence $0 \to F_0' \to F_0 \to M \to 0$. If $n > 0$, let $K_0 = \ker \varepsilon$. If F is a f.g. free module with $P_0 \oplus F$ free, then we have exactness of

$$0 \to P_n \to \cdots \to P_2 \overset{d_2'}{\to} P_1 \oplus F \xrightarrow{d_1 \oplus 1_F} P_0 \oplus F \overset{\varepsilon'}{\to} M \to 0,$$

where d_2': $p_2 \mapsto (d_2 p_2, 0)$ and ε': $(p_0, f) \mapsto \varepsilon(p_0)$. Since $\ker \varepsilon' = K_0 \oplus F$ has a stably free resolution with $n - 1$ terms, the proof is completed by induction. ∎

Theorem 9.41:   *Let R be left noetherian and $0 \to M' \to M \to M'' \to 0$ an exact sequence of left R-modules. If two of the modules have FFR, then so does the third one.*

Proof: Since two of the modules have FFR, they are f.g.; since R is noetherian, the third module is also f.g. That R is noetherian also provides free resolutions of M' and M'' each of whose terms is f.g. (Corollary 4.4). Use the Horseshoe Lemma 6.20 to insert a resolution with f.g. free modules in the middle. For each n, there is an exact sequence of syzygies

$$0 \to K_n' \to K_n \to K_n'' \to 0.$$

Assuming two of $\{M', M, M''\}$ have FFR of length $\leq n$, Exercise 9.16 shows two of $\{K_n', K_n, K_n''\}$ are stably free. If one of these is K_n'', then the sequence of syzygies splits and the third syzygy is also stably free; Lemma 9.40 now gives the result. If K_n' and K_n are stably free, then

$$0 \to K_n' \to K_n \to F_{n-1}'' \to \cdots \to F_0'' \to M'' \to 0$$

is a resolution of M'' by stably free modules, and Lemma 9.40 gives the result in this case as well. ∎

Definition: A **family** $\mathfrak{F}$ is a nonempty subclass of $_R\mathfrak{M}$ such that whenever $0 \to M' \to M \to M'' \to 0$ is exact and two of the modules lie in $\mathfrak{F}$, then the third one lies in $\mathfrak{F}$ as well.

Theorem 9.41 states that the class of all FFR modules over a left noetherian ring is a family.

There are some easy observations. Every family $\mathfrak{F}$ contains the zero module: if $M \in \mathfrak{F}$, then $0 \to M \xrightarrow{1_M} M \to 0 \to 0$ is exact. From this it follows that $M' \cong M$ and $M \in \mathfrak{F}$ implies $M' \in \mathfrak{F}$: families are closed under isomorphism.

It is easy to see that any intersection of families is again a family. Thus, every subclass $\mathfrak{S}$ of $_R\mathfrak{M}$ generates a family.

If $\mathfrak{S}$ is a subclass of $_R\mathfrak{M}$ containing the zero module, define $E(\mathfrak{S})$ as the subclass of all modules contained in a short exact sequence having two terms in $\mathfrak{S}$. Clearly $E(\mathfrak{S})$ is a subclass containing $\mathfrak{S}$. Define an ascending chain of subclasses:

$$E^0(\mathfrak{S}) = \mathfrak{S}; \qquad E^{n+1}(\mathfrak{S}) = E(E^n(\mathfrak{S})).$$

Lemma 9.42: *If $\mathfrak{S}$ is a subclass of $_R\mathfrak{M}$ containing the zero module, then $\bigcup_{n=0}^{\infty} E^n(\mathfrak{S})$ is the family generated by $\mathfrak{S}$.*

Proof: It is clear that any family containing $\mathfrak{S}$ must contain each $E^n(\mathfrak{S})$, hence the union of them. On the other hand, $\bigcup E^n(\mathfrak{S})$ is a family containing $\mathfrak{S}$: if $0 \to M' \to M \to M'' \to 0$ is exact with two terms in $\bigcup E^n(\mathfrak{S})$, then these two terms lie in some $E^n(\mathfrak{S})$, and so the third lies in $E(E^n(\mathfrak{S})) = E^{n+1}(\mathfrak{S})$. ∎

Corollary 9.43: *Let $\mathfrak{S}$ be a class of modules over a left noetherian ring, each member of which has FFR. Then every module in $\mathfrak{F}$, the family generated by $\mathfrak{S}$, has FFR.*

Proof: If $M \in \mathfrak{F}$, there is a least n with $M \in E^n(\mathfrak{S})$; we do an induction on n. If $n = 0$, then $M \in \mathfrak{S}$ and has FFR by hypothesis. If $n > 0$, then M is contained in a short exact sequence whose other two terms lie in $E^{n-1}(\mathfrak{S})$ and hence have FFR, by induction. By Theorem 9.41, M has FFR. ∎

Theorem 9.44: *Let R be commutative noetherian and assume every f.g. R-module has FFR; then every f.g. $R[t]$-module has FFR.*

Before proving Theorem 9.44, let us deduce Serre's theorem from it.

Theorem 9.45 (Serre): *If k is a field, then every f.g. projective $k[t_1, \ldots, t_n]$-module is stably free.*

Proof: We first do an induction on n that every f.g. $k[t_1, \ldots, t_n]$-module has FFR. If $n = 1$, then $k[t]$ is a principal ideal domain and every f.g. module has FFR of length ≤ 1. If $n > 1$, let $R = k[t_1, \ldots, t_{n-1}]$, which is noetherian by the Hilbert Basis Theorem. By induction, every f.g. R-module has FFR, so that Theorem 9.44 gives every f.g. $R[t_n]$-module has FFR. In particular, every f.g. projective module over $R[t_n] = k[t_1, \ldots, t_n]$ has FFR, hence is stably free by Exercise 9.15. ∎

To prove Theorem 9.44, we use two simple lemmas, the first of which is a form of Zorn's lemma available in noetherian rings.

Lemma 9.46: *If R is a left noetherian ring, then every nonempty set $\mathfrak{A}$ of left ideals of R contains a maximal member.*

Proof: Assume $\mathfrak{A}$ has no maximal member. Choose $I_1 \in \mathfrak{A}$; since I_1 is not maximal, there is $I_2 \in \mathfrak{A}$ with $I_1 \subsetneqq I_2$. This procedure may be iterated indefinitely, so the axiom of choice provides a strictly ascending chain that violates ACC.   ∎

Remark: One may show that the property in Lemma 9.46 characterizes left noetherian rings.

Definition: If M is a left R-module, then its **annihilator** is

$$\text{ann}(M) = \{r \in R : rM = 0\}.$$

One defines the annihilator of an element $x \in M$ by

$$\text{ann}(x) = \text{ann}(Rx).$$

It is clear that $\text{ann}(M)$ is a left ideal in R.

Lemma 9.47:   *Let R be commutative noetherian and M a nonzero R-module. If I is an ideal maximal among $\{\text{ann}(x) : x \in M, x \neq 0\}$, then I is a prime ideal.*

Proof: Choose $x \in M$ with $I = \text{ann}(x)$. Let $ab \in I$ and $b \notin I$, i.e., $abx = 0$ and $bx \neq 0$. Then

$$\text{ann}(bx) \supset I + Ra \supset I.$$

If $a \notin I$, the maximality of I is violated. Therefore, I is prime.   ∎

Proof of 9.44:   For any noetherian ring R (not necessarily enjoying the FFR hypothesis), we shall prove the family $\mathfrak{F}$ in $_{R[t]}\mathfrak{M}$ generated by the class $\mathfrak{S}$ of all (extended) modules of the form $R[t] \otimes N$, where N is a f.g. R-module, consists of all f.g. $R[t]$-modules. If, in addition, every f.g. R-module has FFR, then Corollary 9.43 applies, for N has FFR implies $R[t] \otimes N$ has FFR (because $R[t]$ is R-flat), and we are done.

We shall normalize the problem so we may assume that if M' is a f.g. $R[t]$-module with $\text{ann}(M') \cap R \neq 0$, then $M' \in \mathfrak{F}$. Define $I_0 = 0$ and $\mathfrak{F}_0 = \mathfrak{F}$. Construct an ascending chain of ideals in R, $I_0 \subset I_1 \subset I_2 \subset \cdots$, as follows. Assume I_n has been defined and that $\mathfrak{F}_n$ is the family generated by $\mathfrak{S}_n$, the class of all $R_n[t]$-modules extended from f.g. R_n-modules, where R_n denotes R/I_n. If there are f.g. $R_n[t]$-modules M not in $\mathfrak{F}_n$, Lemma 9.46 provides an ideal I_{n+1}/I_n maximal among $\{\text{ann}(M) \cap R_n : M \notin \mathfrak{F}_n\}$: if no such M exist, define $I_{n+1} = I_n$. Since R is noetherian, this chain stops, say, $I_m = I_{m+1} = \cdots$. Observe that $M \notin \mathfrak{F}_m$ implies

$$\text{ann}(M) \cap R_m \subset I_{m+1}/I_m = 0.$$

We prove, by induction on the minimal such $m \geq 0$, that $\mathfrak{F}$ consists of all f.g. $R[t]$-modules. Momentarily deferring the case $m = 0$, we prove the inductive step. Choose a f.g. $R[t]$-module $M \notin \mathfrak{F}$ with $\text{ann}(M) \cap R = I_1$. Since $I_1 M = 0$, M is a f.g. $R_1[t]$-module. By induction, $M \in \mathfrak{F}_1$, the family generated by $\mathfrak{S}_1$. Since

$$R_1[t] \otimes N \cong R[t] \otimes (R_1 \otimes N),$$

it follows that $\mathfrak{S}_1 \subset \mathfrak{S}$ and $\mathfrak{F}_1 \subset \mathfrak{F}$. Therefore $M \in \mathfrak{F}$, a contradiction.

It remains to prove the case $m = 0$: we have normalized so that if M' is a f.g. $R[t]$-module with $\text{ann}(M') \cap R \neq 0$, then $M' \in \mathfrak{F}$. Choose a f.g. module M (over the noetherian ring $R[t]$) with $M \notin \mathfrak{F}$. We may apply Lemma 9.46 to find a nonzero $x_1 \in M$ with $\text{ann}(x_1)$ maximal among $\{\text{ann}(x): x \in M, x \neq 0\}$. Iterating, there are elements $x_1, x_2, \ldots$ in M and submodules $M_0 = 0$ and $M_i = \langle x_1, x_2, \ldots, x_i \rangle$ with

$$0 = M_0 \subset M_1 \subset M_2 \subset \cdots.$$

By construction, $M_i/M_{i-1} \cong R[t]/P_i$, where $P_i = \text{ann}(x_i + M_{i-1})$ is a prime ideal (Lemma 9.47). By Corollary 4.7, $R[t]$ noetherian and M f.g. imply M has ACC; therefore, $M = M_r$ for some r. By induction on r, using the fact that $\mathfrak{F}$ is a family, $M \in \mathfrak{F}$ if $R[t]/P \in \mathfrak{F}$ for every prime ideal P in $R[t]$ (and this is the desired contradiction).

Now $P \cap R \subset \text{ann}(R[t]/P) \cap R$, so the normalization gives $R[t]/P \in \mathfrak{F}$ if $P \cap R \neq 0$. We may thus assume $P \cap R = 0$; as $P \cap R$ is always a prime ideal in R, we may assume R is a domain; let Q be its quotient field. Let $P = (g_1, \ldots, g_n)$ and choose $f \in P$ a nonzero polynomial of minimal degree. We claim $\text{ann}(P/(f)) \cap R \neq 0$. For each i, $1 \leq i \leq n$, the division algorithm in $Q[t]$ yields

$$g_i = fq_i + r_i,$$

where q_i and $r_i \in Q[t]$ and degree $r_i <$ degree f. Clearing denominators, there are nonzero $a_i \in R$ with

$$a_i g_i = fq_i' + r_i',$$

where q_i' and $r_i' \in R[t]$ and degree $r_i' <$ degree f. By minimality of degree f, $a_i g_i = fq_i' \in (f)$ for all i. If $a = \prod a_i$, then $a \neq 0$ and $aP/(f) = 0$. Hence $a \in \text{ann}(P/(f)) \cap R$, and the normalization gives $P/(f) \in \mathfrak{F}$. Exactness of

$$0 \to (f) \to P \to P/(f) \to 0$$

together with $(f) \cong R[t] \in \mathfrak{F}$ imply $P \in \mathfrak{F}$, for $\mathfrak{F}$ is a family. Finally, exactness of

$$0 \to P \to R[t] \to R[t]/P \to 0$$

shows $R[t]/P \in \mathfrak{F}$, as desired. ∎

We mention a useful tool; for some applications, see [Kaplansky, 1970, §4.3].

Exercises: 9.17.   Assume R is a ring with IBN and $0 \to F_n \to \cdots \to F_0 \to M \to 0$ is R-exact, with each F_i f.g. free. Prove that $\chi(M) = \sum_{i=0}^{n}(-1)^i \text{rank}(F_i)$ is independent of the choice of FFR. (Hint: Schanuel.) The integer $\chi(M)$ is called the **Euler characteristic** of M. Compare Exercise 3.6.

9.18.   Assume R has IBN and $0 \to M' \to M \to M'' \to 0$ is an exact sequence of left R-modules, two of which have an Euler characteristic. Then the third module also has an Euler characteristic, and

$$\chi(M) = \chi(M') + \chi(M'').$$

(Hint: Use Lemma 3.63.)

MIXED IDENTITIES

Various natural isomorphisms involving tensor and Hom can be extended to isomorphisms involving Tor and Ext.

Theorem 9.48:   *Let R and S be rings, and consider the situation $(A_R, {}_R B_S, {}_S C)$. If A is flat, there are isomorphisms*

$$\operatorname{Tor}_n^S(A \otimes_R B, C) \cong A \otimes_R \operatorname{Tor}_n^S(B, C).$$

Proof: Let $\mathbf{P}_C$ be a deleted projective resolution of C. Naturality of the associativity isomorphisms for tensor imply an isomorphism of complexes

$$(A \otimes B) \otimes_S \mathbf{P}_C \cong A \otimes_R (B \otimes \mathbf{P}_C)$$

and hence isomorphisms, for all $n \geq 0$,

$$H_n((A \otimes B) \otimes_S \mathbf{P}_C) \cong H_n(A \otimes_R (B \otimes \mathbf{P}_C)).$$

The left hand side is thus $\operatorname{Tor}_n^S(A \otimes_R B, C)$. To evaluate the right hand side, flatness of A means the functor $A \otimes_R$ is exact. By Exercise 6.4,

$$H_n(A \otimes_R (B \otimes \mathbf{P}_C)) \cong A \otimes_R H_n(B \otimes \mathbf{P}_C) = A \otimes_R \operatorname{Tor}_n^S(B, C). \quad \blacksquare$$

There is a very useful variant of this result.

Theorem 9.49:   *Let R be commutative, S a subset of R, and A, B R-modules. There are isomorphisms, for all $n \geq 0$,*

$$S^{-1} \operatorname{Tor}_n^R(A, B) \cong \operatorname{Tor}_n^{S^{-1}R}(S^{-1}A, S^{-1}B).$$

Proof: Recall that $S^{-1}A = S^{-1}R \otimes_R A$. By Lemma 3.77, there is a natural isomorphism

$$S^{-1}R \otimes_R (A \otimes_R B) \cong S^{-1}A \otimes_{S^{-1}R} S^{-1}B.$$

If $\mathbf{P}_B$ is a deleted projective resolution of B, then exactness of localization may be used to show that $S^{-1}\mathbf{P}_B$ is a deleted $S^{-1}R$-projective resolution of $S^{-1}B$. Naturality provides an isomorphism of complexes

$$S^{-1}R \otimes_R (A \otimes_R \mathbf{P}_B) \cong S^{-1}A \otimes_{S^{-1}R} S^{-1}\mathbf{P}_B,$$

and their homology modules are thus isomorphic. Since $S^{-1}R$ is a flat R-module, Theorem 9.48 shows

$$H_n(S^{-1}R \otimes_R (A \otimes_R \mathbf{P}_B)) \cong S^{-1}R \otimes_R \operatorname{Tor}_n^R(A, B) = S^{-1}\operatorname{Tor}_n^R(A, B).$$

On the other hand,

$$H_n(S^{-1}A \otimes_{S^{-1}R} S^{-1}\mathbf{P}_B) = \operatorname{Tor}_n^{S^{-1}R}(S^{-1}A, S^{-1}B). \quad \blacksquare$$

There is a similar result for Ext if we add some hypotheses.

Theorem 9.50: *Let R be commutative noetherian with subset S, and let A, B be R-modules with A f.g. There are isomorphisms, for all $n \geq 0$,*

$$S^{-1}\operatorname{Ext}_R^n(A, B) \cong \operatorname{Ext}_{S^{-1}R}^n(S^{-1}A, S^{-1}B).$$

Proof: Since R is noetherian and A is f.g., Lemma 9.20 shows there is a deleted projective resolution $\mathbf{P}_A$ each of whose terms is f.g. But now the natural isomorphism of Theorem 3.84 gives an isomorphism of complexes

$$S^{-1}R \otimes_R \operatorname{Hom}(\mathbf{P}_A, B) \cong \operatorname{Hom}_{S^{-1}R}(S^{-1}\mathbf{P}_A, S^{-1}B).$$

Taking homology of the left side gives $S^{-1}R \otimes_R \operatorname{Ext}_R^n(A, B)$ (since $S^{-1}R$ is a flat R-module), while homology of the right side is $\operatorname{Ext}_{S^{-1}R}^n(S^{-1}A, S^{-1}B)$. $\quad \blacksquare$

Remark: One can weaken the hypotheses to R a commutative ring for which localizations of injectives are injective (see the remark after Theorem 3.85) and A finitely related (see the proof of Theorem 11.58).

Theorem 9.51: *Let R and S be rings with R left noetherian. Consider the situation $({}_RA, {}_RB_S, C_S)$ with A f.g. and C injective. Then there are isomorphisms, for $n \geq 0$,*

$$\operatorname{Tor}_n^R(\operatorname{Hom}_S(B, C), A) \cong \operatorname{Hom}_S(\operatorname{Ext}_R^n(A, B), C).$$

Proof: Since R is left noetherian, A f.g. implies A finitely related. The hypotheses are precisely those of Lemma 3.60, which provides a natural isomorphism

$$\operatorname{Hom}_S(B, C) \otimes_R A \cong \operatorname{Hom}_S(\operatorname{Hom}_R(A, B), C).$$

Replace A by a deleted projective resolution $\mathbf{P}_A$ each of whose terms is f.g. (Lemma 9.20) to obtain isomorphic complexes; taking homology gives isomorphisms for $n \geq 0$,

$$H_n(\operatorname{Hom}_S(B, C) \otimes_R \mathbf{P}_A) \cong \operatorname{Hom}_S(H^n(\operatorname{Hom}_R(\mathbf{P}_A, B)), C),$$

for H_n commutes with the contravariant exact functor $\operatorname{Hom}_S(\ , C)$ (C is injective). The left side is $\operatorname{Tor}_n^R(\operatorname{Hom}_S(B, C), A)$; the right side is $\operatorname{Hom}_S(\operatorname{Ext}_R^n(A, B), C)$. $\quad \blacksquare$

Remark: The same result holds with no noetherian hypothesis if one assumes A has a projective resolution each of whose terms is f.g.

Exercises: 9.19. Use Theorem 9.51 to give another proof that R left noetherian implies $lD(R) = wD(R)$. (Hint: If A is f.g. and $n \le pd(A)$, then $Ext_R^n(A, B) \ne 0$ for some B; therefore there exists a Z-injective C with $Hom_Z(Ext_R^n(A, B), C) \ne 0$.)

9.20. In the situation $({}_R A, {}_S B_R, {}_S C)$ with A projective, use the adjoint isomorphism to obtain isomorphisms

$$Ext_S^n(B \otimes_R A, C) \cong Hom_R(A, Ext_S^n(B, C)).$$

9.21. In the situation $({}_R A, {}_S B_R, {}_S C)$ with B R-projective, use the adjoint isomorphism to obtain isomorphisms

$$Ext_S^n(B \otimes_R A, C) \cong Ext_R^n(A, Hom_S(B, C)).$$

9.22. Recall that if m is a maximal (even a prime) ideal in a commutative ring R, then $R_m = S^{-1}R$, where $S = R - m$, the complement of m. If R is commutative noetherian and A is a f.g. R-module, then prove A is projective if and only if A_m is R_m-projective for every maximal ideal m of R. (Hint: Theorems 9.50 and 3.80.)

9.23. If R is commutative noetherian, then an R-module B is injective if and only if B_m is an injective R_m-module for every maximal ideal m of R. (Hint: Theorem 9.50 and Lemma 9.11.)

9.24. If I is a nonzero ideal in a Dedekind ring R, then R/I is quasi-Frobenius. (Hint: Use Theorem 4.35 and the fact that R Dedekind implies R_m a principal ideal domain.)

9.25. If R is commutative noetherian with $D(R) = n$ and if $A \in {}_R\mathfrak{M}$, then $Ext_R^n(A, M) \cong M \otimes_R Ext_R^n(A, R)$ for all f.g. R-modules M. (Hint: Use Corollary 3.34.)

9.26. The following are equivalent for a (commutative) domain R: (i) $wD(R) \le 1$; (ii) R is a Prüfer ring; (iii) R_m is a valuation ring for every maximal ideal m in R. (Hint: Use the result of Endo [1962] that f.g. flat modules over a domain are projective, as well as Theorems 9.49 and 3.80 and the fact that local Bézout rings are valuation rings.)

COMMUTATIVE NOETHERIAN LOCAL RINGS

Although homological algebra had admirers from its birth, many mathematicians dismissed it as "merely a language".[1] In 1958 and 1959, Auslander, Buchsbaum, and Serre used homological algebra to solve two

---

[1] To say something is "merely a language" is not to say it is worthless. Indeed, even a mere notation may be valuable: try to deal with polynomials without the usual notation for the variable (due to Viète).

open questions about local rings; the language gave forth poetry, and became more widely accepted thereafter.

The problem of computing the global dimension of a commutative noetherian ring reduces to computing global dimension of its localizations.

Theorem 9.52: *If R is a commutative noetherian ring, then*

$$D(R) = \sup D(R_{\mathfrak{m}})$$

where $\mathfrak{m}$ ranges over all maximal ideals in R.

Proof: First of all, we prove $D(R) \geq D(R_{\mathfrak{m}})$ for every maximal $\mathfrak{m}$. If $D(R) = \infty$, we are done, so assume $D(R) = n < \infty$. Let A be an $R_{\mathfrak{m}}$-module; we may regard A as an R-module and $A \cong A_{\mathfrak{m}} = R_{\mathfrak{m}} \otimes_R A$ (Lemma 3.75). There is an R-projective resolution

$$0 \to P_n \to \cdots \to P_0 \to A \to 0$$

since $D(R) = n$. Since $R_{\mathfrak{m}}$ is a flat R-module and $R_{\mathfrak{m}} \otimes P_i$ are projective $R_{\mathfrak{m}}$-modules,

$$0 \to R_{\mathfrak{m}} \otimes P_n \to \cdots \to R_{\mathfrak{m}} \otimes P_0 \to R_{\mathfrak{m}} \otimes A \to 0$$

is an $R_{\mathfrak{m}}$-projective resolution of $R_{\mathfrak{m}} \otimes A \cong A$ showing $\mathrm{pd}_{R_{\mathfrak{m}}}(A) \leq n$. It follows that $D(R_{\mathfrak{m}}) \leq n = D(R)$.

For the reverse inequality, it suffices to assume $\sup D(R_{\mathfrak{m}}) = n < \infty$. Since R is noetherian, Theorem 9.22 says $D(R) = wD(R)$, so it is enough to prove that

$$\mathrm{Tor}_{n+1}^R(A, B) = 0$$

for all R-modules A and B (Theorem 9.14). However, Theorem 9.49 provides an isomorphism

$$(\mathrm{Tor}_{n+1}^R(A, B))_{\mathfrak{m}} \cong \mathrm{Tor}_{n+1}^{R_{\mathfrak{m}}}(A_{\mathfrak{m}}, B_{\mathfrak{m}}) = 0,$$

for $wD(R_{\mathfrak{m}}) \leq D(R_{\mathfrak{m}}) = n$ (Theorem 9.16; in fact one knows (Example 3 of Chapter 4) that $R_{\mathfrak{m}}$ is noetherian, whence $wD(R_{\mathfrak{m}}) = D(R_{\mathfrak{m}})$). But now Theorem 3.80(i) applies to show $\mathrm{Tor}_{n+1}^R(A, B) = 0$. ∎

If R is commutative noetherian, and if $\mathfrak{m}$ is a prime ideal in R, then $R_{\mathfrak{m}}$ is commutative, noetherian, and local (Chapter 4, Examples 3 and 36). In view of Theorem 9.52, we shall now modify our definition of local ring.

Definition: A commutative ring R is **local** if it is noetherian and if it has a unique maximal ideal $\mathfrak{m}$; the quotient $R/\mathfrak{m}$ is called the **residue field** and is denoted k.

Lemma 9.53: *Let R be local and A a f.g. R-module. Then*

$$\text{pd}(A) \le n \quad \text{if and only if} \quad \text{Tor}^R_{n+1}(A, k) = 0.$$

Proof: If $\text{pd}(A) \le n$, then $\text{fd}(A) \le n$ (see proof of Theorem 9.16), and hence $\text{Tor}_{n+1}(A, k) = 0$ (indeed, $\text{Tor}_{n+1}(A, B) = 0$ for every B).

We prove the converse by induction on n. Assume $n = 0$ and $\text{Tor}_1(A, k) = 0$. Consider the projective cover

$$0 \to N \xrightarrow{i} F \xrightarrow{\varphi} A \to 0$$

(in elementary terms, let $\{a_1, \ldots, a_t\}$ be a minimal set of generators of A, let F be free on $\{x_1, \ldots, x_t\}$, and let $\varphi(x_j) = a_j$, all j; in Lemma 4.43, it is shown that $iN \subset \mathfrak{m}F$). There is an exact sequence

$$0 \to N \otimes k \xrightarrow{i \otimes 1} F \otimes k \xrightarrow{\varphi \otimes 1} A \otimes k \to 0$$

since $\text{Tor}_1(A, k) = 0$. We claim $i \otimes 1 = 0$. Since $iN \subset \mathfrak{m}F$, $y \in N$ implies $iy = \sum r_j x_j$, $r_j \in \mathfrak{m}$. If $\lambda \in k$,

$$i \otimes 1(y \otimes \lambda) = \sum r_j x_j \otimes \lambda = \sum x_j \otimes r_j \lambda = 0$$

because $r_j \lambda = 0$, all j ($\mathfrak{m}$ annihilates $k = R/\mathfrak{m}$). Thus

$$0 = \text{im}(i \otimes 1) = \ker(\varphi \otimes 1) \cong N \otimes k.$$

It is easily seen that $N \otimes k = N \otimes (R/\mathfrak{m}) \cong N/\mathfrak{m}N$. Therefore $N = \mathfrak{m}N$, so Nakayama's lemma gives $N = 0$ (R noetherian gives N f.g.). Thus $A \cong F$ is free and $\text{pd}(A) \le 0$.

For the inductive step, assume $\text{Tor}_{n+1}(A, k) = 0$, and take a projective resolution of A with $(n - 1)$st syzygy K_{n-1}. By Corollary 6.13,

$$\text{Tor}_{n+1}(A, k) \cong \text{Tor}_1(K_{n-1}, k).$$

The case $n = 0$ shows K_{n-1} is free, whence A has a projective resolution of length n, and so $\text{pd}(A) \le n$. ∎

Lemma 9.54: *If R is local, then $\text{D}(R) \le n$ if and only if $\text{Tor}^R_{n+1}(k, k) = 0$.*

Proof: Only sufficiency needs proof. If $\text{Tor}_{n+1}(k, k) = 0$, then $\text{pd}(k) \le n$ (Lemma 9.53); hence, $\text{Tor}_{n+1}(A, k) = 0$ for all A (apply $A \otimes_R$ to a projective resolution of k of length $\le n$); in particular, $\text{Tor}_{n+1}(A, k) = 0$ when A is f.g., so Lemma 9.53 gives $\text{pd}(A) \le n$. Since $\text{D}(R)$ may be computed from projective dimensions of f.g. modules (indeed, cyclic modules suffice), we have shown $\text{D}(R) \le n$. ∎

Corollary 9.55: *If R is local, then*

$$\text{D}(R) = \text{pd}(k).$$

Proof: Immediate from Lemma 9.54. ∎

If R is a local ring, there is a longest chain of prime ideals $\mathfrak{p}_0 \supset \mathfrak{p}_1 \supset \cdots \supset \mathfrak{p}_d$, and the number d is called the **(Krull) dimension** of R and is denoted $\mathrm{Dim}(R)$. A second number one may associate to R is $V(R)$, the minimal number of generators of $\mathfrak{m}/\mathfrak{m}^2$. Since $\mathfrak{m}/\mathfrak{m}^2$ is a vector space over k (being an R-module annihilated by $\mathfrak{m}$), $V(R) = \dim_k \mathfrak{m}/\mathfrak{m}^2$. There is always an inequality

$$\mathrm{Dim}(R) \leq V(R).$$

Definition: A local ring R is **regular** if $\mathrm{Dim}(R) = V(R)$.

Regular local rings have been studied for some time, especially in connection with nonsingular points on algebraic varieties.

Definition: An R-**sequence** in a local ring R is an ordered sequence $\{x_1, x_2, \ldots, x_n\}$ in $\mathfrak{m}$ such that x_1 is not a zero divisor and, for $i > 1$, each x_i is not a zero divisor on the module $R/(x_1, \ldots, x_{i-1})$, i.e., multiplication by x_i is monic.

If R is regular, then $\mathfrak{m}$ can be generated by an R-sequence $\{x_1, \ldots, x_d\}$ with $d = \mathrm{Dim}(R) = V(R)$ [Kaplansky, 1970, p. 119].

Lemma 9.56: *Let R be local and A a f.g. R-module. If $\mathrm{pd}(A) = r < \infty$ and if $x \in \mathfrak{m}$ is not a zero divisor on A, then*

$$\mathrm{pd}(A/xA) = r + 1.$$

Proof: By hypothesis, there is an exact sequence

$$0 \to A \xrightarrow{x} A \to A/xA \to 0,$$

where the first map is multiplication by x. Consider the long exact sequence, first for $i > r + 1$:

$$0 = \mathrm{Tor}_i(A, k) \to \mathrm{Tor}_i(A/xA, k) \to \mathrm{Tor}_{i-1}(A, k) = 0$$

(one shows the ends vanish by using Lemma 9.53 and the fact that $\mathrm{pd}(A) = r$). Thus, $\mathrm{Tor}_i(A/xA, k) = 0$ for $i > r + 1$, which says $\mathrm{pd}(A/xA) \leq r + 1$.

Now consider the long exact sequence when $i = r + 1$:

$$0 = \mathrm{Tor}_{r+1}(A, k) \to \mathrm{Tor}_{r+1}(A/xA, k) \to \mathrm{Tor}_r(A, k) \xrightarrow{x} \mathrm{Tor}_r(A, k).$$

Since $x \in \mathfrak{m}$, multiplication by x annihilates k, and hence multiplication by x is the zero map on $\mathrm{Tor}_r(A, k)$. Exactness gives $\mathrm{Tor}_{r+1}(A/xA, k) \to \mathrm{Tor}_r(A, k)$ an isomorphism. But $\mathrm{pd}(A) = r$ gives $\mathrm{Tor}_r(A, k) \neq 0$; hence $\mathrm{Tor}_{r+1}(A/xA, k) \neq 0$. It follows that $\mathrm{pd}(A/xA) = r + 1$. ∎

Theorem 9.57: *If R is a regular local ring, then*

$$D(R) = \text{Dim}(R) = V(R).$$

Proof: Since R is regular, $\mathfrak{m} = (x_1, \ldots, x_d)$, where $\{x_1, \ldots, x_d\}$ is an R-sequence. Applying Lemma 9.56 repeatedly to the modules R, $R/(x_1)$, $R/(x_1, x_2), \ldots, R/(x_1, \ldots, x_d) = R/\mathfrak{m} = k$, we deduce that $\text{pd}(k) = d$. But we have cited the result that regularity implies $d = \text{Dim}(R) = V(R)$, while we have proved (Corollary 9.55) that $D(R) = \text{pd}(k)$. ∎

The converse of Theorem 9.57 is also true (though more difficult).

Theorem 9.58 (Serre): *A local ring R is regular if and only if $D(R) < \infty$; moreover, in this case*

$$D(R) = \text{Dim}(R) = V(R).$$

Proof: [Northcott, 1960, pp. 202–208] or [Matsumura, 1970, pp. 132–139]. ∎

As a consequence of Serre's characterization, there is a purely ring theoretic result.

Theorem 9.59 (Serre): *If R is a regular local ring and $\mathfrak{p}$ is a prime ideal in R, then $R_\mathfrak{p}$ is also a regular local ring.*

Proof: It is always true that R local implies $R_\mathfrak{p}$ local. If $D(R) < \infty$, then Theorem 9.52 gives $D(R_\mathfrak{p}) \leq D(R) < \infty$. ∎

Theorem 9.60: *A commutative noetherian ring R has $D(R) < \infty$ if and only if every localization $R_\mathfrak{m}$, where $\mathfrak{m}$ is a maximal ideal of R, is a regular local ring.*

Proof: Theorems 9.52 and 9.58. ∎

As a consequence of Theorem 9.60, rings of finite global dimension are often called **regular**. Note that this cannot be confused with von Neumann regular rings: for example, a commutative noetherian ring R is von Neumann regular if and only it it is semisimple, whence $D(R) = 0$.

One may prove that regular local rings R are domains, and it had been conjectured that they are UFDs (unique factorization domains). This is easy to prove when $D(R) \leq 2$. Moreover, Nagata [1958] had proved the inductive step: if R is a UFD when $D(R) = 3$, then every regular local ring is a UFD. That 3-dimensional regular local rings are UFDs was proved by Auslander and Buchsbaum in 1959, and we present their proof. There are now several proofs of this theorem [Kaplansky, 1970, p. 130].

We begin by generalizing the definition of R-sequence; we are still assuming R is a local ring with maximal ideal $\mathfrak{m}$.

Definition: If A is an R-module, then an A-**sequence** is an ordered sequence $\{x_1, x_2, \ldots, x_n\}$ in $\mathfrak{m}$ such that x_1 is not a zero divisor on A (i.e., multiplication by x_1 is monic) and, for $i > 1$, each x_i is not a zero divisor on $A/(x_1, \ldots, x_{i-1})A$.

Of course, if $A = R$, we have the earlier definition of R-sequence.

Definition: If A is a f.g. R-module, then $\operatorname{cod}(A) = $ **codimension** A is the length of a longest A-sequence.

Codimension is also called "depth" or "grade".

We cite the following result, explaining the name.

Theorem 9.61   (Auslander–Buchsbaum):   If A is a f.g. R-module, then

$$\operatorname{pd}(A) + \operatorname{cod}(A) = D(R).$$

Proof: [Northcott, 1960, p. 209] or [Kaplansky, 1970, p. 125].   ∎

Lemma 9.62:   If $D(R) = 3$ and $\mathfrak{p}$ is a prime ideal properly contained in $\mathfrak{m}$, then $\operatorname{pd}(\mathfrak{p}) \leq 1$.

Proof: Let $A = R/\mathfrak{p}$. By hypothesis, there exists $x \in \mathfrak{m} - \mathfrak{p}$, and we claim x is not a zero divisor on A: if $x(r + \mathfrak{p}) = \mathfrak{p}$, then $xr \in \mathfrak{p}$ and $r \in \mathfrak{p}$ (since $x \notin \mathfrak{p}$ and $\mathfrak{p}$ is prime). Therefore $\operatorname{cod}(R/\mathfrak{p}) \geq 1$ and, by Theorem 9.61, $\operatorname{pd}(R/\mathfrak{p}) \leq 2$. Exactness of $0 \to \mathfrak{p} \to R \to R/\mathfrak{p} \to 0$ shows that $\operatorname{pd}(\mathfrak{p}) \leq 1$.   ∎

It is a standard result that a domain R is a UFD if every nonzero minimal prime ideal $\mathfrak{p}$ is principal ($\mathfrak{p}$ is "minimal" if there is no nonzero prime ideal $\mathfrak{q}$ with $\mathfrak{p} \supsetneqq \mathfrak{q}$). Indeed, it suffices to show in our case that such an ideal $\mathfrak{p}$ is projective, for $\mathfrak{p}$ is then free (R is local), and free ideals in (commutative) domains are principal. (One could also argue by invoking Theorem 4.28.) In other words, we know $\operatorname{pd}(\mathfrak{p}) \leq 1$, and we must show $\operatorname{pd}(\mathfrak{p}) = 0$.

Definition: If I is an ideal in a commutative ring R, and $y \in R$, then

$$I : y = \{r \in R : ry \in I\}.$$

It is easy to verify that $I : y$ is an ideal.

Lemma 9.63:   If $\mathfrak{p}$ is a minimal prime ideal in a regular local ring R, then there exists $x \in \mathfrak{p} - \mathfrak{m}\mathfrak{p}$ and $y \in \mathfrak{m} - \mathfrak{p}$ such that

$$\mathfrak{p} = (x) : y.$$

Proof: This is contained in the proof of Auslander–Buchsbaum [1959, Corollary 2] and is a brief argument using localization and the primary decomposition of an ideal.   ∎

We cite one more elementary result of commutative algebra: if A is a f.g. R-module and $x \in A - \mathfrak{m}A$, then x is part of a minimal generating set of A.

Theorem 9.64 (Auslander–Buchsbaum–Nagata): *Every regular local ring R is a unique factorization domain.*

Proof: By Nagata's reduction, we may assume $D(R) = 3$. If $\mathfrak{p}$ is a minimal prime ideal, Lemmas 9.62 and 9.63 allow us to assume:

  (i) $\mathrm{pd}(\mathfrak{p}) \leq 1$;
  (ii) $\mathfrak{p} = (x):y$ for some $x \in \mathfrak{p} - \mathfrak{m}\mathfrak{p}$, some $y \in R$.

We show that $\mathfrak{p} = (x)$, which will complete the proof.

Assume $\mathfrak{p}$ is not principal, so there is a minimal generating set $\{x, a_1, \ldots, a_n\}$ of $\mathfrak{p}$. There is an exact sequence

$$0 \to K \to F \overset{\varphi}{\to} \mathfrak{p} \to 0,$$

where F is free of rank $n + 1$ and $\varphi(r_0, r_1, \ldots, r_n) = r_0 x + \sum r_i a_i$. This is a projective cover, so that $K \subset \mathfrak{m}F$. Since $\mathrm{pd}(\mathfrak{p}) \leq 1$, K is projective, hence free (R is local); moreover, $K \neq 0$ lest $\mathfrak{p}$ be free, hence principal. Choose a basis $\{t^1, \ldots, t^q\}$ of K, where $t^i = (\lambda_0^i, \lambda_1^i, \ldots, \lambda_n^i)$. We have $\lambda_j^i \in \mathfrak{m}$, all i, j, because $K \subset \mathfrak{m}F$.

Now $a_1 \in \mathfrak{p} = (x):y$ implies there is $v \in R$ with

$$a_1 y = vx.$$

There are thus two obvious elements in K:

$$\alpha = (v, -y, 0, \ldots, 0) \qquad \text{and} \qquad \beta = (a_1, -x, 0, \ldots, 0);$$

moreover,

$$x\alpha = y\beta.$$

Now $\alpha = \sum r_j t^j$ and $\beta = \sum s_j t^j$, so that $x\alpha = y\beta$ implies

$$\sum x r_j t^j = \sum y s_j t^j;$$

it follows that $x r_j = y s_j$ for all j (for $\{t^1, \ldots, t^q\}$ is a basis of K). Therefore,

$$s_j \in (x):y = \mathfrak{p}, \qquad \text{all } j.$$

But

$$\beta = (a_1, -x, 0, \ldots, 0) = \sum s_j t^j = \sum s_j(\lambda_0^j, \lambda_1^j, \ldots, \lambda_n^j),$$

so that

$$-x = \sum s_j \lambda_1^j \in \mathfrak{m}\mathfrak{p},$$

and this contradicts the choice of x. ∎

10 The Return of Cohomology of Groups

In Chapter 5, we observed that certain group-theoretical questions led to the construction of a free $\mathbf{Z}G$-resolution of the trivial G-module $\mathbf{Z}$, application of functors $\text{Hom}_G(\ ,A)$ or $\otimes_G A$, and taking homology. The following definition is thus reasonable.

Definition: Let G be a group, A a left G-module, and $\mathbf{Z}$ the integers considered as a trivial G-module. Define

$$H^n(G, A) = \text{Ext}^n_{\mathbf{Z}G}(\mathbf{Z}, A),$$
$$H_n(G, A) = \text{Tor}^{\mathbf{Z}G}_n(\mathbf{Z}, A).$$

The groups H^n are the **cohomology groups** of G (with coefficients A); the groups H_n are the **homology groups** of G.

For the remainder of this chapter, we use the notation H^n and H_n instead of Ext^n and Tor_n; in particular, the long exact sequences will be so written. Observe that we only have long exact sequences in the second variable: if G_1 and G_2 are distinct groups, then $H^n(G_i, \ )$ and $H_n(G_i, \ )$, $i = 1, 2$, deal with modules over distinct rings $\mathbf{Z}G_1$ and $\mathbf{Z}G_2$.

Our aim is not to give a survey of this large subject. We shall give some interpretations of low-dimensional groups, some techniques of calculation, and some purely group-theoretical applications.

HOMOLOGY GROUPS

We remind the reader that we have already proved (Theorem 5.17 and Corollary 5.20) that for a G-module A,

$$H_0(G, A) = A_G = A/\mathfrak{g}A = \mathbf{Z} \otimes_G A,$$

where A_G is the maximal quotient module of A that is G-trivial and $\mathfrak{g}$ is the augmentation ideal of G. In particular, if A is G-trivial, then $A = A_G$ and $H_0(G, A) = A$.

Theorem 10.1:   *If $\mathfrak{g}$ is the augmentation ideal of G, then*

$$H_1(G, \mathbf{Z}) \cong \mathfrak{g}/\mathfrak{g}^2,$$

where $\mathbf{Z}$ is G-trivial (as usual).

Proof: The augmentation sequence $0 \to \mathfrak{g} \to \mathbf{Z}G \overset{\varepsilon}{\to} \mathbf{Z} \to 0$ induces exactness of

$$H_1(G, \mathbf{Z}G) \to H_1(G, \mathbf{Z}) \overset{\partial}{\to} H_0(G, \mathfrak{g}) \to H_0(G, \mathbf{Z}G) \overset{\varepsilon_*}{\to} H_0(G, \mathbf{Z}) \to 0.$$

Since $\mathbf{Z}$ is G-trivial, $H_0(G, \mathbf{Z}) \cong \mathbf{Z}$; also, $H_0(G, \mathbf{Z}G) = \mathbf{Z} \otimes_{\mathbf{Z}G} \mathbf{Z}G \cong \mathbf{Z}$. But any endomorphism of $\mathbf{Z}$ is either 0 or monic; since ε_* is epic, $\varepsilon_* \neq 0$, and hence ε_* is monic. Exactness of the sequence gives ∂ epic. Finally, $H_1(G, \mathbf{Z}G) = 0$ since $\mathbf{Z}G$ is projective (remember, $H_1(G, \mathbf{Z}G) = \text{Tor}_1^{\mathbf{Z}G}(\mathbf{Z}, \mathbf{Z}G)$). Therefore $\partial : H_1(G, \mathbf{Z}) \to H_0(G, \mathfrak{g})$ is an isomorphism. As $H_0(G, A) \cong A/\mathfrak{g}A$ for any G-module A, it follows that $H_0(G, \mathfrak{g}) \cong \mathfrak{g}/\mathfrak{g}^2$.   ∎

Theorem 10.2:   *For any group G, the additive group $\mathfrak{g}/\mathfrak{g}^2$ is isomorphic to the multiplicative group G/G', where G' denotes the commutator subgroup of G.*

Proof: Define $\theta : G \to \mathfrak{g}/\mathfrak{g}^2$ by $x \mapsto (x - 1) + \mathfrak{g}^2$. Note that θ is a homomorphism, for

$$(xy - 1) - (x - 1) - (y - 1) = (x - 1)(y - 1) \in \mathfrak{g}^2.$$

Since $\mathfrak{g}/\mathfrak{g}^2$ is abelian, $G' \subset \ker \theta$, and so θ induces a map $\bar{\theta} : G/G' \to \mathfrak{g}/\mathfrak{g}^2$, namely, $xG' \mapsto (x - 1) + \mathfrak{g}^2$.

We construct an inverse to $\bar{\theta}$. Recall (Theorem 5.19) that $\mathfrak{g}$ is a free abelian group with basis $\{x - 1 : x \in G\}$. Define $\varphi : \mathfrak{g} \to G/G'$ by $\varphi(x - 1) = xG'$. It suffices to prove $\mathfrak{g}^2 \subset \ker \varphi$, for then φ induces a map $\mathfrak{g}/\mathfrak{g}^2 \to G/G'$ that is visibly inverse to $\bar{\theta}$.

If $u \in \mathfrak{g}^2$, then

$$u = (\sum m_i(x_i - 1))(\sum n_j(y_j - 1))$$

$$= \sum_{i,j} m_i n_j (x_i - 1)(y_j - 1)$$

$$= \sum_{i,j} m_i n_j [(x_i y_j - 1) - (x_i - 1) - (y_j - 1)].$$

Therefore

$$\varphi(u) = \prod_{i,j} (x_i y_j x_i^{-1} y_j^{-1})^{m_i n_j} G' = G',$$

and $u \in \ker \varphi$ as desired. ∎

Corollary 10.3: *For any group G,*

$$H_1(G, \mathbf{Z}) \cong G/G'.$$

Exercises: 10.1. Assume G and H are groups with isomorphic group rings, say $\beta: \mathbf{Z}G \overset{\sim}{\to} \mathbf{Z}H$. If $\varepsilon: \mathbf{Z}G \to \mathbf{Z}$ and $\eta: \mathbf{Z}H \to \mathbf{Z}$ are augmentations, prove there exists a ring isomorphism $\beta': \mathbf{Z}G \to \mathbf{Z}H$ with $\eta\beta' = \varepsilon$. Conclude that G and H have the same homology groups and the same cohomology groups. (Hint: Define $\beta' = \beta\alpha$, where $\alpha: \mathbf{Z}G \to \mathbf{Z}G$ is given by $\sum m_x x \mapsto \sum m_x [\eta\beta(x)]x$.)

10.2. If G and H are groups with $\mathbf{Z}G \cong \mathbf{Z}H$, then $G/G' \cong H/H'$. In particular, if G and H are abelian, then $G \cong H$. (Whitcomb has shown that $\mathbf{Z}G \cong \mathbf{Z}H$ implies $G \cong H$ if both groups are metabelian, but the general problem is unsolved.)

10.3. Show that $\mathbf{Z}G$ may be identified with its opposite ring by the anti-isomorphism $\sum m_x x \mapsto \sum m_x x^{-1}$. Conclude that if we regard a right G-module A as a left G-module by $x \cdot a = a \cdot x^{-1}$, and if we regard a left G-module B as a right G-module by $b \cdot x = x^{-1} \cdot b$, then, for all $n \geq 0$,

$$\operatorname{Tor}_n^{\mathbf{Z}G}(A, B) \cong \operatorname{Tor}_n^{\mathbf{Z}G}(B, A).$$

(Compare Theorem 8.5.)

10.4. If G is finite, then every f.g. G-module is f.g. as an abelian group. Conclude that $\mathbf{Z}G$ is left and right noetherian and, for every n and every f.g. G-module A, $H^n(G, A)$ and $H_n(G, A)$ are f.g. abelian groups.

10.5. For an integer m, regard $\mathbf{Z}/m\mathbf{Z}$ as a trivial G-module. Prove that $H_1(G, \mathbf{Z}/m\mathbf{Z}) \cong G/G'G^m$, where G^m is the subgroup of G generated by all mth powers. (Hint: Consider the exact sequence of trivial G-modules $0 \to \mathbf{Z} \overset{m}{\to} \mathbf{Z} \to \mathbf{Z}/m\mathbf{Z} \to 0$.)

Remark: The **Frattini subgroup** $\Phi(G)$ of a group G is defined as the intersection of all maximal subgroups of G. If G is a finite p-group, where p is prime, one may show $\Phi(G) = G'G^p$ and that $G/\Phi(G)$ is a vector space over $\mathbf{Z}/p\mathbf{Z}$ whose dimension is the cardinal of a minimal set of generators of G **(Burnside Basis Theorem)** [Rotman, 1973, p. 126].

Definition: The **(Schur) multiplier** of G is $H_2(G, \mathbf{Z})$.

The multiplier is often called the "multiplicator" (transliterating from German). Our immediate aim is a series of results leading to a formula for $H_2(G, \mathbf{Z})$, due to Hopf, that bears upon free presentations of G.

Recall that a (not necessarily abelian) group G is **free** with **basis** X if, for every function $\varphi : X \to H$, H any group, there exists a unique homo-

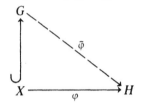

morphism $\tilde{\varphi} : G \to H$ extending φ. When G is free on X, each $g \in G$ $(g \neq 1)$ has a unique factorization

$$g = x_1^{e_1} \cdots x_n^{e_n},$$

where $x_i \in X$, $e_i = \pm 1$, and $x_{i+1}^{e_{i+1}} \neq x_i^{-e_i}$. Clearly G is not abelian if $\mathrm{card}(X) > 1$. (See [Rotman, 1973, pp. 238–239].)

Theorem 10.4:   *If G is a free group with basis X, then $\mathfrak{g}$ is a free G-module with basis $X - 1 = \{x - 1 : x \in X\}$.*

Remark: We know, by Theorem 5.19, that $\mathfrak{g}$ is a free abelian group with basis $G - 1 = \{g - 1 : g \in G\}$.

Proof: The formulas

$$xy - 1 = (x - 1) + x(y - 1) \qquad \text{and} \qquad x^{-1} - 1 = -x^{-1}(x - 1)$$

show that if $g = x_1^{e_1} \cdots x_n^{e_n}$, then $g - 1$ is a G-linear combination of $X - 1$. Therefore, $X - 1$ generates $\mathfrak{g}$ as a G-module.

To see that $\mathfrak{g}$ is freely generated by $X - 1$, we must complete the diagram

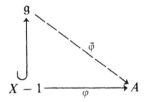

where A is a G-module, φ is a function, and $\tilde{\varphi}$ is a G-map (uniqueness of such a $\tilde{\varphi}$ follows from $\mathfrak{g}$ being generated by $X - 1$). Since $\text{Hom}_G(\mathfrak{g}, A) \cong \text{Der}(G, A)$ (Theorem 5.21), let us seek a derivation. Consider the (necessarily) split extension $0 \to A \to E \xrightarrow{\pi} G \to 1$, so that E consists of all ordered pairs $(a, g) \in A \times G$ and $\pi(a, g) = g$. The given function φ defines a lifting λ of the generators X of G, namely, $\lambda x = (\varphi(x - 1), x)$. Since G is free on X, the function $\lambda: X \to E$ extends to a homomorphism $\tilde{\lambda}: G \to E$. By Exercise 5.10, if $g \in G$, then $\tilde{\lambda}(g) = (\langle g \rangle, g)$, where $\langle \ \rangle : G \to A$ is a derivation. Now the isomorphism of Theorem 5.21 yields a G-map $\tilde{\varphi}: \mathfrak{g} \to A$, namely, $\tilde{\varphi}(g - 1) = \langle g \rangle$. Since $\lambda x = \tilde{\lambda} x = (\langle x \rangle, x)$, we have $\tilde{\varphi}(x - 1) = \varphi(x - 1)$, which shows $\tilde{\varphi}$ extends φ. ∎

Corollary 10.5:   *If G is a free group, then*
$$H_n(G, A) = 0 = H^n(G, A)$$
for every G-module A and every $n > 1$.

Proof: The augmentation sequence $0 \to \mathfrak{g} \to \mathbf{Z}G \to \mathbf{Z} \to 0$ is a G-free resolution of $\mathbf{Z}$. ∎

Exercise: 10.6.   If $G = \{1\}$, then $H_n(G, A) = 0 = H^n(G, A)$ for every G-module A and every $n > 0$.

Let G be a group, and let F be a free group (with basis X) mapping onto G, say, via $\pi: F \to G$. If $R = \ker \pi$, then $R \lhd F$ and $F/R \cong G$. As every subgroup of a free group is itself free **(Nielsen-Schreier theorem)**, R is also free, say, with basis Y.

Every group has its own group ring. Clearly $\mathbf{Z}R$ is a subring of $\mathbf{Z}F$, while π induces a ring map $\bar{\pi}: \mathbf{Z}F \to \mathbf{Z}G$ that is onto, namely, $\sum m_i f_i \mapsto \sum m_i \pi(f_i)$. If we denote $\ker \bar{\pi}$ by $\mathfrak{r}$, then $\mathfrak{r}$ is a two-sided ideal in $\mathbf{Z}F$ and $\mathbf{Z}F/\mathfrak{r} \cong \mathbf{Z}G$. Beware! In this case only, we have deviated from our usual notation: $\mathfrak{r}$ is *not* the augmentation ideal of R, for that is merely an ideal in $\mathbf{Z}R$. Actually, as we see from the next result, $\mathfrak{r}$ is the two-sided ideal in $\mathbf{Z}F$ generated by the augmentation ideal of R.

Lemma 10.6:   *With the notation of the paragraphs above, $\mathfrak{r}$ is a free (left or right) F-module with basis $Y - 1 = \{y - 1 : y \in Y\}$.*

Proof: Clearly $Y - 1$ is contained in $\ker \bar{\pi} = \mathfrak{r}$. Choose a complete set of left coset representatives of R in F:
$$F = \bigcup t_i R.$$
If $\alpha \in \mathbf{Z}F$, then $\alpha = \sum m_{ij} t_i r_j$, where $t_i r_j \in t_i R$ and $m_{ij} \in \mathbf{Z}$. If $\alpha \in \mathfrak{r}$, then
$$0 = \bar{\pi}\alpha = \sum_{i,j} m_{ij} \pi(t_i),$$

and the $\pi(t_i)$ are distinct elements of G. By Theorem 5.19, $\sum_j m_{ij} = 0$ for each i. Therefore

$$\alpha = \alpha - 0 = \sum_i \left(\sum_j m_{ij} t_i (r_j - 1) \right)$$

is an F-linear combination of elements of the form $r - 1$, $r \in R$. However, the proof of Theorem 10.4 shows each $r - 1$ is an R-linear (*a fortiori*, an F-linear) combination of $Y - 1$. Thus, $\mathfrak{r}$ is generated by $Y - 1$.

Assume $\sum \alpha_i (y_i - 1) = 0$, where $\alpha_i \in \mathbf{Z}F$. It is easy to see that $\alpha_i = \sum_j t_j \beta_{ji}$, where $\beta_{ji} \in \mathbf{Z}R$. Now the coset representatives t_j are independent over $\mathbf{Z}R$ ($0 = \sum_{j,k} t_j m_{jk} r_k$ implies each $m_{jk} = 0$ since all $t_j r_k$ are distinct), from which it follows that $\sum_i \beta_{ji}(y_i - 1) = 0$ for each j. The problem has been reduced to Theorem 10.4, for R is free with basis Y.

We have just proved $\mathfrak{r}$ is free as a left F-module. To prove that $\mathfrak{r}$ is free as a right F-module, just repeat the above proof using right coset representatives instead of left. ∎

Lemma 10.7:   *If M is a free left F-module with basis W, then $M/\mathfrak{r}M$ is a free left G-module with basis $\{\bar{w} = w + \mathfrak{r}M : w \in W\}$. A similar statement holds for $M/M\mathfrak{r}$ when M is a free right F-module.*

Proof: First of all, let G act on $M/\mathfrak{r}M$ by

$$g \cdot (m + \mathfrak{r}M) = f \cdot m + \mathfrak{r}M, \qquad \text{where} \quad \pi(f) = g.$$

This is well defined, for if $\pi(f) = \pi(f_1)$, then $f - f_1 \in \ker \bar{\pi} = \mathfrak{r}$ and $(f - f_1) \cdot m \in \mathfrak{r}M$.

Since $M = \coprod_{w \in W} (\mathbf{Z}F)w$ and $\mathfrak{r}M = \coprod \mathfrak{r}w$, it follows that

$$M/\mathfrak{r}M \cong \coprod_w (\mathbf{Z}Fw/\mathfrak{r}w) \cong \coprod (\mathbf{Z}F/\mathfrak{r})\bar{w},$$

and the last module is G-free since $\mathbf{Z}F/\mathfrak{r} \cong \mathbf{Z}G$. ∎

Corollary 10.8:   *If M is a free F-module with basis W, then $M/\mathfrak{f}M$ (or $M/M\mathfrak{f}$ when M is a right F-module) is a free abelian group with basis $\{\bar{w} = w + \mathfrak{f}M : w \in W\}$ (or $\{w + M\mathfrak{f} : w \in W\}$).*

Proof: Take $G = \{1\}$. The trivial map $\pi : F \to G = \{1\}$ induces a ring map $\mathbf{Z}F \to \mathbf{Z}G = \mathbf{Z}$ whose kernel is just $\mathfrak{f}$, the augmentation ideal of F. This corollary is thus a restatement of Lemma 10.7, for $\mathfrak{f}$ now plays the role of $\mathfrak{r}$. ∎

Let us make two elementary observations before using these lemmas.

(i)   If $\mathfrak{a}$ and $\mathfrak{b}$ are two-sided ideals in $\mathbf{Z}F$, free on A and B respectively, then $\mathfrak{a}\mathfrak{b}$ is free with basis $AB = \{ab : a \in A, b \in B\}$.

(ii) If $M_1 \subset M_2 \subset M$, then there is an epimorphism $M/M_1 \to M/M_2$ given by $m + M_1 \mapsto m + M_2$ (of course, this is the map of the Third Iso-morphism Theorem, and its kernel is M_2/M_1). If $M \subset M'$, then $M/M_2 \subset M'/M_2$, and the composite

$$M/M_1 \to M/M_2 \hookrightarrow M'/M_2$$

is called **enlargement of coset**.

Theorem 10.9   (Gruenberg):   *Let* $1 \to R \to F \overset{\pi}{\to} G \to 1$ *be an exact sequence of groups, with* F *free on* X *and* R *free on* Y. *There is a* G-*free resolution of* $\mathbf{Z}$,

$$\cdots \to P_2 \overset{d_2}{\to} P_1 \overset{d_1}{\to} \mathbf{Z}G \overset{\varepsilon}{\to} \mathbf{Z} \to 0,$$

where

$P_{2n} = \mathfrak{r}^n/\mathfrak{r}^{n+1}$, *the* G-*free module with basis*
$$\{(y_1 - 1)\cdots(y_n - 1) + \mathfrak{r}^{n+1} : y_i \in Y\},$$

$P_{2n-1} = \mathfrak{r}^{n-1}\mathfrak{f}/\mathfrak{r}^n\mathfrak{f}$, *the* G-*free module with basis*
$$\{(y_1 - 1)\cdots(y_{n-1} - 1)(x - 1) + \mathfrak{r}^n\mathfrak{f} : y_i \in Y, x \in X\},$$

and the maps $d_k : P_k \to P_{k-1}$ *are enlargements of coset.*

Proof: Since $\mathfrak{f}$ is F-free on $X - 1$ and $\mathfrak{r}$ is F-free on $Y - 1$, iterated use of observation (i) shows the left F-modules $\mathfrak{r}^{n-1}\mathfrak{f}$ and $\mathfrak{r}^n$ are free with bases $\{(y_1 - 1)\cdots(y_{n-1} - 1)(x - 1) : y_i \in Y, x \in X\}$ and $\{(y_1 - 1)\cdots(y_n - 1) : y_i \in Y\}$, respectively. Lemma 10.7 now applies, showing that P_{2n-1} and P_{2n} are free left G-modules on the corresponding cosets.

Let us describe the maps in more detail. Since $\mathfrak{r}^{n+1} \subset \mathfrak{r}^n\mathfrak{f}$, $d_{2n} : P_{2n} \to P_{2n-1}$ is the enlargement of coset map

$$\mathfrak{r}^n/\mathfrak{r}^{n+1} \to \mathfrak{r}^n/\mathfrak{r}^n\mathfrak{f} \hookrightarrow \mathfrak{r}^{n-1}\mathfrak{f}/\mathfrak{r}^n\mathfrak{f}.$$

The Third Isomorphism Theorem gives

$$\operatorname{im} d_{2n} = \mathfrak{r}^n/\mathfrak{r}^n\mathfrak{f} \qquad \text{and} \qquad \ker d_{2n} = \mathfrak{r}^n\mathfrak{f}/\mathfrak{r}^{n+1}.$$

Since $\mathfrak{r}^{n+1}\mathfrak{f} \subset \mathfrak{r}^{n+1}$, $d_{2n+1} : P_{2n+1} \to P_{2n}$ is the enlargement of coset map

$$\mathfrak{r}^n\mathfrak{f}/\mathfrak{r}^{n+1}\mathfrak{f} \to \mathfrak{r}^n\mathfrak{f}/\mathfrak{r}^{n+1} \hookrightarrow \mathfrak{r}^n/\mathfrak{r}^{n+1};$$

hence

$$\operatorname{im} d_{2n+1} = \mathfrak{r}^n\mathfrak{f}/\mathfrak{r}^{n+1} \qquad \text{and} \qquad \ker d_{2n+1} = \mathfrak{r}^{n+1}/\mathfrak{r}^{n+1}\mathfrak{f}.$$

Our calculation of images and kernels shows exactness of the sequence, with the possible exception of

$$\mathfrak{f}/\mathfrak{r}\mathfrak{f} = P_1 \overset{d_1}{\to} \mathbf{Z}G \overset{\varepsilon}{\to} \mathbf{Z}.$$

If we interpret $\mathfrak{r}^0$ as $\mathbf{Z}F$, then $P_0 = \mathfrak{r}^0/\mathfrak{r}^1 = \mathbf{Z}F/\mathfrak{r} \xrightarrow{\sim} \mathbf{Z}G$, and $d_1 : x - 1 + \mathfrak{r}\mathfrak{f}$ $\mapsto x - 1 + \mathfrak{r} \mapsto \pi x - 1$. Thus, $\operatorname{im} d_1$ is the augmentation ideal $\mathfrak{g} = \ker \varepsilon$, and the proof is complete. ∎

The resolution above is called **Gruenberg's resolution** with respect to the given presentation of G as F/R.

In Theorem 10.2, we saw that $\mathfrak{g}/\mathfrak{g}^2 \cong G/G'$. There is thus a relation between group ring constructions and group constructions. The next result reveals further such.

Lemma 10.10:   *With the notation of Theorem* 10.9, *there are isomorphisms of abelian groups*

$$\mathfrak{r}/\mathfrak{r}\mathfrak{f} \xrightarrow{\sim} R/R' \qquad and \qquad (\mathfrak{r}\mathfrak{f} + \mathfrak{f}\mathfrak{r})/\mathfrak{r}\mathfrak{f} \xrightarrow{\sim} [F, R]/R'.$$

Remark: $[F, R] = [R, F]$ is the subgroup of F generated by all commutators $[f, r] = frf^{-1}r^{-1}$, where $f \in F$ and $r \in R$. Actually, $R \lhd F$ implies that $[F, R]$ is a subgroup of R, even a normal subgroup.

Proof: By Corollary 10.8, $\mathfrak{r}/\mathfrak{r}\mathfrak{f}$ is free abelian on $\{y - 1 + \mathfrak{r}\mathfrak{f} : y \in Y\}$, while R/R' is visibly free abelian on $\{yR' : y \in Y\}$. There is thus an isomorphism of abelian groups

$$\theta : \mathfrak{r}/\mathfrak{r}\mathfrak{f} \to R/R'$$

defined by

$$y - 1 + \mathfrak{r}\mathfrak{f} \mapsto yR'.$$

If $r = y_1^{e_1} \cdots y_n^{e_n} \in R$, what is $\theta(r - 1 + \mathfrak{r}\mathfrak{f})$? The identities

$$(uv - 1) - (u - 1) - (v - 1) = (u - 1)(v - 1)$$

and

$$-(u^{-1} - 1) - (u - 1) = (u - 1)(u^{-1} - 1),$$

together with $\mathfrak{r}^2 \subset \mathfrak{r}\mathfrak{f}$, show that $\theta(r - 1 + \mathfrak{r}\mathfrak{f}) = rR'$.

Now restrict the isomorphism θ to $(\mathfrak{r}\mathfrak{f} + \mathfrak{f}\mathfrak{r})/\mathfrak{r}\mathfrak{f}$, the subgroup generated by all $(f - 1)(r - 1) + \mathfrak{r}\mathfrak{f}$, where $f \in F$ and $r \in R$. The identity

$$(f - 1)(r - 1) = ([f, r] - 1) + ([f, r] - 1)(rf - 1) + (r - 1)(f - 1)$$

shows that

$$(f - 1)(r - 1) + \mathfrak{r}\mathfrak{f} = [f, r] - 1 + \mathfrak{r}\mathfrak{f},$$

and so

$$\theta((f - 1)(r - 1) + \mathfrak{r}\mathfrak{f}) = [f, r]R'. ∎$$

Remark: Actually, more is true. The multiplicative abelian group R/R' is a G-module if one defines

$$g \cdot rR' = frf^{-1}R', \qquad \text{where} \quad \pi f = g.$$

The isomorphism θ is now a G-isomorphism, for

$$
\begin{aligned}
\theta(g \cdot (r - 1 + \mathfrak{r}\mathfrak{f})) &= \theta(f(r - 1) + \mathfrak{r}\mathfrak{f}) \qquad \text{(see Lemma 10.7)} \\
&= \theta(f(r - 1) + f(r - 1)(f^{-1} - 1) + \mathfrak{r}\mathfrak{f}) \\
&= \theta(frf^{-1} - 1 + \mathfrak{r}\mathfrak{f}) \\
&= frf^{-1}R' = g \cdot rR' = g\theta(r - 1 + \mathfrak{r}\mathfrak{f}).
\end{aligned}
$$

Exercises: 10.7. Let Π be a group with normal subgroup N, and let B and A be Π-modules. Prove that $\operatorname{Hom}_N(B, A)$ is a Π/N-module if one defines, for $b \in B$ and $f \in \operatorname{Hom}_N(B, A)$,

$$(\bar{x} \cdot f)(b) = x \cdot f(x^{-1}b),$$

where $x \in \Pi$ has coset $\bar{x} \in \Pi/N$.

10.8. With the same notation as in Exercise 10.7, prove that $B \otimes_{\mathbf{Z}N} A$ is a Π/N-module if one defines

$$\bar{x}(b \otimes a) = bx^{-1} \otimes xa.$$

10.9. Let N be a subgroup (not necessarily normal) of a group Π, and let T be a complete set of right coset representatives, i.e., one element from each right coset Nt in Π. Show there is an isomorphism of N-modules

$$\mathbf{Z}\Pi \cong \coprod_{t \in T} (\mathbf{Z}N)t.$$

Conclude that every projective Π-module is a projective N-module.

We need one more elementary lemma.

Lemma 10.11: *If M is a left F-module, then*

$$\mathbf{Z} \otimes_G (M/\mathfrak{r}M) \cong M/\mathfrak{f}M.$$

Proof: A left G-module A may be regarded as a left F-module annihilated by $\mathfrak{r}$; moreover, if $\pi f = g$ (where $\pi: F \to G$), then $g \cdot a = f \cdot a$ for all $a \in A$. Therefore $(g - 1) \cdot a = (f - 1) \cdot a$ which implies

$$\mathfrak{g}A = \mathfrak{f}A.$$

In particular,

$$\mathfrak{g}(M/\mathfrak{r}M) = \mathfrak{f}(M/\mathfrak{r}M) = \mathfrak{f}M/\mathfrak{r}M.$$

By Theorem 5.18,

$$\mathbf{Z} \otimes_G (M/\mathfrak{r}M) \cong (M/\mathfrak{r}M)/\mathfrak{g}(M/\mathfrak{r}M)$$
$$= (M/\mathfrak{r}M)/(\mathfrak{f}M/\mathfrak{r}M) \cong M/\mathfrak{f}M. \quad \blacksquare$$

Theorem 10.12   (Hopf's formula):   *For any group G,*

$$H_2(G, \mathbf{Z}) \cong (R \cap F')/[F, R],$$

where F is free and $F/R \cong G$.

Proof: Apply $\mathbf{Z} \otimes_G$ to the Gruenberg resolution

$$\cdots \to \mathfrak{r}\mathfrak{f}/\mathfrak{r}^2\mathfrak{f} \xrightarrow{d_3} \mathfrak{r}/\mathfrak{r}^2 \xrightarrow{d_2} \mathfrak{f}/\mathfrak{r}\mathfrak{f} \to \cdots$$

to obtain the complex (using Lemma 10.11):

$$\cdots \to \mathfrak{r}\mathfrak{f}/\mathfrak{f}\mathfrak{r}\mathfrak{f} \xrightarrow{\Delta_3} \mathfrak{r}/\mathfrak{f}\mathfrak{r} \xrightarrow{\Delta_2} \mathfrak{f}/\mathfrak{f}^2 \to \cdots;$$

the maps Δ_i are still enlargements of coset. By Exercise 10.3,

$$H_2(G, \mathbf{Z}) = \ker \Delta_2/\operatorname{im} \Delta_3 = (\mathfrak{r} \cap \mathfrak{f}^2)/(\mathfrak{f}\mathfrak{r} + \mathfrak{r}\mathfrak{f})$$
$$= \ker(\mathfrak{r}/(\mathfrak{f}\mathfrak{r} + \mathfrak{r}\mathfrak{f}) \to \mathfrak{f}/\mathfrak{f}^2),$$

this last arrow being enlargement of coset. But $\mathfrak{f}/\mathfrak{f}^2 \cong F/F'$ and, by Lemma 10.10,

$$\mathfrak{r}/(\mathfrak{f}\mathfrak{r} + \mathfrak{r}\mathfrak{f}) \cong (\mathfrak{r}/\mathfrak{r}\mathfrak{f})/[(\mathfrak{f}\mathfrak{r} + \mathfrak{r}\mathfrak{f})/\mathfrak{r}\mathfrak{f}]$$
$$\cong (R/R')/([F, R]/R') \cong R/[F, R].$$

We conclude that

$$H_2(G, \mathbf{Z}) \cong \ker(R/[F, R] \to F/F') = (R \cap F')/[F, R],$$

for the reader may check commutativity of the diagram

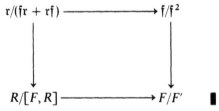

Corollary 10.13:   *If $G \cong F/R$, where F is free, then $(R \cap F')/[F, R]$ depends only on G (and not on the choice of F and R).*

Let us now give a minor group-theoretic application of Hopf's formula, so the reader may appreciate it.

Every group G may be described as a quotient of a free group (in many ways): $G \cong F/R$. If X is a basis of F, then X is called a set of **generators** of G and R is the normal subgroup of **relations**. The ordered pair $(X|R)$ obviously describes G to isomorphism. Of course, there is no need to list every element of R: if Y is a basis of R (which always exists), then G may be described by $(X|Y)$. But we can be even more efficient. Since $R \triangleleft F$, it may be possible to describe R by a set of smaller cardinal than that of Y, namely, a set W that generates R as a normal subgroup. Thus, R is the subgroup of F generated by all conjugates fwf^{-1} of elements $w \in W$ by elements $f \in F$. The ordered pair $(X|W)$ is called a **presentation** of G; this says that $G \cong F/R$, where F is free with basis X and R is the normal subgroup of F generated by W.

As an example, let G be the 4-group $\mathbf{Z}/2\mathbf{Z} \oplus \mathbf{Z}/2\mathbf{Z}$. One presentation of G is

$$(x_1, x_2 | x_1^2, x_2^2, [x_1, x_2]).$$

Here, F is the free group with basis $\{x_1, x_2\}$; the normal subgroup R generated by the relators turns out to have a basis of five elements, three of which are the displayed relators. (In contrast to free abelian groups, a basis of a subgroup of a free group F may have larger cardinal than that of a basis of F.)

Let us use Hopf's formula to show $H_2(G, \mathbf{Z}) \cong \mathbf{Z}/2\mathbf{Z}$. Set $K = F/[R, F]$. Since F/R is abelian, $F' \subset R$ and $R \cap F' = F'$. Therefore,

$$K' = F'/[R, F] = (R \cap F')/[R, F] \cong H_2(G, \mathbf{Z}).$$

Define $a = x[R, F]$ and $b = y[R, F]$, so that $K = \langle a, b \rangle$. If $L = \langle [a, b] \rangle$, then clearly $L \subset K'$; on the other hand, $L \triangleleft K$ and K/L is abelian, being generated by two commuting elements, so that $L \supset K'$. Therefore $K' = L$ is cyclic with generator $[a, b]$.

We claim $[a, b]^2 = 1$. Observe first that $F' \subset R$ implies $[F', F] \subset [R, F]$, from which it follows that $K' \subset Z(K)$, the center of K; furthermore $b^2 \in Z(K)$ (for $y^2 \in R$ and $[y^2, f] \in [R, F]$ for all $f \in F$). Hence

$$b^2 = ab^2 a^{-1} = (aba^{-1})^2 = ([a, b]b)^2$$
$$= [a, b]^2 b^2, \qquad \text{since} \quad [a, b] \in Z(K).$$

Canceling b^2 gives $[a, b]^2 = 1$ and $|K'| \leq 2$.

It remains to show $K' \neq \{1\}$, i.e., that $K = F/[R, F]$ is not abelian. Consider the group of quaternions, with presentation

$$Q = (x, y | x^2 = y^2, xyx = y).$$

We record simple properties of Q:

(i) Q is a group of order 8 with generators c and d, where $c = xS$, $d = yS$, and S is the normal subgroup of F generated by x^2y^{-2} and $xyxy^{-1}$;

(ii) $N = \langle c^2 \rangle$ has order 2;

(iii) $Z(Q) = N = Q'$.

If we can show that $[R, F] \subset S$, then $F/[R, F]$ maps onto the nonabelian group $F/S = Q$; this will show that $F/[R, F]$ is not abelian, i.e., $K' \neq \{1\}$. To prove $[R, F] \subset S$, it suffices to prove $[\bar{r}, \bar{f}] = 1$ in Q (bar means coset mod S) for every $f \in F$ and $r \in \{x^2, y^2, [x, y]\}$. But this is true, for $\bar{x}^2 = c^2 = \bar{y}^2$ lie in $N = Z(Q)$, and $[\bar{x}, \bar{y}] \in Q' = N = Z(Q)$.

Using spectral sequences, we shall generalize this result (Theorem 11.48). Let us return to Hopf's formula.

Definition: A group S is **finitely related** if it has a finite presentation

$$(x_1, \ldots, x_n | y_1, \ldots, y_r).$$

Finitely related groups are f.g. (the converse is false). Of course, finite groups and f.g. abelian groups are finitely related. Is there some way to describe the integers n and r in terms of G when G is finitely related? If one chooses F wisely, then n can be described as the cardinal of a minimal generating set of G. The example of the presentation F/R of the 4-group given above shows that the cardinal of a basis of R may be bigger than r (there the cardinal is 5 and not 3); thus, r is more subtle than n.

Notation: If A is a f.g. abelian group, then $A = T \oplus B$, where T is finite and B is f.g. free abelian. Let $d(A)$ denote the smallest cardinal of a generating set of A, and let $\rho(A) = \operatorname{rank} B$.

Exercises: 10.10. If A is a f.g. abelian group, then $\rho(A) \leq d(A)$; moreover, $\rho(A) = d(A)$ if and only if A is free, and $\rho(A) = 0$ if and only if A is torsion.

10.11. If A and A' are f.g. abelian groups with A' free, then $d(A \oplus A') = d(A) + d(A')$. (This is false if A' is not free: $\mathbf{Z}/2\mathbf{Z} \oplus \mathbf{Z}/3\mathbf{Z} \cong \mathbf{Z}/6\mathbf{Z}$.)

10.12. Prove that $\rho(A) = \dim_{\mathbf{Q}} \mathbf{Q} \otimes_{\mathbf{Z}} A$. Conclude that if $0 \to A' \to A \to A'' \to 0$ is an exact sequence of f.g. abelian groups, then

$$\rho(A) = \rho(A') + \rho(A'').$$

Lemma 10.14: *Let G have a presentation $(x_1, \ldots, x_n | y_1, \ldots, y_r)$ (so that $G \cong F/R$, where F is free on $\{x_1, \ldots, x_n\}$ and R is the normal subgroup of F generated by $\{y_1, \ldots, y_r\}$). Then $R/[F, R]$ is a f.g. abelian group and*

$$d(R/[F, R]) \leq r.$$

Proof: Since $R \lhd F$, one sees easily that $[F, R]$ is a normal subgroup of R containing $[R, R] = R'$; therefore, $R/[F, R]$ is an abelian group. It suffices to show $R/[F, R]$ is generated by the cosets of the y's. Every element of R is a product of elements of the form fsf^{-1}, where s lies in the subgroup generated by the y's. But $fsf^{-1}s^{-1} \in [F, R]$, so that $fsf^{-1} \equiv s \bmod [F, R]$ and the result follows. ∎

Theorem 10.15: *If G has a finite presentation $(x_1, \ldots, x_n | y_1, \ldots, y_r)$, then $H_2(G, \mathbf{Z})$ is f.g. with $d(H_2(G, \mathbf{Z})) \leq r$, and*

$$n - r \leq \rho(G/G') - d(H_2(G, \mathbf{Z})).$$

Proof: Let F be free on $\{x_1, \ldots, x_n\}$ and R the normal subgroup generated by $\{y_1, \ldots, y_r\}$. The chain to the left is a descending series of normal subgroups of F. By Lemma 10.14, $R/[F, R]$ is a f.g. abelian group with at most r generators. There is an exact sequence of abelian groups

F

$F'R$

R

$$0 \to (R \cap F')/[F, R] \to R/[F, R] \to R/R \cap F' \to 0.$$

$R \cap F'$

$[F, R]$

By Hopf's formula, $d(H_2(G, \mathbf{Z})) = d((R \cap F')/[F, R]) \leq r$ (Corollary 4.19). Since $R/R \cap F' \cong F'R/F' \subset F/F'$, the group $R/R \cap F'$ is free abelian (because F/F' is), and the sequence splits. By Exercise 10.11,

$$d(R/[F, R]) = d(R/R \cap F') + d((R \cap F')/[F, R])$$
$$= d(F'R/F') + d(H_2(G, \mathbf{Z})).$$

There is another exact sequence

$$0 \to F'R/F' \to F/F' \to F/F'R \to 0,$$

and Exercise 10.12 gives $\rho(F/F') = \rho(F'R/F') + \rho(F/F'R)$. Because F/F' and its subgroup $F'R/F'$ are free, however, we have

$$d(F'R/F') = d(F/F') - \rho(F/F'R) = n - \rho(F/F'R).$$

We conclude that

$$r \geq d(R/[F, R]) = n - \rho(F/F'R) + d(H_2(G, \mathbf{Z})).$$

This completes the proof, for $F/F'R \cong G/G'$ (under the map $F \to G$, the subgroup of F mapping onto G' is $F'R$). ∎

Definition: A finite presentation is **balanced** if $n = r$: it has the same number of generators as relators.

Corollary 10.16: *If a finitely related group G has a balanced presentation, then* $d(H_2(G, \mathbf{Z})) \leq \rho(G/G')$. *In particular, if a finite group G has a balanced presentation, then* $H(G, \mathbf{Z}) = 0$.

Corollary 10.17:   *Let G be a finite group having a presentation with n generators and r relations. Then*

$$d(H_2(G, \mathbf{Z})) \leq r - n.$$

The multiplier must be computed in order to use Corollary 10.16. If G is the 4-group, we have seen that $H_2(G, \mathbf{Z}) \neq 0$, hence G admits no balanced presentation.

A more impressive result is due to Golod and Šafarevič. Let G be a group, p a prime, and $\mathbf{Z}/p\mathbf{Z}$ a trivial G-module. Since multiplication by p is the zero map on $\mathbf{Z}/p\mathbf{Z}$, it follows from Theorem 8.13 that $pH_n(G, \mathbf{Z}/p\mathbf{Z}) = 0$ for all n, i.e., $H_n(G, \mathbf{Z}/p\mathbf{Z})$ is a vector space over $\mathbf{Z}/p\mathbf{Z}$; let d_n denote its dimension. Now Exercise 10.5 and our subsequent remarks about the Burnside Basis Theorem show that if G is a finite p-group, then $d_1(G) = d(G)$. In view of Theorem 10.15, it is reasonable to expect that $d_2(G)$ somehow involves the number of relators of G.

Theorem   (Golod and Šafarevič):   *If G is a finite p-group, then*

$$d_2(G) > \tfrac{1}{4}(d_1(G) - 1)^2.$$

(This result may be improved to: $d_2(G) > \tfrac{1}{4}(d_1(G))^2$ [Gruenberg, 1970, p. 104].)

Thus, a criterion exists to determine whether a p-group G is finite. They also construct a f.g. p-group G violating this inequality, thereby exhibiting an infinite f.g. p-group. Burnside had proved fifty years earlier that a f.g. torsion subgroup of $GL(n, \mathbf{C})$, $n \times n$ nonsingular matrices over the complex numbers $\mathbf{C}$, must be finite. (Thus, the group of Golod and Šafarevič has no faithful finite-dimensional complex representation.) Burnside's problem asked whether every f.g. torsion group is finite; thus, the answer is "no". A much more difficult version of Burnside's problem replaces "torsion" by the stronger condition "there is an integer $e > 0$ with $x^e = 1$ for all $x \in G$." In 1968, Adjan and Novikov proved the answer is negative in this case as well, if e is odd and sufficiently large; in 1982, Ol'šanskiĭ found a more elegant proof. (No homological algebra is used in the last proofs, and the proof of Golod and Šafarevič can now be given without homological algebra [Herstein, 1968].)

There is another context in which multipliers arise.

Definition: A **central extension** of G is an exact sequence of groups

$$1 \to A \to E \to G \to 1$$

in which $A \subset Z(E)$, the center of E. A **universal central extension** of G is a central extension

$$1 \to M \to U \to G \to 1$$

such that, given any central extension, there exists a unique map $U \to E$ making the following diagram commute:

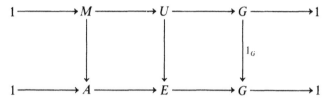

If G is a finite group, then it is has a universal central extension if and only if G is **perfect**, i.e., $G = G'$, in which case $M \cong H_2(G, \mathbf{Z})$ [Milnor, 1971, pp. 43–46]. For example, nonabelian simple groups are perfect. What if G is not perfect?

Definition: A **stem extension** of G is a central extension

$$1 \to A \to E \to G \to 1$$

in which $A \subset E'$ (Schur called such groups E "representation groups").

If G is perfect, every central extension is a stem extension.

Definition: A **stem cover** of G is a stem extension of the form

$$1 \to H_2(G, \mathbf{Z}) \to S \to G \to 1.$$

When G is the 4-group, $H_2(G, \mathbf{Z}) \cong \mathbf{Z}/2\mathbf{Z}$ and the two nonabelian groups of order 8 (dihedral group and quaternions) give stem covers of G. There are two main results:

(a)   Each stem extension is a homomorphic image of a stem cover (this is the analogue of universal central extension when G is not perfect);

(b)   If $1 \to A \to E \to G \to 1$ is a central extension with A a divisible abelian group, then there exists a homomorphism of every stem cover of G into it.

Proofs and discussion of stem extensions and stem covers may be found in [Gruenberg, 1970] and [Stammbach, 1973].

The significance of result (b) is bound up with "projective representations", so this is a convenient place to switch from homology to cohomology.

COHOMOLOGY GROUPS

Before continuing the story of stem covers and multipliers, let us recall some items from Chapter 5. Using homological language that was not available then, we saw in Theorem 5.15 that

$$H^0(G, A) = \operatorname{Hom}_G(\mathbf{Z}, A) = A^G,$$

where A^G is the maximal G-trivial submodule of A; in particular, if A is G-trivial, then $H^0(G, A) = A$.

Using Theorem 5.16,

$$H^1(G, A) = \operatorname{stab}(G, A) = \operatorname{Der}(G, A)/\operatorname{PDer}(G, A).$$

In case A is G-trivial, then $\operatorname{PDer}(G, A) = 0$ and every derivation is a homomorphism. Therefore

$$H^1(G, A) = \operatorname{Hom}(G, A)$$

whenever A is G-trivial.

If $e(G, A) = $ (normalized factor sets)/(normalized coboundaries), then we had almost proved (in Chapter 5) that $e(G, A) \cong H^2(G, A)$. Let us remind the reader of the formulas.

A **normalized factor set** is a function $f : G \times G \to A$ with

(i) $xf(y, z) - f(xy, z) + f(x, yz) - f(x, y) = 0$;
(ii) $f(x, 1) = 0 = f(1, x)$;

a **normalized coboundary** is a function $g : G \times G \to A$ such that

(iii) $g(x, y) = xh(y) - h(xy) + h(x)$,

where $h : G \to A$ is a function with

(iv) $h(1) = 0$.

The gap in obtaining an isomorphism $H^2(G, A) \cong e(G, A)$ was that the resolution of $\mathbf{Z}$ in Theorem 5.16 gave cocycles satisfying condition (i) but not (ii), and coboundaries satisfying condition (iii) but not (iv).

With these formulas before our eyes, let us return to stem covers. Representation theory deals with homomorphisms $G \to \operatorname{GL}(n, k)$, where k is a field. If $k^{\#}$ is the multiplicative group of nonzero elements of k, then $k^{\#}$ is isomorphic to the center of $GL(n, k)$, all nonzero $n \times n$ scalar matrices.

Definition: The **projective general linear group** is

$$PGL(n, k) = GL(n, k)/\text{scalars}.$$

A **projective representation** of G is a homomorphism $G \to PGL(n, k)$.

Now one really prefers a representation of G, but he may only have a projective representation. The next theorem says, when k is algebraically closed, that one can exchange a projective representation of G if he pays a price: G must be replaced by a stem cover (i.e., a representation group) S of G. (A more complete account of this material can be found in [Isaacs, 1976].)

Theorem   (Schur):   *If* $\tau: G \to PGL(n, k)$, *where* k *is algebraically closed, and if* $1 \to M \to S \xrightarrow{\pi} G \to 1$ *is a stem cover, then there exists a commutative diagram*

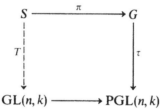

i.e., τ *"comes from" an ordinary representation* T *of* S.

Sketch of proof: First form the pullback

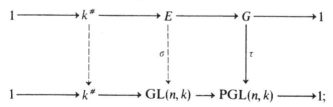

the top row is a central extension. Since k is algebraically closed, $k^{\#}$ is divisible (for each n, every element has an nth root), so result (b) on p. 279 provides a homomorphism of any stem cover (hence a map $\rho: S \to E$) into the top row. Then $T = \sigma\rho: S \to GL(n, k)$ is the desired representation. ∎

The next exercises show that projective representations give rise to a cohomology group.

Exercises: 10.13.   Consider the diagram

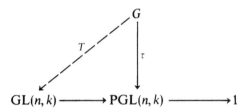

where τ is a projective representation. For each $x \in G$, define $T(x) \in GL(n,k)$ as some lifting of $\tau(x)$. Prove that the function T satisfies

$$T(xy) = f(x, y)T(x)T(y),$$

where $f: G \times G \to k^{*}$. If the multiplicative group k^{*} is considered as a trivial G-module, prove that f is a factor set (equation (i) written multiplicatively).

10.14.   If, in the situation of Exercise 10.13, one chooses other liftings $T'(x)$ of $\tau(x)$, so that $T'(xy) = f'(x, y)T'(x)T'(y)$, show that $f^{-1}f'$ is a coboundary (equation (iii) written multiplicatively).

The exercises above show $H^{2}(G, k^{*}) = $ factor sets/coboundaries, so that cohomology occurs in this context (as well as in the context of extensions). Surely this group $H^{2}(G, k^{*})$ deserves the name of "multiplier", and we shall prove (Theorem 10.31) that $H^{2}(G, k^{*}) \cong H_{2}(G, \mathbf{Z})$ when G is finite and k is an algebraically closed field of characteristic zero.

Our present task is to establish an isomorphism $H^{2}(G, A) \cong e(G, A)$, and there is no choice but to write down some G-projective resolutions of $\mathbf{Z}$. Before taking the plunge, we alert the reader that a topological interpretation of cohomology groups will appear.

Definition: Let $G^{(n)}$ denote the cartesian product of n copies of G; let P_n be the free abelian group on $G^{(n+1)}$ made into a G-module by $x(x_0, x_1, \ldots, x_n) = (xx_0, xx_1, \ldots, xx_n)$. Define maps $\partial_n: P_n \to P_{n-1}$, whenever $n \geq 1$, by

$$(x_0, \ldots, x_n) \mapsto \sum_{i=0}^{n} (-1)^i (x_0, \ldots, \hat{x}_i, \ldots, x_n),$$

where $\hat{}$ indicates deletion.

Theorem 10.18: $\mathbf{P} = \cdots \to P_1 \overset{\partial_1}{\to} P_0 \overset{\varepsilon}{\to} \mathbf{Z} \to 0$ *is a G-free resolution of* $\mathbf{Z}$.

Proof: Observe that $P_0 = \mathbf{Z}G$; by definition, ε is the augmentation map. We let the reader check that P_n is G-free (a basis consists of all $(n+1)$-tuples of the form $(1, x_1, \ldots, x_n)$). Next, $\mathbf{P}$ is a complex, for the boundary homomorphism is precisely that arising in topology (Theorem 1.19). To prove exactness of **P**, it suffices to construct a contracting homotopy (Exercise 6.19); we want maps

$$\cdots \leftarrow P_2 \overset{s_1}{\leftarrow} P_1 \overset{s_0}{\leftarrow} P_0 \overset{s_{-1}}{\longleftarrow} \mathbf{Z}$$

with

$$\partial_{n+1} s_n + s_{n-1} \partial_n = 1_{P_n}, \qquad n \geq 0 \qquad (\partial_0 = \varepsilon) \qquad \text{and} \qquad \varepsilon s_{-1} = 1_{\mathbf{Z}}.$$

Define $s_{-1} : \mathbf{Z} \to P_0$ by $1 \mapsto (1)$ (the 1 in parentheses is the identity of G); if $n \geq 0$, define s_n by

$$(x_0, \ldots, x_n) \mapsto (1, x_0, \ldots, x_n).$$

Note that the maps s_n are only **Z**-maps (which is all we need: the conclusion is that **P** is exact as a complex of abelian groups; but, by the definition of exactness, **P** is then exact as a complex of G-modules). Here are the simple computations.

(i) $\varepsilon s_{-1}(1) = \varepsilon(1) = 1$;

(ii) $(\partial_{n+1} s_n + s_{n-1} \partial_n)(x_0, \ldots, x_n)$

$$= \partial_{n+1}(1, x_0, \ldots, x_n) + s_{n-1} \sum_{i=0}^{n} (-1)^i (x_0, \ldots, \hat{x}_i, \ldots, x_n)$$

$$= (x_0, \ldots, x_n) + \sum_{k=0}^{n} (-1)^{k+1}(1, x_0, \ldots, \hat{x}_k, \ldots, x_n)$$

$$+ \sum_{i=0}^{n} (-1)^i (1, x_0, \ldots, \hat{x}_i, \ldots, x_n)$$

$$= (x_0, \ldots, x_n). \qquad \blacksquare$$

Exercise: 10.15. Prove that P_n is isomorphic (as a G-module) to the tensor product (over **Z**) of **Z**G with itself $n + 1$ times.

The resolution **P** of "homogeneous" $(n + 1)$-tuples is quite simple, but it does not resemble the formulas that arose in Chapter 5. We now construct a resolution of "inhomogeneous" n-tuples.

Definition: For $n > 0$, define Q_n as the free G-module with basis all n-tuples $[x_1, \ldots, x_n]$ of elements of G; define Q_0 as the free G-module on the single generator $[\ ]$.

For each $n \geq 0$, $P_n \cong Q_n$, for both are free G-modules on a set in one-to-one correspondence with $G^{(n)}$; we give a specific isomorphism.

Define $\tau : P_n \to Q_n$ by

$$\tau(x_0, \ldots, x_n) = x_0[x_0^{-1}x_1, x_1^{-1}x_2, \ldots, x_{n-1}^{-1}x_n];$$

define $\sigma : Q_n \to P_n$ by

$$\sigma[x_1, \ldots, x_n] = (1, x_1, x_1 x_2, x_1 x_2 x_3, \ldots, x_1 x_2 \cdots x_n).$$

It is easy to check that σ and τ are inverse to one another. There are thus unique maps $d_n : Q_n \to Q_{n-1}$, $n \geq 1$, making the following diagrams commute:

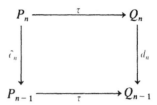

Since $\tau^{-1} = \sigma$, we have $d_n = \tau \hat{\partial}_n \sigma$. The formula is

$$d_n[x_1, \ldots, x_n] = x_1[x_2, x_3, \ldots, x_n] + \sum_{i=1}^{n-1} (-1)^i[x_1, \ldots, x_i x_{i+1}, \ldots, x_n]$$

$$+ (-1)^n[x_1, \ldots, x_{n-1}].$$

In particular, we have

$$d_1[x] = x[\,] - [\,],$$
$$d_2[x, y] = x[y] - [xy] + [x],$$
$$d_3[x, y, z] = x[y, z] - [xy, z] + [x, yz] - [x, y].$$

Thus $\mathbf{Q}$ is the resolution of $\mathbf{Z}$ we began building in Chapter 5, and it is called the **(unnormalized) standard resolution** of $\mathbf{Z}$.

Theorem 10.19: $\mathbf{Q} = \cdots \to Q_1 \overset{d_1}{\to} Q_0 \overset{\varepsilon}{\to} \mathbf{Z} \to 0$ *is a G-free resolution of* $\mathbf{Z}$.

Proof: First, $\mathbf{Q}$ is a complex, for $d_{n-1}d_n = \tau \hat{\partial}_{n-1} \hat{\partial}_n \sigma = 0$ and $\varepsilon d_1 = 0$. Next, $\tau : \mathbf{P} \to \mathbf{Q}$ is an isomorphism of complexes, hence it induces isomorphisms $H_n(\mathbf{P}) \to H_n(\mathbf{Q})$, all n. Since $\mathbf{P}$ is exact, each $H_n(\mathbf{P}) = 0$; therefore each $H_n(\mathbf{Q}) = 0$ and $\mathbf{Q}$ is exact.   ∎

The complex $\mathbf{P}$ suggests a connection between group cohomology and cohomology of topological spaces.

Definition: Let X be a topological space, and let $\text{Aut}(X)$ be the group of all homeomorphisms of X with itself. A group G **operates** on X if there is a homomorphism $G \to \text{Aut}(X)$.

If G operates on X, we may regard each $g \in G$ as a homeomorphism of X; moreover, there are identities

$$g_1(g_2 x) = (g_1 g_2)x, \qquad 1x = x,$$

where $g_1, g_2, 1 \in G$ and $x \in X$.

Definition: A group G **operates without fixed points** on a space X if $gx = x$ for some $x \in X$ implies $g = 1$.

Theorem 10.20: *If a group G operates on a topological space X, then the singular complex $\cdots \to S_n(X) \xrightarrow{\partial} S_{n-1}(X) \to \cdots$ is a complex of G-modules. Moreover, if G operates without fixed points, each $S_n(X)$ is a free G-module.*

Proof: Recall that $S_n(X)$ is the free abelian group with basis all continuous $T: \Delta_n \to X$. If $g \in G$, we may regard g as a homeomorphism $X \to X$, so that gT is again an n-simplex. The identities above show that $S_n(X)$ is a G-module. Also (using the notation of Chapter 1), if $T^i = Te_i$ is the ith face of T, then $(gT)^i = gTe_i = g(T^i)$; it follows that ∂ is a G-map.

If $x \in X$, its **orbit** is the set $\{gx : g \in G\}$; clearly the orbits partition X. Let $X_0 \subset X$ consist of one element from each orbit. Call a simplex T **basic** if $Tv_0 \in X_0$ (where $\Delta_n = [v_0, \ldots, v_n]$); we claim $S_n(X)$ is free on all basic simplexes. They do generate, for if $S: \Delta_n \to X$, then $Sv_0 = gx_0$ for some $g \in G$, $x_0 \in X_0$; hence $g^{-1}S$ is basic and $S = g(g^{-1}S)$. Finally, suppose $\sum \gamma_i T_i = 0$, where $\gamma_i = \sum m_{ik} g_k \in \mathbf{Z}G$, and the T_i are distinct. Then

$$\sum_{i,k} m_{ik}(g_k T_i) = 0.$$

We claim that if $gT_1 = hT_2$, where T_1, T_2 are basic, then $g = h$ and $T_1 = T_2$. Suppose $T_1 v_0 \neq T_2 v_0$; then, as these elements lie in distinct orbits, $gT_1 v_0 \neq hT_2 v_0$, a contradiction; therefore $T_1 v_0 = T_2 v_0 = x$. But $gx = hx$ implies $g^{-1}h$ fixes x; since G operates without fixed points, $g = h$. Finally, since g is a homeomorphism, $T_1 = T_2$. It follows that all the simplexes $g_k T_i$ are distinct, so that each $m_{ik} = 0$, hence each $\gamma_i = 0$. ∎

Definition: A topological space X is **acyclic** if $H_0(X) \cong \mathbf{Z}$ and $H_n(X) = 0$ for $n > 0$.

Theorem 10.21: *If a group G operates without fixed points on an acyclic space X, then the singular complex of X is a deleted G-free resolution of $\mathbf{Z}$.*

Proof: We already know the singular complex is a complex of G-free modules. The condition that X is acyclic gives an exact sequence $\cdots \to S_1(X) \to S_0(X) \to H_0(X) \to 0$, and $H_0(X) \cong \mathbf{Z}$. ∎

Let A be a G-module. Now $H^n(X; A)$ is the homology of the complex $\operatorname{Hom}_{\mathbf{Z}}(S(X), A)$. If X is acyclic and G operates on X without fixed points, then $H^n(G, A)$ is the homology of the complex $\operatorname{Hom}_G(S(X), A)$, by Theorem 10.21. To complete this discussion, we assume some knowledge of topology. Suppose G operates **properly** on X, i.e., each $x \in X$ lies in an open set U with $gU \cap U = \varnothing$, all $g \in G$, $g \neq 1$ (this implies that G operates without fixed points). One can give an isomorphism of complexes [MacLane, 1963, pp. 135–136]

$$\operatorname{Hom}_{\mathbf{Z}}(S(X/G), A) \cong \operatorname{Hom}_G(S(X), A),$$

where A is G-trivial and X/G is the orbit space of X, and this induces isomorphisms for all $n \geq 0$

$$H^n(X/G; A) \cong H^n(G, A).$$

The next step is to exhibit a space X as above. Given a group G, there exists an **Eilenberg–MacLane space** $Y = K(G, 1)$: a path connected, "aspherical" space (i.e., the nth homotopy groups $\pi_n(Y) = 0$ for $n > 1$) having fundamental group $\pi_1(Y) \cong G$. If one defines $X = \tilde{Y}$, the universal covering space of Y, then X is acyclic, G acts properly on X, and $X/G \approx Y$. It follows that

$$H^n(K(G, 1); A) \cong H^n(G, A):$$

the cohomology of an abstract group G (with G-trivial coefficients) coincides with the cohomology of a certain topological space $Y = K(G, 1)$.

It is also true that there are isomorphisms in homology: if $Y = K(G, 1)$ and A is G-trivial, then

$$H_n(K(G, 1); A) \cong H_n(G, A)$$

(a proof using spectral sequences may be found in [MacLane, 1963, p. 344]).

Theorem 10.22 (Universal Coefficient Theorem): *If G is a group and A is G-trivial, then*

$$H^n(G, A) \cong \operatorname{Hom}_{\mathbf{Z}}(H_n(G, \mathbf{Z}), A) \oplus \operatorname{Ext}_{\mathbf{Z}}^1(H_{n-1}(G, \mathbf{Z}), A).$$

Proof: The Universal Coefficient Theorem 8.27 gives such an isomorphism for any topological space Y. Choose $Y = K(G, 1)$ and use the fact that $H^n(Y; A) \cong H^n(G, A)$ and $H_n(Y; A) \cong H_n(G, A)$ for all n. $\blacksquare$

Remark: A purely algebraic proof of Theorem 10.22 may be found in [Gruenberg, 1970, p. 49].

There is also a universal coefficient theorem for homology, using Corollary 8.23: if A is G-trivial,

$$H_n(G, A) \cong H_n(G, \mathbf{Z}) \otimes_{\mathbf{Z}} A \oplus \mathrm{Tor}_1^{\mathbf{Z}}(H_{n-1}(G, \mathbf{Z}), A).$$

The fact that one may realize homology groups of G as homology groups of a topological space "explains" Theorem 10.3. If Y is a path connected space having fundamental group π, then the **Hurewicz theorem** states that

$$H_1(Y; \mathbf{Z}) \cong \pi/\pi'.$$

Setting $Y = K(\pi, 1)$ shows $H_1(Y; \mathbf{Z}) \cong H_1(\pi, \mathbf{Z})$, which gives a topological proof of Theorem 10.3.

We return to algebra, still seeking to prove $H^2(G, A) \cong e(G, A)$.

Definition: The **bar resolution** (or **standard resolution** or **normalized resolution**) is

$$\mathbf{B} = \cdots \to B_1 \xrightarrow{d_1} B_0 \xrightarrow{\varepsilon} \mathbf{Z} \to 0,$$

where B_n is the free G-module on all $[x_1, \ldots, x_n]$ with $x_i \in G$ and $x_i \neq 1$, and the formula for $d_n : B_n \to B_{n-1}$ is the same as in $\mathbf{Q}$. (B_0 is free on the single generator $[\ ]$.)

Remarks: 1. In order that d_n be defined, we agree that $[x_1, \ldots, x_n] = 0$ whenever some $x_i = 1$.

2. It is not obvious that $\mathbf{B}$ is a complex (this does not follow immediately from $d_{n-1} d_n = 0$ in $\mathbf{Q}$, for here we are making some of the terms in the formula equal to 0).

3. $\mathbf{B}$ is called the bar resolution because the original notation for $[x_1, \ldots, x_n]$ was $[x_1 | x_2 | \cdots | x_n]$.

Theorem 10.23: *The bar resolution $\mathbf{B}$ is a G-free resolution of $\mathbf{Z}$.*

Proof: Again, we do not yet know $\mathbf{B}$ is even a complex (so far, all we know is that each B_n is G-free and the d's are G-maps). First, we construct a contracting homotopy

$$\cdots \leftarrow B_1 \xleftarrow{s_0} B_0 \xleftarrow{s_{-1}} \mathbf{Z},$$

where each s_n is a $\mathbf{Z}$-map. Define

$$s_{-1} : \mathbf{Z} \to B_0 \qquad \text{by} \quad 1 \mapsto [\ ]$$

and

$$s_n : B_n \to B_{n+1} \qquad \text{by} \quad x[x_1, \ldots, x_n] \mapsto [x, x_1, \ldots, x_n]$$

(since s_n is only a $\mathbf{Z}$-map, it must be defined on a $\mathbf{Z}$-free set of generators, not merely on the G-free generators $[x_1, \ldots, x_n]$). It is easy to check that this is a contracting homotopy, i.e.,

$$d_{n+1}s_n + s_{n-1}d_n = 1_{B_n} \qquad \text{(where} \quad d_0 = \varepsilon\text{)}$$

and

$$\varepsilon s_{-1} = 1_{\mathbf{Z}}.$$

If we show $\mathbf{B}$ is a complex, then Exercise 6.19 completes the proof. Now B_{n+1} is generated as a G-module by the subgroup im s_n, so that it suffices to show $d_n d_{n+1} = 0$ on this subgroup. We do an induction on n, noting that $0 = \varepsilon d_1 = s_{-1}d_1$. For the inductive step,

$$\begin{aligned} d_n d_{n+1} s_n &= d_n(1 - s_{n-1}d_n) \qquad \text{(contracting homotopy)} \\ &= d_n - (1 - s_{n-2}d_{n-1})d_n \qquad \text{(ditto)} \\ &= d_n - d_n - s_{n-2}d_{n-1}d_n = 0. \quad \blacksquare \end{aligned}$$

Theorem 10.24: $H^2(G, A) \cong e(G, A)$.

Proof: Applying $\text{Hom}_G(\ , A)$ to $\mathbf{B}$ gives the complex

$$\cdots \to \text{Hom}_G(B_1, A) \overset{d_2^*}{\to} \text{Hom}_G(B_2, A) \overset{d_3^*}{\to} \text{Hom}_G(B_3, A) \to \cdots$$

and

$$H^2(G, A) = \ker d_3^* / \text{im } d_2^*.$$

Suppose $f: B_2 \to A$ is a cocycle. Then $0 = d_3^* f = f d_3$, so that

$$0 = f d_3[x, y, z] = xf(y, z) - f(xy, z) + f(x, yz) - f(x, y)$$

(that f is a G-map is used to get $f(x[y, z]) = xf(y, z)$). Moreover, $f(x, 1) = 0 = f(1, x)$, since $[x, 1] = 0 = [1, x]$, so that f is a normalized factor set as needed in the definition of $e(G, A)$.

Suppose $g: B_2 \to A$ lies in im d_2^*, i.e., there is a map $h: B_1 \to A$ with $g = d_2^* h = h d_2$. Since $h: B_1 \to A$, we have $h(1) = 0$. Also, $g(x, y) = h d_2[x, y] = xh(y) - h(xy) + h(x)$ (that h is a G-map is used to get $h(x[y]) = xh(y)$). Therefore, g is a normalized coboundary as needed in the definition of $e(G, A)$. $\blacksquare$

Theorem 10.24 follows immediately from the fact (Theorem 10.23) that the bar resolution $\mathbf{B}$ is a G-free resolution of $\mathbf{Z}$. The proof given above is the proof in the original paper of Eilenberg and MacLane (and is credited there to Weil).

We now give a second proof of Theorem 10.23, using the Gruenberg resolution of the trivial G-module $\mathbf{Z}$ corresponding to a suitable free presentation F/R of G. In order to proceed, we must know a basis of R; fortunately, the Nielsen–Schreier theorem actually exhibits one [Rotman, 1973, pp. 260–261].

A **left transversal** T of a (not necessarily normal) subgroup R of F is a complete set of coset representatives, i.e., one element from each left coset tR in F. A **Schreier transversal** of R in F, where F is free with basis X, is a transversal T such that, whenever $t = x_1^{e_1} \cdots x_n^{e_n} \in T$, where $x_i \in X$ and $e_i = \pm 1$, then every terminal segment of $t : x_2^{e_2} \cdots x_n^{e_n}, \ldots, x_{n-1}^{e_{n-1}} x_n^{e_n}, x_n^{e_n}, 1$, also lies in T. It may be shown that Schreier transversals always exist (most accounts of this theory deal with right transversals of R in F; of course, inverting is the way to change one account to the other). If $t \in T$ and $x \in X$, then $xtR = uR$ for some (unique) $u \in T$ (perhaps $u = xt$, perhaps not). Define

$$y_{x,t} = u^{-1}xt.$$

The Nielsen–Schreier theorem states that if T is a Schreier transversal of R in F, and if X is a basis of F, then

$$\{y_{x,t} \neq 1 : x \in X, t \in T\}$$

is a basis of R.

Let F be free on $X = \{s_x : x \in G, x \neq 1\}$, define $\pi : F \to G$ by $s_x \mapsto x$, and let $R = \ker \pi$. For notational convenience, set $s_1 = 1$. It is clear that $T = \{s_x : x \in G\}$ is a Schreier transversal of R in F; moreover, if $x, y \in G$, then $s_{xy}R = s_x s_y R$ and $s_x s_y \neq s_{xy}$ (when $x \neq 1$ and $y \neq 1$). Thus, R has a basis

$$\{r_{x,y} = s_{xy}^{-1} s_x s_y : x, y \in G, x, y \neq 1\}.$$

Theorem 10.25 ($=$ Theorem 10.23): *The bar resolution $\mathbf{B}$ is a G-free resolution of $\mathbf{Z}$.*

Proof: Let us use the presentation of G given in the paragraph above. Since R is free on all elements $r_{x,y} = s_{xy}^{-1} s_x s_y$, $x, y \neq 1$, the ideal $\mathfrak{r}$ is free on all such $r_{x,y} - 1$. As s_{xy} is a unit in $\mathbf{Z}F$, we may modify this basis to obtain a basis of $\mathfrak{r}$ comprised of all elements (x, y), where

$$(x, y) = s_{xy}(r_{x,y} - 1) = s_x s_y - s_{xy}, \qquad x, y \in G, \quad x, y \neq 1.$$

Theorem 10.9 gives an explicit description of the corresponding Gruenberg resolution $\cdots \to P_2 \xrightarrow{d_2} P_1 \xrightarrow{d_1} P_0 \xrightarrow{\varepsilon} \mathbf{Z} \to 0$:

$P_{2n} = \mathfrak{r}^n / \mathfrak{r}^{n+1}$ is G-free with basis all elements

$$[x_1, \ldots, x_{2n}] = (x_1, x_2) \cdots (x_{2n-1}, x_{2n}) + \mathfrak{r}^{n+1};$$

$P_{2n-1} = \mathfrak{r}^{n-1} \mathfrak{f} / \mathfrak{r}^n \mathfrak{f}$ is G-free with basis all elements

$$[x_1, \ldots, x_{2n-1}] = (x_1, x_2) \cdots (x_{2n-3}, x_{2n-2})(s_{x_{2n-1}} - 1) + \mathfrak{r}^n \mathfrak{f};$$

the maps $d_k: P_k \to P_{k-1}$ are enlargements of coset. If we regard $P_0 \cong \mathbf{Z}G$ as the G-free module with single generator $[\ ]$, then the end of the proof of Theorem 10.9 shows $d_1: [x] \mapsto (x-1)[\ ] = x[\ ] - [\ ]$.

Here are two notational conventions. For $x \in G$, denote $s_x[x_1, \ldots, x_k]$ by $x[x_1, \ldots, x_k]$. Also, allow an entry in $[x_1, \ldots, x_k]$ to be 1 by defining such a bracket to be 0. This last convention allows us to state the formula whose validity we shall prove by induction on k:

$$d_k[x_1, \ldots, x_k] = x_1[x_2, \ldots, x_k] + \sum_{i=1}^{k-1} (-1)^i [x_1, \ldots, x_i x_{i+1}, \ldots, x_k]$$
$$+ (-1)^k [x_1, \ldots, x_{k-1}].$$

Because the definitions of bracket depend on the parity of k, it is no surprise that the induction reflects this fact.

If $k = 2$, use the identity

$$(x, y) = s_x s_y - s_{xy} = s_x(s_y - 1) - (s_{xy} - 1) + (s_x - 1).$$

In bracket notation, this is the desired formula

$$d_2[x, y] = x[y] - [xy] + [x].$$

The case $k = 3$ is trickier. Evaluate $s_x s_y s_z$ in two ways, using the defining equation $s_{ab} r_{a,b} = s_a s_b$. On the one hand, $(s_x s_y) s_z = s_{xy} r_{x,y} s_z$; on the other,

$$s_x(s_y s_z) = s_x s_{yz} r_{y,z} = s_{xyz} r_{x,yz} r_{y,z} = s_{xy} s_z r_{xy,z}^{-1} r_{x,yz} r_{y,z}.$$

Therefore,

$$(x, y)(s_z - 1) = s_{xy}(r_{x,y} - 1)s_z - (x, y)$$
$$= s_{xy} s_z r_{xy,z}^{-1} r_{x,yz} r_{y,z} - s_{xy} s_z - (x, y)$$
$$= s_{xy} s_z (r_{xy,z}^{-1} r_{x,yz} r_{y,z} - 1) - (x, y).$$

The identities

$$ab - 1 = (a - 1) + (b - 1) + (a - 1)(b - 1);$$
$$a^{-1} - 1 = -(a - 1) - (a^{-1} - 1)(a - 1),$$

yield congruences mod $\mathfrak{r}^2$:

$$(x, y)(s_z - 1) \equiv s_{xy} s_z\{-(r_{xy,z} - 1) + (r_{x,yz} - 1) + (r_{y,z} - 1)\} - (x, y)$$
$$\equiv s_{xy} s_z\{-s_{xyz}^{-1}(xy, z) + s_{xyz}^{-1}(x, yz) + s_{yz}^{-1}(y, z)\} - (x, y).$$

Clearly $s_a s_b s_{ab}^{-1} \in R$ and $s_a s_b s_{ab}^{-1} - 1 \in \mathfrak{r}$; hence, if $\alpha \in \mathfrak{r}$, then

$$s_a s_b s_{ab}^{-1} \alpha \equiv \alpha \bmod \mathfrak{r}^2.$$

In particular,

$$(x, y)(s_z - 1) \equiv -(xy, z) + (x, yz) + s_x(y, z) - (x, y) \bmod \mathfrak{r}^2,$$

and this completes the case $k = 3$.

The inductive step assumes the result for all integers $<k$; there are two cases, depending on whether k is even or odd.

Assume $k = 2n$. We must show $(x_1, x_2) \cdots (x_{2n-1}, x_{2n})$ is congruent to the alternating sum formula mod $\mathfrak{r}^n\overline{\mathfrak{f}}$. Denote $s_x - 1$ by (x), and set $x_{k-1} = y$ and $x_k = z$. Modifying slightly the identity occurring in the case $k = 2$,

$$(x_1, x_2) \cdots (y, z) = (x_1, \ldots, x_{k-2})\{-(yz) + (y) + (z) + (y)(z)\}.$$

Expanding gives a sum of four terms; examine the last term $(x_1, \ldots, x_{k-2})(y)(z)$. By induction, $(x_1, \ldots, x_{k-2})(y)$, a term of odd length, can be written

$$(x_1, \ldots, x_{k-2})(y) = x_1(x_2, \ldots, y) + \sum_{i=1}^{k-2} (-1)^i(x_1, \ldots, x_i x_{i+1}, \ldots, y)$$

$$+ (-1)^{k-1}(x_1, \ldots, x_{k-2}) + u,$$

where $u \in \mathfrak{r}^n$ (induction takes place in $P_{2n-2} = P_{k-2} = \mathfrak{r}^{n-1}/\mathfrak{r}^n$). Multiplying on the right by (z) leaves us with two offensive terms: $(-1)^{k-1}(x_1, \ldots, x_{k-2})(z)$ and $u(z)$. As $k = 2n$ is even, $(-1)^{k-1} = -1$ and the signed term cancels against the third term of the original four terms. Finally, $u(z) \in \mathfrak{r}^n\overline{\mathfrak{f}}$, so that we have the desired congruence.

Assume $k = 2n + 1$. We must show $(x_1, x_2) \cdots (x_{2n-2}, x_{2n-1})(x_{2n})$ is congruent to the alternating sum formula mod $\mathfrak{r}^{n+1}$. Set $x_{2n-2} = x, x_{2n-1} = y$, and $x_{2n} = z$. Modifying slightly the congruence occurring in the case $k = 3$, there is a five-term expansion

$$(x_1, \ldots, x_{k-3})(x, y)(z) \equiv (x_1, \ldots, x_{k-3})\{-(xy, z) + (x, yz)$$
$$+ (x)(y, z) + (y, z) - (x, y)\},$$

the congruence mod $\mathfrak{r}^{n+1}$ (the term $(x_1, \ldots, x_{k-3}) \in \mathfrak{r}^{n-1}$ since $k = 2n + 1$). By induction, we may rewrite part of the term involving $(x)(y, z)$:

$$(x_1, \ldots, x_{k-3})(x) \equiv x_1(x_2, \ldots, x) + \sum_{i=1}^{k-3} (-1)^i(x_1, \ldots, x_i x_{i+1}, \ldots, x)$$

$$+ (-1)^{k-2}(x_1, \ldots, x_{k-3}) \bmod \mathfrak{r}^n.$$

Multiplying on the right by (y, z) yields a congruence mod $\mathfrak{r}^{n+1}$. Since k, hence $k - 2$, is odd, the sign of the last term is negative, hence cancels against term four of the original five terms. What survives is the desired formula. This completes the induction. ∎

COMPUTATIONS AND APPLICATIONS

We have given some interpretations of low-dimensional homology and cohomology groups, but are we in a stronger position having done so? In order to use homology, some computations are necessary.

Theorem 10.26:   *Let G be a finite group of order m. For every G-module A and every $n > 0$,*

$$mH^n(G, A) = 0 = mH_n(G, A).$$

Proof: We use the unnormalized standard resolution $\mathbf{Q}$. If $f : Q_n \to A$, define $g : Q_{n-1} \to A$ by

$$g(x_1, \ldots, x_{n-1}) = \sum_{x \in G} f(x_1, \ldots, x_{n-1}, x).$$

Now sum the coboundary formula

$$(df)(x_1, \ldots, x_{n+1}) = x_1 f(x_2, \ldots, x) + \sum_{i=1}^{n-2} (-1)^i f(x_1, \ldots, x_i x_{i+1}, \ldots, x)$$

$$+ (-1)^{n-1} f(x_1, \ldots, x_n x) + (-1)^n f(x_1, \ldots, x_n)$$

over all $x = x_{n+1}$ in G. In the next to last term, as x varies over G, so does $x_n x$. Therefore, if f is a cocycle, then $df = 0$ and

$$0 = x_1 g(x_2, \ldots, x_n) + \sum_{i=1}^{n-2} (-1)^i g(x_1, \ldots, x_i x_{i+1}, \ldots, x_n)$$

$$+ (-1)^{n-1} g(x_1, \ldots, x_{n-1}) + m(-1)^n f(x_1, \ldots, x_n)$$

(the last term is independent of x). Hence

$$0 = dg + m(-1)^n f,$$

and mf is a coboundary.

The same proof works for homology, and we merely set up notation. If $f(x_1, \ldots, x_n, a) = [x_1, \ldots, x_n] \otimes a$, where $a \in A$, then define

$$g(x_1, \ldots, x_{n-1}, a) = \sum_{x_n \in G} f(x_1, \ldots, x_n, a).$$

The proof proceeds as above, but remember that one begins with an element of the form $\sum_{j=1}^p f(x_1^j, \ldots, x_n^j, a^j)$.   ∎

Remark: If e is a positive integer, then a group G has **exponent** e if $x^e = 1$ for all $x \in G$. (Often, one defines the exponent of G as the minimal such e.) One might ask whether Theorem 10.26 can be improved for finite groups G

by replacing the order m of G by its exponent e, for clearly $e \le m$. In general, the answer is "no": the group of quaternions Q has order 8 and exponent 4, but it may be shown that $H^4(Q, \mathbf{Z}) \cong \mathbf{Z}/8\mathbf{Z}$ [Cartan–Eilenberg, 1956, pp. 253–254]. On the other hand, if G is finite abelian and A is a trivial G-module, then one may replace m by e.

⋅ This seemingly harmless result, Theorem 10.26, has many consequences.

Theorem 10.27 **(Schur–Zassenhaus):** *Let E be a finite group having a normal abelian subgroup A with $(|A|, |E/A|) = 1$ (where $|A|$ denotes the order of A). Then E is a semidirect product of A by E/A, and any two complements of A in E are conjugate.*

Proof: Let $E/A = G$ have order m and let $|A| = k$. By Theorem 5.2, we may consider A as a G-module. If $n > 0$, Theorem 10.26 gives $mH^n(G, A) = 0$. On the other hand, $kA = 0$ implies $kH^n(G, A) = 0$ (Exercise 7.6). Since $(m, k) = 1$, it follows that $H^n(G, A) = 0$ for all $n > 0$. When $n = 2$, this says that E is a semidirect product of A by G (Corollary 5.10); when $n = 1$, this says that any two complements of A in E are conjugate (Theorem 5.13). ∎

Remark: The hypothesis that A is abelian may be removed ([Rotman, 1973, p. 149] for the first statement; conjugacy is true also, but much harder).

Theorem 10.28 **(Maschke):** *Let G be a finite group of order m, and let k be a field whose characteristic does not divide m; then the group ring kG is semisimple.*

Proof: By Theorem 5.14, it suffices to prove $H^1(G, A) = 0$ for every kG-module A. Now $1/m \in k$, since k is a field whose characteristic does not divide m. Therefore, multiplication by m is an automorphism of A (whose inverse is multiplication by $1/m$) which induces an automorphism of $H^1(G, A)$ that is also multiplication by m (Theorem 7.16). It follows that $mH^1(G, A) = H^1(G, A)$. On the other hand, $mH^1(G, A) = 0$, by Theorem 10.26, whence $H^1(G, A) = 0$. ∎

Theorem 10.29: *If G is a finite group and A is a f.g. G-module, then $H^n(G, A)$ and $H_n(G, A)$ are finite for all $n > 0$.*

Proof: By Exercise 10.4, $H^n(G, A)$ and $H_n(G, A)$ are f.g. abelian groups for all $n \ge 0$. If $n > 0$, these groups are annihilated by $|G|$, whence the result. ∎

There is a remarkable connection between cohomology and homology groups of G. For a G-module B, we write B^* for the character module (with G-action as in Theorem 1.15):

$$B^* = \mathrm{Hom}_{\mathbf{Z}}(B, \mathbf{Q}/\mathbf{Z}).$$

Theorem 10.30   (Duality Theorem):   *For every finite group G and every G-module B, there are isomorphisms for all $n \geq 0$*

$$(H^n(G, B))^* \cong H_n(G, B^*).$$

Proof: Recall the mixed identity, Theorem 9.51: Let R and S be rings with R left noetherian, and consider the situation $(_R A, _R B_S, C_S)$ in which A is f.g. and C is injective. Then there are isomorphisms for all $n \geq 0$,

$$\text{Tor}_n^R(\text{Hom}_S(B, C), A) \cong \text{Hom}_S(\text{Ext}_R^n(A, B), C).$$

Set $S = \mathbf{Z}$ and $R = \mathbf{Z}G$ (which is left noetherian, by Exercise 10.4); take $A = \mathbf{Z}$ and $C = \mathbf{Q}/\mathbf{Z}$. The isomorphism becomes

$$\text{Tor}_n^{\mathbf{Z}G}(B^*, \mathbf{Z}) \cong (\text{Ext}_{\mathbf{Z}G}^n(\mathbf{Z}, B))^*.$$

By Exercise 10.3, $\text{Tor}_n^{\mathbf{Z}G}(B^*, \mathbf{Z}) \cong \text{Tor}_n^{\mathbf{Z}G}(\mathbf{Z}, B^*) = H_n(G, B^*)$, while the right hand side is $(H^n(G, B))^*$, as desired.   ∎

Remark: Using the remark after the proof of Theorem 9.51, one may prove the Duality Theorem 10.30 for any group G for which the trivial G-module $\mathbf{Z}$ has a G-projective resolution each of whose terms is f.g.

The next theorem explains why $H_2(G, \mathbf{Z})$ is called the multiplier of G when one recalls the discussion of projective representations.

Theorem 10.31:   *Let G be a finite group, and let k be an algebraically closed field of characteristic 0. If $k^{\#}$ is the multiplicative group of nonzero elements of k considered as a trivial G-module, then*

$$H_2(G, \mathbf{Z}) \cong H^2(G, k^{\#}).$$

Proof: Since k is algebraically closed, every element in $k^{\#}$ has an nth root for every $n > 0$. This says that the multiplicative abelian group $k^{\#}$ is divisible. The torsion subgroup of $k^{\#}$, consisting of the roots of unity, is isomorphic to $\mathbf{Q}/\mathbf{Z}$ because k has characteristic 0 (consider $k = \mathbf{C}$ to convince yourself). Divisibility, hence injectivity, of $\mathbf{Q}/\mathbf{Z}$ gives

$$k^{\#} = \mathbf{Q}/\mathbf{Z} \oplus V,$$

where V is a vector space over $\mathbf{Q}$ (Exercise 3.20); thus,

$$H^2(G, k^{\#}) \cong H^2(G, \mathbf{Q}/\mathbf{Z}) \oplus H^2(G, V).$$

However, the argument of Theorem 10.28 gives $H^2(G, V) = 0$, whence

$$H^2(G, k^{\#}) \cong H^2(G, \mathbf{Q}/\mathbf{Z}).$$

Exactness of $0 \to \mathbf{Z} \to \mathbf{Q} \to \mathbf{Q}/\mathbf{Z} \to 0$ gives exactness of

$$H^2(G, \mathbf{Q}) \to H^2(G, \mathbf{Q}/\mathbf{Z}) \to H^3(G, \mathbf{Z}) \to H^3(G, \mathbf{Q}),$$

and the outside terms are 0 by the Theorem 10.28 argument. We conclude that

$$H^2(G, k^\#) \cong H^2(G, \mathbf{Q}/\mathbf{Z}) \cong H^3(G, \mathbf{Z}).$$

A similar argument gives an isomorphism in homology

$$H_3(G, \mathbf{Q}/\mathbf{Z}) \cong H_2(G, \mathbf{Z}).$$

Applying the Duality Theorem 10.30 yields an isomorphism

$$(H^3(G, \mathbf{Z}))^* \cong H_3(G, \mathbf{Z}^*).$$

By Theorem 10.29, $H^3(G, \mathbf{Z})$ is finite, whence $(H^3(G, \mathbf{Z}))^* \cong H^3(G, \mathbf{Z})$. Also, $\mathbf{Z}^* \cong \mathbf{Q}/\mathbf{Z}$, so that $H^3(G, \mathbf{Z}) \cong H_3(G, \mathbf{Q}/\mathbf{Z})$. Assembling all the isomorphisms gives $H^2(G, k^\#) \cong H_2(G, \mathbf{Z})$, as desired. ∎

Observe that H_3 and H^3 were useful to us, even though we have not given them down-to-earth interpretations.

Definition: A group G has **cohomological dimension** $\leq n$, denoted $\mathrm{cd}(G) \leq n$, if $H^k(G, A) = 0$ for all G-modules A and all $k > n$. If n is the least such integer, one defines $\mathrm{cd}(G) = n$; if no such integer n exists, define $\mathrm{cd}(G) = \infty$.

Exercises: 10.16.   Prove that $\mathrm{cd}(G) = 0$ if and only if $G = \{1\}$.
    10.17.   If G is free of rank > 0, then $\mathrm{cd}(G) = 1$.
    10.18.   If $H^{n+1}(G, A) = 0$ for all G-modules A, then $H^k(G, A) = 0$ for all $k > n$ and all G-modules A.
    10.19.   Let G be free abelian with basis $S = \{x_1, \ldots, x_n\}$. Prove that $\mathbf{Z}G \cong S^{-1}\mathbf{Z}[x_1, \ldots, x_n]$; use Hilbert's Syzygy Theorem to prove $\mathrm{cd}(G) \leq n + 1$.

Remarks: 1.   The ring of **Laurent polynomials** in x, coefficients in $\mathbf{Z}$, consists of all formal sums

$$\sum_{i=k}^{n} m_i x^i, \qquad m_i \in \mathbf{Z},$$

k, n (possibly negative) integers, with obvious addition and multiplication. One may easily generalize this definition to several (commuting) variables, and observe, using Exercise 10.19, that Laurent polynomials in n variables, coefficients in $\mathbf{Z}$, is $\mathbf{Z}G$, where G is free abelian of rank n.
    2.   If G is free abelian of rank n, then $\mathrm{cd}(G) = n$ (Corollary 11.50). The reason the bound in Exercise 10.19 is too high is that global dimension $\leq n$

demands vanishing of $\text{Ext}^k_{ZG}(B, A)$, all $k > n$ and all pairs of G-modules B, A, whereas $\text{cd}(G) \leq n$ only demands such vanishing in the special case $B = Z$.

Is there a relation between $\text{cd}(S)$ and $\text{cd}(G)$ when S is a subgroup of G?

Theorem 10.32 (Shapiro's Lemma): *If S is a subgroup of G and A is an S-module, then, for all $n \geq 0$,*

$$H^n(S, A) \cong H^n(G, \text{Hom}_S(ZG, A)).$$

Proof: First of all, the right side makes sense, for $\text{Hom}_S(ZG, A)$ may be regarded as a G-module (as in Theorem 1.15, i.e., as in the adjoint isomorphism). A mixed identity, Exercise 9.23, arising from the adjoint isomorphism, gives

$$\text{Ext}^n_{ZS}(ZG \otimes_G Z, A) \cong \text{Ext}^n_{ZG}(Z, \text{Hom}_S(ZG, A)).$$

Since $ZG \otimes_G Z \cong Z$, this is the desired isomorphism

$$H^n(S, A) \cong H^n(G, \text{Hom}_S(ZG, A)). \quad \blacksquare$$

Corollary 10.33: *If S is a subgroup of G, then $\text{cd}(S) \leq \text{cd}(G)$.*

Proof: Without loss of generality, we may assume $\text{cd}(G) = n < \infty$. If $k > n$ and A is an S-module, then Shapiro's lemma gives

$$H^k(S, A) \cong H^k(G, A^*) = 0,$$

where $A^* = \text{Hom}_S(ZG, A)$. $\quad \blacksquare$

Lemma 10.34: *Let G be a finite cyclic group of order k and with generator x. Define elements of ZG:*

$$D = x - 1 \qquad \text{and} \qquad N = 1 + x + x^2 + \cdots + x^{k-1}.$$

Then the following is a G-free resolution of Z:

$$\cdots \to ZG \xrightarrow{D} ZG \xrightarrow{N} ZG \xrightarrow{D} ZG \xrightarrow{\varepsilon} Z \to 0,$$

where D and N denote multiplication by D and N, respectively.

Proof: First, ZG is commutative, so multiplications are G-maps. Since $ND = DN = x^k - 1 = 0$, we do have a complex.

Since $\ker \varepsilon = \mathfrak{g} = \text{im } D$ (Theorem 5.19), there is exactness at the first step.

Suppose $u = \sum_{i=0}^{k-1} m_i x^i \in \ker D$. Computing gives $m_0 = m_1 = \cdots = m_{k-1}$, so that $u = m_0 N \in \text{im } N$.

Finally, if $u \in \ker N$, then $0 = uN = \sum_{j=0}^{k-1} (\sum_{i=0}^{k-1} m_i) x^j$, so that $\sum m_i = 0$, and

$$u = -D[m_0 1 + (m_0 + m_1)x + \cdots + (m_0 + \cdots + m_{k-1})x^{k-1}] \in \text{im } D. \quad \blacksquare$$

Exercises: 10.20. Prove Lemma 10.34 using Gruenberg's resolution with respect to the presentation $(x\,|\,x^k)$ of G.

  10.21.   If G is finite cyclic, and A is a G-module, then $DA = \mathfrak{g}A$.

Theorem 10.35: *Let G be a finite cyclic group of order k. If A is a G-module, define* $_NA = \{a \in A : Na = 0\}$. *Then*

$$H^0(G, A) = A^G, \qquad H^{2n-1}(G, A) = {}_NA/DA, \qquad H^{2n}(G, A) = A^G/NA.$$

Proof: Apply $\operatorname{Hom}_G(\ , A)$ to the resolution of Lemma 10.34. ∎

Corollary 10.36: *Let G be a finite cyclic group of order k and let A be a trivial G-module. Then*

$$H^0(G, A) = A,$$
$$H^{2n-1}(G, A) = {}_kA = \{a \in A : ka = 0\},$$
$$H^{2n}(G, A) = A/kA.$$

In particular,

$$H^0(G, \mathbf{Z}) = \mathbf{Z}, \qquad H^{2n-1}(G, \mathbf{Z}) = 0, \qquad and \qquad H^{2n}(G, \mathbf{Z}) = \mathbf{Z}/k\mathbf{Z}.$$

Corollary 10.37: *If G is finite cyclic (of order $\neq 1$), then* $\operatorname{cd}(G) = \infty$.

Proof: Take $A = \mathbf{Z}$ considered as a trivial G-module; by Corollary 10.36, $H^{2n}(G, \mathbf{Z}) \neq 0$ for all n. ∎

Corollary 10.38: *If $\operatorname{cd}(G) < \infty$, then G is torsion-free.*

Proof: If G contains an element $x \neq 1$ of finite order, then Corollary 10.33 shows $\operatorname{cd}(\langle x \rangle) \leq \operatorname{cd}(G)$, contradicting Corollary 10.37. ∎

Exercises: 10.22.   If G is finite cyclic, then $H_n(G, A) \cong H^{n+1}(G, A)$ for all G-modules A and all $n \geq 1$. (Compare with the Duality Theorem 10.30.)

  10.23.   If G is finite cyclic, then $\operatorname{pd}(\mathbf{Z}) = \infty$ and $D(\mathbf{Z}G) = \infty$.

  10.24.   If $R = \mathbf{Z}[x]/(x^k - 1)$, then $D(R) = \infty$.

  10.25.   (**Barr–Rinehart**) Define $\tilde{H}^n(G, A) = \operatorname{Ext}^n_{\mathbf{Z}G}(\mathfrak{g}, A)$. Prove that

$$\tilde{H}^0(G, A) \cong \operatorname{Der}(G, A) \qquad and \qquad \tilde{H}^n(G, A) \cong H^{n+1}(G, A), \qquad n \geq 1.$$

(Hint: Theorem 5.21.)

What are the groups of cohomological dimension 1? Stallings [1968] proves the following results.

Theorem. *A finitely related torsion-free group G is a nontrivial free product if and only if $H^1(G, (\mathbf{Z}/2\mathbf{Z})G)$ has more than two elements.*

The hypothesis on G can be considerably relaxed [Cohen, 1972]; the important point here is that Stallings discovered a homological criterion for certain groups to be free products; he used this criterion to prove the next two theorems.

Theorem.   *If G is f.g., then $\mathrm{cd}(G) \leq 1$ if and only if G is free.*

This theorem was a conjecture of Eilenberg and Ganea. Observe that this generalizes the Schreier–Nielsen theorem.

Lemma.   *If S is a subgroup of finite index in G, then*

$$H^1(S, (\mathbf{Z}/2\mathbf{Z})S) \cong H^1(G, (\mathbf{Z}/2\mathbf{Z})G).$$

This is an easy result which we quote to make the next result plausible.

Theorem.   *If G is f.g. torsion-free, and if G has a free subgroup of finite index, then G is free.*

This theorem was a conjecture of Serre who proved that if G is torsion-free and S is a subgroup of finite index, then $\mathrm{cd}(G) = \mathrm{cd}(S)$. Swan [1969], using Stallings's results, was able to remove the hypothesis that G be f.g. in the last two theorems.

We end this chapter by voicing the reader's discontent; one must compute homology in order to use it, and there is still difficulty even in determining whether a homology group is nonzero. A technique is needed, and, in fact, a technique exists: spectral sequences. While it does not solve all problems, it is a powerful tool.

11 Spectral Sequences

Spectral sequences (invented by Leray and Koszul) are certain types of sequences of homology modules; however, one often abuses language by saying spectral sequences is a technique of computing homology. Actually, there are two main techniques. The first gives a useful exact sequence, the "five-term exact sequence of terms of low degree". The second technique involves a notion of convergence of a spectral sequence to a limit module H. We shall be more precise later, but we wish to indicate now to what extent the limit H is determined by the approximating sequence. It is a familiar procedure in group theory to consider a normal series of a group G,

$$G = G_0 \supset G_1 \supset G_2 \supset \cdots,$$

and its corresponding factor groups G_i/G_{i+1}. It is also a familiar procedure to consider a series of submodules of a module H,

$$H = H_0 \supset H_1 \supset H_2 \supset \cdots,$$

now called a "filtration" instead of a normal series, and its factor modules H_i/H_{i+1}. The limit of a spectral sequence is H only in the sense that one obtains the factor modules of some filtration of H. Of course, this usually does not determine H to isomorphism, but it does provide information.

EXACT COUPLES AND FIVE-TERM SEQUENCES

We begin with some simple definitions.

Definition: A **graded module** is a sequence of modules $M = \{M_p : p \in \mathbf{Z}\}$. If $M = \{M_p\}$ and $N = \{N_p\}$ are graded modules and a is a fixed integer, then a sequence of homomorphisms $f = \{f_p : M_p \to N_{p+a}\}$ is a **map** of **degree** a. One writes $f : M \to N$.

A complex $\mathbf{C} = \cdots \to C_p \overset{d_p}{\to} C_{p-1} \to \cdots$ determines a graded module $C = \{C_p : p \in \mathbf{Z}\}$ if one ignores the differentiation $d = \{d_p : p \in \mathbf{Z}\}$. The map $d : C \to C$ has degree -1. If $(\mathbf{C}', d')$ is another complex, a chain map $f : \mathbf{C} \to \mathbf{C}'$ gives a map of degree 0 (with $fd = df'$), while a homotopy is a certain type of map of degree $+1$. A second example of a graded module is the homology of a complex $\mathbf{C} : H_*(\mathbf{C}) = \{H_p(\mathbf{C}) : p \in \mathbf{Z}\}$. One may reverse this procedure: given a graded module $\{A_p : p \in \mathbf{Z}\}$, define a complex

$$\mathbf{A} = \cdots \to A_p \overset{d_p}{\to} A_{p-1} \to \cdots$$

in which each $d_p = 0$; one says $\mathbf{A}$ is a **complex with zero differentiation.**

Exercises: 11.1.   Degrees add under composition: if $f : M \to N$ has degree a and $g : N \to K$ has degree b, then $gf : M \to K$ is a map of degree $a + b$.

11.2.   All graded modules (over a fixed ring) and all maps having a degree comprise a category. Note that

$$\operatorname{Hom}(M, N) = \bigcup_{a \in \mathbf{Z}} \left(\prod_p \operatorname{Hom}(M_p, N_{p+a}) \right).$$

Recall a mnemonic introduced when we first saw long exact sequences: exact triangles. If $0 \to \mathbf{A} \overset{f}{\to} \mathbf{B} \overset{g}{\to} \mathbf{C} \to 0$ is a short exact sequence of complexes, then the long exact sequence may be written

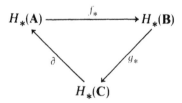

Regarding the vertices as graded modules, the maps f_* and g_* have degree 0 and ∂ has degree -1. Conversely, given any exact triangle, one may write down a long exact sequence if he knows the degrees of the maps.

Definition: A **bigraded module** is a doubly indexed family of modules $M = \{M_{p,q} : (p, q) \in \mathbf{Z} \times \mathbf{Z}\}$. If $M = \{M_{p,q}\}$ and $N = \{N_{p,q}\}$ are bigraded modules and if (a, b) is a fixed ordered pair of integers, then a family of homomorphisms $f = \{f_{p,q} : M_{p,q} \to N_{p+a,\,q+b}\}$ is a **map** of **bidegree** (a, b). One writes $f : M \to N$.

Exercises: 11.3. Bidegrees add under composition: if $f : M \to N$ has bidegree (a, b) and $g : N \to K$ has bidegree (a', b'), then $gf : M \to K$ has bidegree $(a + a', b + b')$.

11.4. All bigraded modules (over a fixed ring) and all maps having a bidegree comprise a category.

In the category of bigraded modules, there are subobjects and quotient objects. If $M_{p,q} \subset N_{p,q}$ for all p, q, then $M = \{M_{p,q}\}$ is a **(bigraded) submodule** of $N = \{N_{p,q}\}$; visibly, the inclusion map $M \to N$ has bidegree $(0, 0)$. Define the **(bigraded) quotient module** N/M as $\{N_{p,q}/M_{p,q}\}$; the natural map $N \to N/M$ also has bidegree $(0, 0)$.

There is a consequence of this elementary definition. Given $f : M \to N$ with bidegree (a, b), im f should be a (bigraded) submodule of N; what is $(\text{im } f)_{p,q}$? Since $(\text{im } f)_{p,q} \subset N_{p,q}$, we are forced to define

$$(\text{im } f)_{p,q} = f_{p-a,\,q-b}(M_{p-a,\,q-b}) = \text{im}(f_{p-a,\,q-b}) \subset N_{p,q}.$$

(Thus, $(\text{im } f)_{p,q}$ is *not* $\text{im}(f_{p,q})$, which lies in $N_{p+a,\,q+b}$.) On the other hand, there is no problem with indices of ker f: if one defines

$$(\text{ker } f)_{p,q} = \text{ker}(f_{p,q}),$$

then ker f is a (bigraded) submodule of M. It is now clear how to define **exactness** of a sequence of bigraded modules.

We defer giving examples of bigraded modules; later, we shall define a "bicomplex" which will determine bigraded modules in much the same way that a complex determines graded modules (ignore differentiation; homology).

Definition: An **exact couple** is a pair of bigraded modules D and E, and maps α, β, γ (each of some bidegree) such that there is exactness at each vertex of the triangle

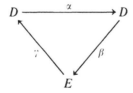

Obviously, an exact couple generalizes the notion of exact triangle (since one does not generalize merely for the sake of generalization, there are not

three distinct bigraded modules at the vertices; in practice, an exact couple is what one encounters). Given an exact couple $D = \{D_{p,q}\}$, $E = \{E_{p,q}\}$, and maps α, β, γ of bidegrees (a, a'), (b, b'), (c, c'), respectively, one may write down a long exact sequence for each fixed p, q:

$$\cdots \xrightarrow{\beta} E_{p-c, q-c'} \xrightarrow{\gamma} D_{p,q} \xrightarrow{\alpha} D_{p+a, q+a'} \xrightarrow{\beta} E_{p+a+b, q+a'+b'} \xrightarrow{\gamma} \cdots .$$

Conversely, given infinitely many long exact sequences as above, they may be assembled into one exact couple.

Let us now use this barrage of notation (actually, a clever organization of a maze of data, due to Massey) to obtain concrete results. The next theorem is due to Kreimer, while the notion of acyclicity when dealing with composite functors was first formulated by Grothendieck.

We maintain our (too restrictive) hypothesis that all functors are additive functors between categories of modules (one may replace module categories by certain more general categories).

Definition: Let $F: \mathfrak{B} \to \mathfrak{C}$ be a functor of either variance. A module B in $\mathfrak{B}$ is **right** F-**acyclic** if $(R^p F)B = 0$ for all $p \geq 1$, where $R^p F$ is the pth right derived functor of F; a module B in $\mathfrak{B}$ is **left** F-**acyclic** if $(L_p F)B = 0$ for all $p \geq 1$, where $L_p F$ is the pth left derived functor of F.

If F is covariant, recall that $(R^p F)B = H^p(F\mathbf{E}_B)$, where $\mathbf{E}_B$ is a deleted injective resolution of B. It follows from Theorem 7.6 that every injective module is right F-acyclic.

Exercises: 11.5. Every projective module is left F-acyclic for any covariant functor F, and is right F-acyclic for any contravariant functor F.
   11.6. If $F = A \otimes_R$, then every flat module $_R B$ is left F-acyclic.

A composite of functors may give an exact couple. Observe first that if $G: \mathfrak{A} \to \mathfrak{B}$, then $(R^n G)A \in \mathfrak{B}$ for every module A in $\mathfrak{A}$ (if $\cdots \to X_n \xrightarrow{d_n} X_{n-1} \to \cdots$ is a deleted resolution in $\mathfrak{A}$, then $\cdots \to GX_n \xrightarrow{Gd_n} GX_{n-1} \to \cdots$ is a complex in $\mathfrak{B}$, whence $\ker Gd_n / \operatorname{im} Gd_{n+1} \in \mathfrak{B}$).

Theorem 11.1: *Let* $G: \mathfrak{A} \to \mathfrak{B}$ *and* $F: \mathfrak{B} \to \mathfrak{C}$ *be functors such that* F *is left exact and whenever* E *is injective in* $\mathfrak{A}$, *then* GE *is right* F-*acyclic. For each module* A *in* $\mathfrak{A}$, *choose an injective resolution* $0 \to A \to E^0 \to E^1 \to \cdots$ *and define*

$$Z^q = \ker(GE^q \to GE^{q+1}).$$

Then there exists an exact couple with

$$E_{p,q} = \begin{cases} (R^p F)(R^q G(A)) & \text{if } p \geq 0, \ q \geq 0, \\ 0 & \text{otherwise}, \end{cases}$$

$$D_{p,q} = \begin{cases} (R^{p+1}F)Z^{q-1} & \text{if } p \geq 0, \ q \geq 1, \\ R^{p+q}(FG)A & \text{if } p = -1, \ q \geq 1, \\ 0 & \text{otherwise}, \end{cases}$$

and maps $\alpha : D \to D$ *of bidegree* $(-1, 1)$, $\beta : D \to E$ *of bidegree* $(1, -1)$, *and* $\gamma : E \to D$ *of bidegree* $(1, 0)$.

Remarks: 1.   Visualize a bigraded module as a family of modules, one sitting on each lattice point in the p-q plane. Thus, E lives in the first quadrant and D lives above the line $q = 1$ and to the right of $p = -1$.

2.   The basic idea is just to assemble the long exact sequences arising from the obvious short exact sequences [1] and [4] below.

Proof: Abbreviate $R^p G$, $R^p F$, $R^p(FG)$ to G^p, F^p, $(FG)^p$, respectively. Our task is to exhibit, for each $q \geq 0$, an exact sequence (since every $D_{p,q}$ occurs)

$$E_{-1, q+1} \to D_{0, q+1} \to D_{-1, q+2} \to E_{0, q+1} \to \cdots$$
$$\to D_{p, q+1} \to D_{p-1, q+2} \to E_{p, q+1} \to \cdots .$$

By definition of E and D, we want exact sequences

$$0 \to F^1 Z^q \to (FG)^{q+1} A \to F(G^{q+1} A) \to \cdots$$
$$\to F^{p+1} Z^q \to F^p Z^{q+1} \to F^p(G^{q+1} A) \to \cdots .$$

Now $G^q A$ is just the qth homology of the complex $\cdots \to GE^q \to GE^{q+1} \to \cdots$. We have already denoted the q-cycles of this complex by Z^q; denote the q-boundaries, $\mathrm{im}(GE^{q-1} \to GE^q)$, by B^q. There are short exact sequences

[1] $0 \to Z^q \to GE^q \overset{\lambda}{\to} B^{q+1} \to 0$

which give rise to exact sequences (since F is left exact and GE^q is right F-acyclic)

[2] $0 \to FZ^q \to FGE^q \to FB^{q+1} \to F^1 Z^q \to 0,$ all $q \geq 0,$

and isomorphisms

[3] $F^p B^{q+1} \overset{\backsim}{\to} F^{p+1} Z^q,$ all $p \geq 1, \ q \geq 0.$

The definition of homology gives short exact sequences

[4] $0 \to B^{q+1} \overset{\mu}{\to} Z^{q+1} \overset{\nu}{\to} G^{q+1} A \to 0,$

which yields long exact sequences

$$0 \to FB^{q+1} \to FZ^{q+1} \to F(G^{q+1}A) \to F^1B^{q+1} \to F^1Z^{q+1} \to F^1(G^{q+1}A) \to \cdots.$$

Using [3] to replace each term F^pB^{q+1} by its isomorphic copy $F^{p+1}Z^q$, for $p \geq 1$, almost leaves us with the desired exact sequences; only the first two terms are not correct.

Consider the diagram

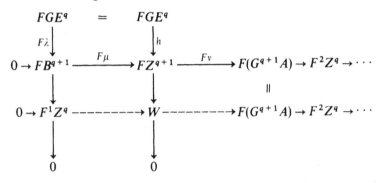

where the first column is the end of the exact sequence [2], the map h is the composite $F\mu F\lambda = F(GE^q \to B^{q+1} \hookrightarrow Z^{q+1})$, and $W = \operatorname{coker} h$. Commutativity of the top square and exactness of the columns provide a unique map $F^1Z^q \to W$ (Exercise 2.7) making the bottom square commute; that $Fv \circ h = 0$ implies the existence of a map $W \to F(G^{q+1}A)$ which makes the remaining square commute. Exactness of the middle row implies, by diagram-chasing, exactness of the bottom row. It remains to identify W with $(FG)^{q+1}A$.

Now $Z^{q+1} = \ker(GE^{q+1} \to GE^{q+2})$, so that left exactness of F gives $FZ^{q+1} = \ker(FGE^{q+1} \to FGE^{q+2})$. Therefore

$$W = \operatorname{coker}(FGE^q \to FZ^{q+1}) = FZ^{q+1}/\operatorname{im}(FGE^q \to FZ^{q+1})$$
$$= \ker(FGE^{q+1} \to FGE^{q+2})/\operatorname{im}(FGE^q \to FZ^{q+1}).$$

Using left exactness of F once again, $Z^{q+1} \hookrightarrow GE^{q+1}$ implies $FZ^{q+1} \hookrightarrow FGE^{q+1}$, and thus $\operatorname{im}(FGE^q \to FZ^{q+1}) = \operatorname{im}(FGE^q \to FGE^{q+1})$. Therefore, $W = H^{q+1}(FG\mathbf{E}_A) = (FG)^{q+1}A$. ∎

Exercise: 11.7. Give explicit formulas for the maps $F^1Z^q \to W$ and $W \to F(G^{q+1}A)$; give explicit formulas for these maps once W has been identified with $(FG)^{q+1}A$.

Theorem 11.2 (Cohomology Five-Term Sequence): *Let* $G: \mathfrak{A} \to \mathfrak{B}$ *and* $F: \mathfrak{B} \to \mathfrak{C}$ *be left exact functors such that E injective in $\mathfrak{A}$ implies GE is right*

F-acyclic. Then there is an exact sequence

$$0 \to (R^1 F)(GA) \to R^1(FG)A \to F(R^1 GA) \to (R^2 F)(GA) \to R^2(FG)A$$

for every module A in $\mathfrak{A}$.

Proof: We return to the abbreviated notation above. Consider the exact sequences in Theorem 11.1 for $q = 0$ and $q = 1$:

$$0 \to F^1 Z^0 \to (FG)^1 A \to F(G^1 A) \to F^2 GA \to F^1 Z^1 \to \cdots$$

and

$$0 \to F^1 Z^1 \to (FG)^2 A \to \cdots.$$

Just splice these two sequences together at $F^1 Z^1$, and remember that $Z^0 = \ker(GE^0 \to GE^1) = GA$, for G is left exact. ∎

Remark: If G is not left exact, one must replace GA by $(R^0 G)A$ in the sequence.

The following variant of Theorem 11.2 is useful.

Theorem 11.3: *Let $G: \mathfrak{A} \to \mathfrak{B}$ and $F: \mathfrak{B} \to \mathfrak{C}$ be left exact functors such that E injective in $\mathfrak{A}$ implies GE is right F-acyclic. If A is a module in $\mathfrak{A}$ with $(R^i G)A = 0$ for $1 \le i < q$, then there is an exact sequence*

$$0 \to (R^q F)(GA) \to R^q(FG)A \to F(R^q GA) \to (R^{q+1} F)(GA) \to R^{q+1}(FG)A.$$

Proof: Let us return once again to the abbreviated notation for derived functors. We shall only prove the special (though most important) case when $G^1 A = 0$ (the rest being an exercise for the reader). Consider the exact sequences in Theorem 11.1 for $q = 0$, 1, and 2:

$$0 \to F^1 Z^0 \to (FG)^1 A \to F(G^1 A) \to F^2 Z^0 \to F^1 Z^1 \to F^1(G^1 A) \to \cdots,$$
$$0 \to F^1 Z^1 \to (FG)^2 A \to F(G^2 A) \to F^2 Z^1 \to F^1 Z^2 \to \cdots,$$

and

$$0 \to F^1 Z^2 \to (FG)^3 A \to \cdots.$$

Splice the last two sequences together at $F^1 Z^2$ to obtain exactness of

$$0 \to F^1 Z^1 \to (FG)^2 A \to F(G^2 A) \to F^2 Z^1 \to (FG)^3 A.$$

Since $G^1 A = 0$, the first sequence ($q = 0$) gives an isomorphism

$$F^2 Z^0 \cong F^1 Z^1.$$

Recalling that $Z^0 = GA$, the sequence now begins with $F^2(GA)$. Let us deal with the fourth term $F^2 Z^1$. Since $G^1 A = Z^1/B^1$, the hypothesis gives $Z^1 = B^1$

and hence $F^2Z^1 = F^2B^1$. But isomorphism [3] in the proof of Theorem 11.1 gives $F^2B^1 \cong F^3Z^0 = F^3GA$. ∎

An important use of this last result occurs in the cohomology of algebras, involving Hilbert's "Theorem 90" and the Brauer group.

It should be clear that the proofs just given dualize; we merely state the results for right exact covariant functors.

Theorem 11.4 (Homology Five-Term Sequence): *Let $G:\mathfrak{A} \to \mathfrak{B}$ and $F:\mathfrak{B} \to \mathfrak{C}$ be right exact functors such that P projective in $\mathfrak{A}$ implies GP is left F-acyclic. Then there is an exact sequence*

$$L_2(FG)A \to (L_2F)(GA) \to F(L_1GA) \to L_1(FG)A \to (L_1F)(GA) \to 0$$

for every module A in $\mathfrak{A}$. Moreover, if $(L_iG)A = 0$ for $1 \leq i < q$, then there is an exact sequence

$$L_{q+1}(FG)A \to (L_{q+1}F)(GA) \to F(L_qGA) \to L_q(FG)A \to (L_qF)(GA) \to 0.$$

Let us illustrate these general results. Assume Π is a group with normal subgroup N. Clearly, every Π-module A may be regarded as an N-module, so that $\mathrm{Hom}_N(\mathbf{Z}, A)$ is defined. Recall

$$\mathrm{Hom}_N(\mathbf{Z}, A) = A^N = \{a \in A : n \cdot a = a \text{ for all } n \in N\}.$$

For a Π-module A, the module A^N is actually a Π/N-module: if $x \in \Pi$ has coset $\bar{x} \in \Pi/N$, define

$$\bar{x} \cdot a = x \cdot a, \qquad a \in A^N$$

(one checks this is well defined). Therefore $\mathrm{Hom}_N(\mathbf{Z}, \ )$, and its derived functors $\mathrm{Ext}^i_{\mathbf{Z}N}(\mathbf{Z}, \ ) = H^i(N, \ )$ are functors from Π-modules to Π/N-modules.

What is the action of Π/N on $H^i(N, A)$ for a Π-module A? Take a Π-projective resolution of $\mathbf{Z}$

$$\mathbf{P} = \cdots \to P_1 \to P_0 \to \mathbf{Z} \to 0$$

(which is automatically N-projective, by Exercise 10.9), and let us compute $H^i(N, A)$ by examining the complex $\mathrm{Hom}_N(\mathbf{P_Z}, A)$. Exercise 10.7 tells us a reasonable way to make each $\mathrm{Hom}_N(P_i, A)$ into a Π/N-module:

$$(\bar{x}f)(b) = x \cdot f(x^{-1}b), \qquad \bar{x} \in \Pi/N, \quad b \in P_i.$$

Hence, $H^i(N, A)$ becomes a Π/N-module by

$$\bar{x}(z_i + B_i) = \bar{x} \cdot z_i + B_i,$$

where $z_i \in \mathrm{Hom}_N(P_i, A)$ is a cycle and B_i is the submodule of boundaries. (Observe that the action of Exercise 10.7 coincides with the action of Π/N on $\mathrm{Hom}_N(\mathbf{Z}, A)$ given above.)

A similar discussion, now using Exercise 10.8, shows how Π/N acts on the homology groups $H_i(N, A)$ when A is Π-module:

$$\bar{x}(b \otimes a) = bx^{-1} \otimes xa, \qquad \bar{x} \in \Pi/N, \quad b \in P_i.$$

Remark: One may prove [Gruenberg, 1970, p. 151] that if $N \subset Z(\Pi)$ and A is Π-trivial, then $H^p(N, A)$ and $H_p(N, A)$ are Π/N-trivial for all p. The proof of this is not difficult, but involves a longish digression. This fact will be used several times in the sequel.

Theorem 11.5: *If N is a normal subgroup of Π and A is a Π-module, then there is an exact sequence*

$$0 \to H^1(\Pi/N, A^N) \to H^1(\Pi, A) \to H^1(N, A)^{\Pi/N} \to H^2(\Pi/N, A^N) \to H^2(\Pi, A).$$

Moreover, if $H^i(N, A) = 0$ for $1 \leq i < q$, there is an exact sequence

$$0 \to H^q(\Pi/N, A^N) \to H^q(\Pi, A) \to H^q(N, A)^{\Pi/N} \to H^{q+1}(\Pi/N, A^N) \to H^{q+1}(\Pi, A).$$

Proof: Let $\mathfrak{A}$ be the category of Π-modules and $\mathfrak{B}$ the category of Π/N-modules; define $G : \mathfrak{A} \to \mathfrak{B}$ by $G = \mathrm{Hom}_N(\mathbf{Z}, \ )$ and $F : \mathfrak{B} \to \mathbf{Ab}$ by $F = \mathrm{Hom}_{\Pi/N}(\mathbf{Z}, \ )$. It is clear that G and F, being Hom's, are left exact.

If E is an injective Π-module, we claim $GE = E^N$ is Π/N-injective (hence right F-acyclic). Consider the diagram

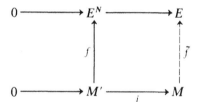

where f and i are Π/N-maps. By change of rings $\mathbf{Z}\Pi \to \mathbf{Z}(\Pi/N)$, every Π/N-module may be regarded as a Π-module and every Π/N-map may be regarded as a Π-map. Since E is Π-injective, there is a Π-map $\tilde{f} : M \to E$ extending f. But $\mathrm{im}\, \tilde{f} \subset E^N : n \cdot \tilde{f}(x) = \tilde{f}(n \cdot x) = \tilde{f}(x)$, since each $x \in M$ is fixed by every $n \in N$. It follows that $\tilde{f} : M \to E^N$ is a Π/N-map, and E^N is Π/N-injective.

The hypotheses of Theorem 11.2 (or 11.3) hold. To apply these theorems, one must evaluate the composite FG. It is a simple matter, however, to see that $(A^N)^{\Pi/N} = A^\Pi$, i.e., $FG = \mathrm{Hom}_\Pi(\mathbf{Z}, \ )$, and we are done. ∎

Remark: One may identify the map $H^i(\Pi/N, A^N) \to H^i(\Pi, A)$ as "inflation" and the map $H^i(\Pi, A) \to H^i(N, A)^{\Pi/N}$ as "restriction", both of which are familiar to specialists. The map $H^1(N, A)^{\Pi/N} \to H^2(\Pi/N, A^N)$ is called "transgression".

There are formulas for these maps in terms of the bar resolution **B**; recall that B_i is the free Π-module on all symbols $[x_1, \ldots, x_i]$, $x_i \in \Pi$, $x_i \neq 1$, and cocycles are certain functions f with domain B_i. It may be shown [Weiss, 1969, §2.5] that if cls $f \in H^i(\Pi/N, A^N)$, then inflation sends cls $f \mapsto$ cls g, where

$$g[x_1, \ldots, x_i] = f[\bar{x}_1, \ldots, \bar{x}_i],$$

$\bar{x}_j$ is the coset of x_j in Π/N, and cls f denotes the homology class of f. If cls $f \in H^i(\Pi, A)$, then restriction sends cls $f \mapsto$ cls h, where

$$h[x_1, \ldots, x_i] = f[x_1, \ldots, x_i], \qquad x_j \in N, \quad j = 1, 2, \ldots, i.$$

Finally, if A is a Π-module, and cls $f \in H^i(N, A)$, then $\bar{t} \in \Pi/N$ acts by

$$(\bar{t} \cdot f)[x_1, \ldots, x_i] = t \cdot f[t^{-1}x_1 t, \ldots, t^{-1}x_i t], \qquad x_j \in N, \quad j = 1, 2, \ldots, i.$$

In particular, this last formula shows that the action of Π/N on $H^i(N, A)$ is trivial when A is Π-trivial and $N \subset Z(\Pi)$.

Here is the corresponding result for homology.

Theorem 11.6: *If N is a normal subgroup of Π and A is a Π-module, then there is an exact sequence*

$$H_2(\Pi, A) \to H_2(\Pi/N, A_N) \to H_1(N, A)_{\Pi/N} \to H_1(\Pi, A) \to H_1(\Pi/N, A_N) \to 0.$$

Moreover, if $H_i(N, A) = 0$ for $1 \leq i < q$, there is an exact sequence

$$H_{q+1}(\Pi, A) \to H_{q+1}(\Pi/N, A_N) \to H_q(N, A)_{\Pi/N} \to H_q(\Pi, A) \to H_q(\Pi/N, A_N) \to 0.$$

Proof: Recall that $A_N = \mathbf{Z} \otimes_N A = A/\mathfrak{n}A$, where $\mathfrak{n}$ is the augmentation ideal of N.

Let $\mathfrak{A}$ be the category of Π-modules and $\mathfrak{B}$ the category of Π/N-modules. Define $G: \mathfrak{A} \to \mathfrak{B}$ by $G = \mathbf{Z} \otimes_N$. Note that $\mathbf{Z} \otimes_N A$ is a Π/N-module if one defines $\bar{g} \cdot (m \otimes a) = m \otimes g \cdot a$, where $\bar{g} \in \Pi/N$ is the coset of $g \in \Pi$, a lies in the Π-module A, and $m \in \mathbf{Z}$. Define $F: \mathfrak{B} \to \mathbf{Ab}$ by $F = \mathbf{Z} \otimes_{\Pi/N}$. It is clear that G and F, being tensors, are right exact.

If P is a projective Π-module, we claim $GP = \mathbf{Z} \otimes_N P$ is Π/N-projective, hence left F-acyclic. If $P \cong \mathbf{Z}\Pi$, then $\mathbf{Z} \otimes_N P \cong P/\mathfrak{n}P \cong \mathbf{Z}\Pi/\mathfrak{n}\mathbf{Z}\Pi \cong \mathbf{Z}(\Pi/N)$. It follows that if P is Π-free or Π-projective, then $P_N = \mathbf{Z} \otimes_N P$ is Π/N-free or Π/N-projective.

The hypotheses of Theorem 11.4 hold. To apply it, one must evaluate the composite FG. It is a simple matter to see that $(A_N)_{\Pi/N} = A_\Pi$, i.e., $FG = \mathbf{Z} \otimes_\Pi$, and the theorem is proved. ∎

Theorem 11.7 (=Theorem 10.12): *For any group G,*

$$H_2(G, \mathbf{Z}) \cong (R \cap F')/[F, R],$$

where F is free and $F/R \cong G$.

Proof: Since R is a normal subgroup of F, Theorem 11.6 applies to give exactness of

$$H_2(F, \mathbf{Z}) \to H_2(G, \mathbf{Z}_R) \to H_1(R, \mathbf{Z})_G \to H_1(F, \mathbf{Z}) \to \cdots.$$

Now $\mathbf{Z}_R = \mathbf{Z}$, since $\mathbf{Z}$ is F-trivial, hence R-trivial, and Corollary 10.5 gives $H_2(F, \mathbf{Z}) = 0$ since F is free. Thus,

$$H_2(G, \mathbf{Z}) = \ker(H_1(R, \mathbf{Z})_G \to H_1(F, \mathbf{Z})).$$

Corollary 10.3 shows $H_1(F, \mathbf{Z}) \cong F/F'$ and $H_1(R, \mathbf{Z}) \cong R/R'$; moreover, $R/R' \cong \mathfrak{r}/\mathfrak{r}\mathfrak{f}$ as G-modules, by Lemma 10.10 and the remark immediately thereafter. With the further aid of Lemma 10.11, we have

$$H_1(R, \mathbf{Z})_G = \mathbf{Z} \otimes_G R/R' \cong \mathbf{Z} \otimes_G \mathfrak{r}/\mathfrak{r}\mathfrak{f} \cong (\mathfrak{r}/\mathfrak{r}\mathfrak{f})/\mathfrak{f}(\mathfrak{r}/\mathfrak{r}\mathfrak{f})$$
$$= (\mathfrak{r}/\mathfrak{r}\mathfrak{f})/(\mathfrak{f}\mathfrak{r} + \mathfrak{r}\mathfrak{f})/\mathfrak{r}\mathfrak{f} \cong \mathfrak{r}/\mathfrak{f}\mathfrak{r} + \mathfrak{r}\mathfrak{f} \cong R/[F, R].$$

Therefore,

$$H_2(G, \mathbf{Z}) = \ker(R/[F, R] \to F/F').$$

We now cheat, asserting the map is just enlargement of coset, from which the result follows. ∎

Our little cheating is just that; it is not difficult to identify the map. Accepting this, the reader may begin to appreciate the power of the five-term sequence.

Here is another application of the five-term sequence. Let Q be the quaternion group of order 8:

$$Q = (x, y \mid x^2 = y^2, xyx = y).$$

We recall the following properties of Q: there is a normal subgroup N of order 2, namely, $\langle x^2 \rangle$, with $Q/N \cong G$, where G is the 4-group; center $Q = N = Q'$, the commutator subgroup of Q; Q has a balanced presentation (exhibited above). It follows from Corollary 10.16 that $H_2(Q, \mathbf{Z}) = 0$.

The five-term sequence with $A = \mathbf{Z}$ is

$$H_2(Q, \mathbf{Z}) \to H_2(G, \mathbf{Z}) \xrightarrow{\alpha} H_1(N, \mathbf{Z})_G \to H_1(Q, \mathbf{Z}) \xrightarrow{\beta} H_1(G, \mathbf{Z}) \to 0.$$

Since $H_2(Q, \mathbf{Z}) = 0$, the map α is monic. By Corollary 10.3, $H_1(Q, \mathbf{Z}) \cong Q/Q' = Q/N \cong G$ and $H_1(G, \mathbf{Z}) \cong G/G' = G$; since both of these groups have order 4 and β is epic, β must be monic. It follows from exactness that α is epic, i.e.,

$$H_2(G, \mathbf{Z}) \cong H_1(N, \mathbf{Z})_G.$$

Now Q acts on $\mathfrak{n} = \{n - 1 : n \in N\}$ by conjugation: if $x \in Q$, then $x \cdot n = xnx^{-1}$; since N is the center of Q, however, $xnx^{-1} = n$. It follows that $G = Q/N$ acts trivially on $\mathfrak{n}$ and hence on $\mathfrak{n}/\mathfrak{n}^2 = H_1(N, \mathbf{Z})$. Therefore,

$H_1(N, \mathbf{Z})_G = H_1(N, \mathbf{Z}) \cong N/N' = N$, and the multiplier of G has order 2 (as we saw in Chapter 10).

We shall return to composite functors, so let us give a simple necessary condition guaranteeing acyclicity.

Theorem 11.8:   *Assume (F, G) is an adjoint pair of functors, where $F: \mathfrak{A} \to \mathfrak{B}$ and $G: \mathfrak{B} \to \mathfrak{A}$.*

(i)   *If G preserves epimorphisms, then F preserves projectives;*

(ii)   *If F preserves monomorphisms, then G preserves injectives.*

Proof: Let P be projective in $\mathfrak{A}$, and consider the diagram in $\mathfrak{B}$

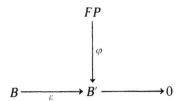

If $\tau: \mathrm{Hom}_{\mathfrak{B}}(F, \ ) \to \mathrm{Hom}_{\mathfrak{A}}(\ , G)$ is the natural bijection, then there is a diagram in $\mathfrak{A}$

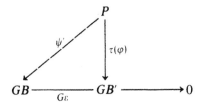

The row is exact, since G preserves epimorphisms, and so projectivity of P provides a map $\psi' \in \mathrm{Hom}(P, GB)$ with

$$(G\varepsilon) \circ \psi' = \tau(\varphi).$$

We claim $\tau^{-1}(\psi'): FP \to B$ completes the original diagram. Naturality of τ gives commutativity of

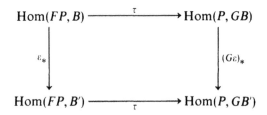

Evaluating on $\tau^{-1}\psi'$, we have $\tau\varepsilon_*(\tau^{-1}\psi') = (G\varepsilon)_*\tau(\tau^{-1}\psi')$. But the left side is $\tau(\varepsilon \circ \tau^{-1}\psi')$, while the right side is $(G\varepsilon)_*\psi' = (G\varepsilon) \circ \psi' = \tau(\varphi)$. Since τ is a bijection,

$$\varepsilon \circ \tau^{-1}\psi' = \varphi,$$

as desired. The dual proof gives the second statement. ∎

Exercises: 11.8. Show that the functor $\mathrm{Hom}_N(Z, \ )$ of Theorem 11.5 satisfies the hypothesis of Theorem 11.8.

11.9. Show that the functor $Z \otimes_N$ of Theorem 11.6 satisfies the hypothesis of Theorem 11.8.

DERIVED COUPLES AND SPECTRAL SEQUENCES

The main observation in this section is that an exact couple is more than a clever organization of long exact sequences; it determines other exact couples.

Consider the exact couple, with maps of the indicated bidegrees:

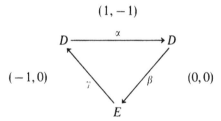

Define $d^1 : E \to E$ by $d^1 = \beta\gamma$ (first γ and then β!). Since $\gamma\beta = 0$, it follows that $d^1 d^1 = 0$ and E has homology groups $H(E, d^1) = \ker d^1/\mathrm{im}\, d^1$. One usually denotes $H(E, d^1)$ by E^2 and regards it as a bigraded module. In more detail,

$$d^1_{p,q} : E_{p,q} \xrightarrow{\gamma} D_{p-1,q} \xrightarrow{\beta} E_{p-1,q},$$

so that d^1 has bidegree $(-1, 0)$. By definition of bigraded quotient module,

$$E^2_{p,q} = \ker d^1_{p,q}/\mathrm{im}\, d^1_{p+1,q}.$$

Define a second bigraded module $D^2 = \mathrm{im}\, \alpha$. Since α has bidegree $(1, -1)$, the definition of bigraded image gives

$$D^2_{p,q} = \alpha_{p-1,q+1}(D_{p-1,q+1}) = \mathrm{im}\, \alpha_{p-1,q+1} \subset D_{p,q}.$$

We next define α^2, β^2, γ^2 as pictured below:

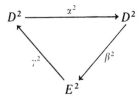

Set α^2 to be the restriction of α (for $D^2 = \operatorname{im} \alpha \subset D$). Since the inclusion map $i: D^2 = \operatorname{im} \alpha \hookrightarrow D$ has bidegree $(0,0)$, the map $\alpha^2 = \alpha \circ i$ has the same bidegree as α, namely, $(1, -1)$. Define $\beta^2: D^2 \to E^2$ by

$$\beta^2 y = [\beta \alpha^{-1} y],$$

where bracket denotes homology class (more details in a moment), and define $\gamma^2: E^2 \to D^2$ by

$$\gamma^2_{p,q}[z_{p,q}] = \gamma_{p,q} z_{p,q} \in D_{p-1,q}.$$

It is a brief exercise that γ^2 is well defined, so that γ^2 has the same bidegree as γ, namely, $(-1, 0)$. Let us return to β^2. If $y \in D^2_{p,q}$, then $y = \alpha_{p-1,q+1}(x_{p-1,q+1})$; define

$$\beta^2_{p,q} = \beta_{p-1,q+1} \alpha^{-1}_{p-1,q+1} : y \mapsto [\beta_{p-1,q+1}(x_{p-1,q+1})] \in E^2_{p-1,q+1}.$$

Again, we let the reader check a map is well defined. Visibly, β^2 has bidegree $(-1, 1)$.

Observe that the bidegrees of α^2, β^2, γ^2 are the same bidegrees occurring in the exact couple of Theorem 11.1. Note also that all definitions make sense if we begin with an exact couple whose maps have any bidegrees.

Theorem 11.9:   *With the definitions above,*

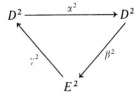

is an exact couple; α^2 has bidegree $(1, -1)$, β^2 has bidegree $(-1, 1)$, and γ^2 has bidegree $(-1, 0)$.

Proof: This is a simple diagram chase. The reader may verify that successive composites are 0, i.e., $\operatorname{im} \subset \ker$, and we shall prove the reverse inclusions. We drop subscripts.

Assume $x \in D^2 = \operatorname{im} \alpha$ and $\alpha^2 x = \alpha x = 0$. Then $x \in \ker \alpha = \operatorname{im} \gamma$, so that $x = \gamma y$ for some $y \in E$. But $x \in \operatorname{im} \alpha = \ker \beta$ implies $0 = \beta x = \beta \gamma y = d^1 y$. Therefore, y is a cycle and $x = \gamma^2 [y]$.

Assume $x \in D^2 = \operatorname{im} \alpha$ and $\beta^2 x = 0$. Thus, $x = \alpha y$ and $\beta y = d^1 w = \beta \gamma w$. Hence $y - \gamma w \in \ker \beta = \operatorname{im} \alpha = D^2$, and $\alpha^2(y - \gamma w) = \alpha(y - \gamma w) = \alpha y = x$ (for $\alpha \gamma = 0$).

Assume $x \in E^2$ and $\gamma^2 x = 0$. Now $x = [y] \in E^2$, and $\gamma^2 x = \gamma y = 0$; hence $y \in \ker \gamma = \operatorname{im} \beta$, so $y = \beta w$. One checks that $\beta^2(\alpha w) = [y] = x$. ∎

Definition: The exact couple $(D^2, E^2, \alpha^2, \beta^2, \gamma^2)$ is called the **derived couple** of $(D, E, \alpha, \beta, \gamma)$.

Of course, we may (and do) iterate this construction to get a sequence of exact couples

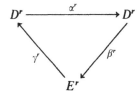

where, by definition, the $(r + 1)$st exact couple is the derived couple of the rth one.

It is useful to have more explicit descriptions of the data of the rth derived couple.

Theorem 11.10: Let $(D, E, \alpha, \beta, \gamma)$ be an exact couple, where α, β, γ have bidegrees $(1, -1)$, $(0, 0)$, and $(-1, 0)$, respectively. If the rth derived couple is $(D^r, E^r, \alpha^r, \beta^r, \gamma^r)$, then

   (i) α^r has bidegree $(1, -1)$, β^r has bidegree $(1 - r, r - 1)$, and γ^r has bidegree $(-1, 0)$;

   (ii) $d^r = \beta^r \gamma^r$ has bidegree $(-r, r - 1)$ and is induced by $\beta \alpha^{-r+1} \gamma$;

   (iii) $E^{r+1}_{p,q} = \ker d^r_{p,q} / \operatorname{im} d^r_{p+r, q-r+1}$.

Proof: An easy induction on r, using the fact that bidegrees add. ∎

Definition: A **spectral sequence** is a sequence $\{E^r, d^r : r \geq 1\}$ of bigraded modules and maps with $d^r d^r = 0$ such that

$$E^{r+1} = H(E^r, d^r)$$

as bigraded modules.

Of course, one may denote the original E by E^1 (as we have just done). Thus, every exact couple determines a spectral sequence.

An elementary definition is useful here.

Definition: A **subquotient** of a module M is a module of the form M'/M'', where $M'' \subset M'$ are submodules of M.

Obviously, one may also speak of subquotients of graded modules or of bigraded modules. In our context, the most obvious example of a subquotient is homology: if $\mathbf{C}$ is a complex, then for each q, $H_q(\mathbf{C})$ is a subquotient of C_q.

Exercises: 11.10. Subquotient is transitive: if B is a subquotient of A and C is a subquotient of B, then C is a subquotient of A.

11.11. If an abelian group G has any of the following properties, then so does every subquotient of G:

(i) $G = 0$; (ii) G finite; (iii) G f.g.; (iv) G torsion; (v) G p-primary; (vi) $mG = 0$ for some $m > 0$; (vii) G cyclic.

Returning to spectral sequences, we see that each E^r is a subquotient of $E = E^1$, or of E^2, indeed, of any earlier term. This remark can be pictured. Write $E^2 = Z^2/B^2$ (actually, $E^2_{p,q} = Z^2_{p,q}/B^2_{p,q}$). Since $E^3 = Z^3/B^3$ is a subquotient of E^2, the third isomorphism theorem allows us to assume

$$0 \subset B^2 \subset B^3 \subset Z^3 \subset Z^2 \subset E^1.$$

Iterating,

$$0 \subset B^2 \subset \cdots \subset B^r \subset B^{r+1} \subset \cdots \subset Z^{r+1} \subset Z^r \subset \cdots \subset Z^2 \subset E^1.$$

Definition: $Z^\infty_{p,q} = \bigcap_r Z^r_{p,q}$; $B^\infty_{p,q} = \bigcup_r B^r_{p,q}$; $E^\infty_{p,q} = Z^\infty_{p,q}/B^\infty_{p,q}$.

The bigraded module E^∞ is called the **limit term** of the spectral sequence $\{E^r\}$, and it is clear that, as r gets large, the terms E^r do "approximate" the limit term.

There are other descriptions of the modules $Z^r_{p,q}, B^r_{p,q}, E^r_{p,q}$, for $r \le \infty$; indeed, many accounts of spectral sequences use these descriptions as their starting point. The appropriate context is that of "*dfg*-modules" (differential, filtered, graded). It is instructive to see this approach, and the reader is advised to look since other viewpoints are valuable [MacLane, 1963, XI.1 and XI.3].

Exercise: 11.12. If $d^r_{p,q} = 0$ for all p, q, then $E^r = E^{r+1}$. If $d^r_{p,q} = 0$ for all p, q and all $r \ge s$, then $E^\infty_{p,q} = E^s_{p,q}$ for all p, q.

FILTRATIONS AND CONVERGENCE

Definition: Let $\mathfrak{A}$ be a category and let A be an object in $\mathfrak{A}$. A **filtration** of A is a family of subobjects of A, $\{F^pA : p \in \mathbf{Z}\}$, such that

$$\cdots \subset F^{p-1}A \subset F^pA \subset F^{p+1}A \subset \cdots.$$

We are especially interested in two special cases: $\mathfrak{A} = $ complexes; $\mathfrak{A} = $ graded modules. Thus, a filtration of a complex $\mathbf{C}$ is a family of sub-complexes $\{F^p\mathbf{C} : p \in \mathbf{Z}\}$ with $F^{p-1}\mathbf{C} \subset F^p\mathbf{C}$ for all p. A filtration of a graded module $H = \{H_n : n \in \mathbf{Z}\}$ is a family of graded submodules $\{F^pH : p \in \mathbf{Z}\}$ with $F^{p-1}H \subset F^pH$ for all p. Note that this last condition merely says that there are (ordinary) submodules F^pH_n of H_n, for each n, such that $F^{p-1}H_n \subset F^pH_n$, all p.

Theorem 11.11: *Every filtration $\{F^p\mathbf{C} : p \in \mathbf{Z}\}$ of a complex $\mathbf{C}$ determines an exact couple*

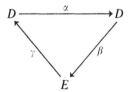

in which α has bidegree $(1, -1)$, β has bidegree $(0, 0)$, and γ has bidegree $(-1, 0)$.

Proof: Let us abbreviate $F^p\mathbf{C}$ to F^p. For each fixed p, there is a short exact sequence of complexes

$$0 \to F^{p-1} \to F^p \to F^p/F^{p-1} \to 0.$$

The corresponding long exact homology sequence is

$$\cdots \to H_{p+q}(F^{p-1}) \xrightarrow{\alpha} H_{p+q}(F^p) \xrightarrow{\beta} H_{p+q}(F^p/F^{p-1}) \xrightarrow{\gamma} H_{p+q-1}(F^{p-1}) \to \cdots.$$

Observe that there are only two types of terms: homology of some F^p or homology of some quotient F^p/F^{p-1}. Accordingly, define

$$D_{p,q} = H_{p+q}(F^p) \qquad \text{and} \qquad E_{p,q} = H_{p+q}(F^p/F^{p-1}).$$

The long exact sequence now reads

$$\cdots \to D_{p-1, q+1} \xrightarrow{\alpha} D_{p,q} \xrightarrow{\beta} E_{p,q} \xrightarrow{\gamma} D_{p-1, q} \to \cdots.$$

The reader may now equip the maps α, β, γ with bidegrees, namely, $(1, -1)$, $(0, 0)$, and $(-1, 0)$, respectively. All has been verified. ∎

It is at this point the reader may begin to feel that he could have invented exact couples; he also sees why the bidegrees before Theorem 11.9 were so chosen.

Corollary 11.12:   *Every filtration $\{F^pC\}$ of a complex $\mathbf{C}$ determines a spectral sequence.*

Proof: The filtration determines an exact couple with maps of the proper bidegrees; Theorem 11.9 gives a sequence of derived exact couples whose terms $E^2, E^3, \ldots, E^r, \ldots$ form a spectral sequence. ∎

It would be pleasant to know the E^2 term of this spectral sequence. Let us be content for the moment with exhibiting the differential $d^1_{p,q}:E_{p,q} \to E_{p-1,q}$. By definition, $d^1 = \beta\gamma$; with subscripts,

$$d^1_{p,q} = \beta_{p-1,q}\gamma_{p,q}:H_{p+q}(F^p/F^{p-1}) \to H_{p+q-1}(F^{p-1}/F^{p-2});$$

moreover, $\gamma_{p,q}$ is the connecting homomorphism $H_{p+q}(F^p/F^{p-1}) \to H_{p+q-1}(F^{p-1})$, while $\beta_{p-1,q}$ is the map induced by the natural map $F^{p-1} \to F^{p-1}/F^{p-2}$.

Exercise: 11.13.   Consider the short exact sequence of complexes

$$0 \to F^{p-1}/F^{p-2} \to F^p/F^{p-2} \to F^p/F^{p-1} \to 0.$$

Show that the map $d^1_{p,q}$ is the connecting homomorphism $H_{p+q}(F^p/F^{p-1}) \to H_{p+q-1}(F^{p-1}/F^{p-2})$. Hint: Use naturality of the connecting homomorphism in the situation

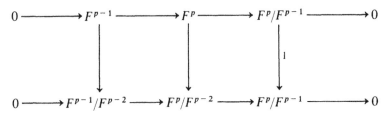

A filtration $\{F^pC\}$ of a complex $\mathbf{C}$ defines a filtration of the (graded) homology module $H_*(\mathbf{C})$ in the following obvious way. If $i_p:F^pC \to \mathbf{C}$ is the inclusion map, then there is the induced map $H_n(i_p):H_n(F^pC) \to H_n(\mathbf{C})$ for each n. Clearly im $H_n(i_{p-1}) \subset$ im $H_n(i_p)$ for all p, i.e., the (graded) submodules $\{$im $H(i_p)\}$ form a filtration of $H_*(\mathbf{C})$. Wouldn't it be nice if the spectral sequence of $\{F^pC\}$, whose E^1 term is $H_*(F^p/F^{p-1})$, determined the factor modules of this filtration of $H_*(\mathbf{C})$? If so, the spectral sequence would allow us to commute homology and filtration, better, homology and factor modules.

Definition: Let H be a graded module. A spectral sequence $\{E^r\}$ **roughly converges** to H, denoted $E^2_{p,q} \underset{p}{\to} H_{p+q}$, if there is some filtration $\{\Phi^p H\}$ of H such that

$$E^\infty_{p,q} \cong \Phi^p H_{p+q}/\Phi^{p-1} H_{p+q}$$

for all p, q.

There is a universal notational agreement in this subject that

$$n = p + q.$$

Hence, the isomorphisms in the definition of rough convergence are written $E^\infty_{p,q} \cong \Phi^p H_n/\Phi^{p-1} H_n$. The index p written under the arrow $E^2_{p,q} \underset{p}{\to} H$ is called the **filtration degree**; it tells which of the two subscripts index the filtration of H, and this is necessary because $E^\infty_{p,q}$ and $E^\infty_{q,p}$ may not be isomorphic.

Let us state some difficulties implicit in the notion of rough convergence by considering a filtration $\{F^p C\}$ of a complex C and its corresponding spectral sequence $\{E^r\}$. First, we need some criterion to guarantee rough convergence. Next, even if we have rough convergence, it may be useless. For example, define $F^p C = 0$, the zero subcomplex, for all p. Clearly $0 = E = E^2 = E^3 = \cdots$, so that $E^\infty_{p,q} = 0$ for all p, q. Moreover, the "obvious" filtration $\{\text{im } H(i_p)\}$ of $H_*(C)$ consists of zero submodules, hence has zero factor modules. Thus, we do have rough convergence, and the information it conveys is $0 = 0$. Plainly, some hypothesis is needed on the filtration $\{F^p C\}$. In most applications, the following condition does hold.

Definition: A filtration $\{F^p H\}$ of a graded module H is **bounded** if, for each n, there exist integers $s = s(n)$ and $t = t(n)$ such that

$$F^s H_n = 0 \qquad \text{and} \qquad F^t H_n = H_n.$$

Since a complex C is a graded module if we forget its differentiation, it makes sense to say that a filtration $\{F^p C\}$ of C is bounded.

In any filtration, $F^{p-1} H \subset F^p H$ for all p. In particular, if $\{F^p H\}$ is bounded, then for each n, $F^p H_n = 0$ for all $p \le s$, $F^p H_n = H_n$ for all $p \ge t$, and there is a finite chain

$$0 = F^s H_n \subset F^{s+1} H_n \subset \cdots \subset F^t H_n = H_n.$$

Definition: Let H be a graded module. A spectral sequence $\{E^r\}$ **converges** to H, denoted $E^2_{p,q} \underset{p}{\Rightarrow} H_n$, if there is some *bounded* filtration $\{\Phi^p H\}$ of H such that

$$E^\infty_{p,q} \cong \Phi^p H_n/\Phi^{p-1} H_n$$

for all p, q (remember, $n = p + q$).

The previous notion of rough convergence may be forgotten; convergence is the important notion.

Theorem 11.13:   *Assume* $\{F^pC\}$ *is a bounded filtration of a complex* C, *and let* $\{E^r\}$ *be the spectral sequence it determines. Then*

(i)   *for each* p, q, *we have* $E_{p,q}^{\infty} = E_{p,q}^r$ *for large* r *(depending on* p, q*)*;

(ii) $E_{p,q}^2 \underset{p}{\Rightarrow} H_n(C)$.

Proof: (i)   How can we say anything at all about $E_{p,q}^r$? The one thing we do know is that its differentiation d^r has bidegree $(-r, r-1)$, Choose p large, i.e., $p > t(n)$. For any such p, $F^{p-1} = F^p$ and $F^p/F^{p-1} = 0$. Hence $E_{p,q} = H_{p+q}(F^p/F^{p-1}) = 0$, and so $E_{p,q}^r = 0$ for all r, all such p (for $E_{p,q}^r$ is a subquotient of $E_{p,q}$). Similarly, choosing p small, i.e., $p < s(n)$, gives $E_{p,q}^r = 0$ for all r.

Now choose any p, q. Since d^r has bidegree $(-r, r-1)$, $d^r(E_{p,q}^r) \subset E_{p-r, q+r-1}^r$. For large r, the index $p - r$ is small and $E_{p-r, q+r-1}^r = 0$. Hence $E_{p,q}^r = \ker d_{p,q}^r$ for large r. Let us now compute $E_{p,q}^{r+1} = \ker d_{p,q}^r / \operatorname{im} d_{p+r, q-r+1}^r$. When r is large, the index $p + r$ is large, and so $\operatorname{im} d_{p+r, q-r+1}^r = d^r(E_{p+r, q-r+1}^r) = 0$. Therefore, $E_{p,q}^{r+1} = \ker d_{p,q}^r = E_{p,q}^r$ for large r (depending on p and q). The spectral sequence is thus ultimately constant, whence $E_{p,q}^{\infty} = E_{p,q}^r$ for large r.

(ii)   Equip $H_*(C)$ with the obvious filtration

$$\Phi^p H_n(C) = \operatorname{im}(H_n(F^pC) \to H_n(C)),$$

where the map is induced by the inclusion $F^pC \to C$. That F is bounded implies Φ is bounded:

$$0 = \Phi^s H_n(C) \subset \Phi^{s+1} H_n(C) \subset \cdots \subset \Phi^t H_n(C) = H_n(C).$$

Consider the exact sequence arising from the rth derived couple (the indices are determined by the bidegrees of the maps, which were calculated in Theorem 11.10):

(*) $\cdots \to D_{p+r-2, q-r+2}^r \overset{\alpha^r}{\to} D_{p+r-1, q-r+1}^r \overset{\beta^r}{\to} E_{p,q}^r \overset{\gamma^r}{\to} D_{p-1, q}^r \to \cdots$.

Recall that $D^r = \alpha \circ \cdots \circ \alpha D$ ($r - 1$ factors equal to α) and that α has bidegree $(1, -1)$. It follows that

$$D_{p+r-1, q-r+1}^r = \alpha \circ \cdots \circ \alpha D_{p,q} \qquad (r - 1 \text{ factors of } \alpha)$$
$$= \operatorname{im}(H_{p+q}(F^p) \to H_{p+q}(F^{p+r-1})),$$

for $\alpha_{p,q}$ is just the homology map induced by inclusion $F^p \to F^{p+1}$. For fixed p, q, boundedness of the filtration gives $F^{p+r-1}C = C$ for large r. We conclude that for large r

$$D^r_{p+r-1, q-r+1} = \Phi^p H_{p+q}(\mathbf{C}).$$

A similar calculation shows that, for large r, the first term in the exact sequence $(*)$ is $\Phi^{p-1} H_{p+q}(\mathbf{C})$. Substituting, exact sequence $(*)$ becomes

$$\cdots \to \Phi^{p-1} H_{p+q}(\mathbf{C}) \to \Phi^p H_{p+q}(\mathbf{C}) \to E^r_{p,q} \to D^r_{p-1, q} \to \cdots.$$

If we can show that $D^r_{p-1, q} = 0$ for large r, then we have

$$\Phi^p H_{p+q}(\mathbf{C})/\Phi^{p-1} H_{p+q}(\mathbf{C}) \cong E^r_{p,q} = E^\infty_{p,q},$$

and we are done. But this is essentially the same calculation:

$$\begin{aligned}
D^r_{p-1, q} &= \alpha \circ \cdots \circ \alpha D_{p-r, q+r-1} &&(r-1 \text{ factors of } \alpha) \\
&= \alpha \circ \cdots \circ \alpha H_{p+q-1}(F^{p-r}\mathbf{C}) &&(r-1 \text{ factors}) \\
&= 0, &&\text{if } p - r < s. \quad \blacksquare
\end{aligned}$$

In practice, the vast majority of spectral sequences arise from a bounded filtration $\{F^p\mathbf{C}\}$ of a complex $\mathbf{C}$; moreover, the typical application involves a certain complex (the "total complex") associated to a "bicomplex". For this reason, we proceed to bicomplexes. (The reader finding a more general spectral sequence in his hands is referred to any other textbook treating spectral sequences; we maintain that an understanding of the special case suffices for an understanding of the general.)

BICOMPLEXES

Definition: A **bicomplex** (or **double complex**) is a bigraded module $M = \{M_{p,q}\}$ with maps $d', d'': M \to M$ of bidegree $(-1, 0)$ and $(0, -1)$, respectively, such that

$$d'd' = 0, \qquad d''d'' = 0, \qquad \text{and} \qquad d'd'' + d''d' = 0.$$

As usual, picture a bigraded module as a family of modules in the p-q plane. The maps $d'_{p,q}: M_{p,q} \to M_{p-1, q}$ go one step left; the maps $d''_{p,q}: M_{p,q} \to M_{p, q-1}$ go one step down. The first two conditions $d'd' = 0$ and $d''d'' = 0$ say that each row and each column is a complex. Before giving examples of bicomplexes, let us associate a complex to M in order to explain the anti-commutativity equation $d'd'' + d''d' = 0$.

Definition: If M is a bicomplex, its **total complex** $\mathrm{Tot}(M)$ is the complex defined by

$$\mathrm{Tot}(M)_n = \coprod_{p+q=n} M_{p,q},$$

and

$$d_n : \mathrm{Tot}(M)_n \to \mathrm{Tot}(M)_{n-1}$$

is given by

$$d_n = \sum_{p+q=n} d'_{p,q} + d''_{p,q}.$$

Lemma 11.14:   *If M is a bicomplex, then $\mathrm{Tot}(M)$ is a complex.*

Proof: Begin by drawing a picture.

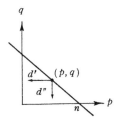

$\mathrm{Tot}(M)_n$ is the sum of all $M_{p,q}$ lying on the $45°$ line $p + q = n$. Note that if $p + q = n$, then $\mathrm{im}\, d'_{p,q} \subset M_{p-1,q}$ and $\mathrm{im}\, d''_{p,q} \subset M_{p,q-1}$; in either case, the sum of the indices is $p + q - 1 = n - 1$, so that $\mathrm{im}\, d_n \subset \mathrm{Tot}(M)_{n-1}$. Finally, we must show that $dd = 0$. But

$$dd = \sum(d' + d'')(d' + d'') = \sum d'd' + \sum(d'd'' + d''d') + \sum d''d'',$$

and each of the summands is zero, by definition of bicomplex.   ∎

The first example of a bicomplex is very general, and we state it as a very trivial lemma.

Lemma 11.15   (Sign Lemma):   *Let M be a bigraded module with maps $d', d'' : M \to M$ of bidegree $(-1, 0)$ and $(0, -1)$, respectively; assume $d'd' = 0$, $d''d'' = 0$, and the diagram commutes. If $d''_{p,q}$ is replaced by $\Delta''_{p,q} = (-1)^p d''_{p,q}$, then (M, d', Δ'') is a bicomplex.*

Proof: Changing sign does not alter kernel or image, so that $\Delta''\Delta'' = 0$. For the remaining identity,

$$d'_{p, q-1}\Delta''_{p,q} = (-1)^p d'_{p, q-1} d''_{p,q} = (-1)^p d''_{p-1, q} d'_{p,q}$$
$$= (-1)^p (-1)^{p-1}\Delta''_{p-1, q} d'_{p,q} = -\Delta''_{p-1, q} d'_{p,q}.$$ ∎

In short, a big commutative diagram whose rows and columns are complexes can be modified by a simple sign change to be a bicomplex.

Examples: 1. Let A be a right R-module and B a left R-module; let

$$\mathbf{P}_A = \cdots \to P_p \xrightarrow{\Delta_p'} P_{p-1} \to \cdots \to P_0 \to 0$$

be a deleted projective resolution of A, and let

$$\mathbf{Q}_B = \cdots \to Q_q \xrightarrow{\Delta_q''} Q_{q-1} \to \cdots \to Q_0 \to 0$$

be a deleted projective resolution of B.

Define a bigraded module M by $M_{p,q} = P_p \otimes Q_q$; the reader may check that M becomes a bicomplex if one defines

$$d_{p,q}' = \Delta_p' \otimes 1_q \qquad \text{and} \qquad d_{p,q}'' = (-1)^p 1_p \otimes \Delta_q'',$$

where 1_q and 1_p are identity maps on Q_q and P_p, respectively.

Observe that $M_{p,q} = 0$ if $p < 0$ or $q < 0$; such a bicomplex is an example of a **first quadrant bicomplex**.

2. Nowhere in the construction just given did we use the hypothesis that the complexes were deleted projective resolutions. Thus, if $\mathbf{A}$ and $\mathbf{B}$ are complexes, we may form a bicomplex M as above. Tot(M) is usually called the **tensor product of the complexes** $\mathbf{A}$ and $\mathbf{B}$, and one defines

$$\mathbf{A} \otimes \mathbf{B} = \text{Tot}(M).$$

Thus if $\mathbf{A} = (\mathbf{A}, \Delta')$ and $\mathbf{B} = (\mathbf{B}, \Delta'')$, then

$$(\mathbf{A} \otimes \mathbf{B})_n = \coprod_{p+q=n} A_p \otimes B_q,$$

and

$$d_n : (\mathbf{A} \otimes \mathbf{B})_n \to (\mathbf{A} \otimes \mathbf{B})_{n-1}$$

is given by

$$a_p \otimes b_q \mapsto \Delta' a_p \otimes b_q + (-1)^p a_p \otimes \Delta'' b_q,$$

where $a_p \in A_p$ and $b_q \in B_q$.

3. The **Eilenberg–Zilber theorem** [MacLane, 1963, VIII.8] states that if X and Y are topological spaces, then $H_n(X \times Y) \cong H_n(\mathbf{S}(X) \otimes \mathbf{S}(Y))$, where $\mathbf{S}(X)$ is the singular complex of X.

4. Let A and B be left R-modules, let

$$\mathbf{P}_A = \cdots \to P_p \xrightarrow{\Delta_p'} P_{p-1} \to \cdots \to P_0 \to 0$$

be a deleted projective resolution of A, and let

$$\mathbf{E}_B = 0 \to E^0 \to \cdots \to E^q \xrightarrow{\Delta_q''} E^{q+1} \to \cdots$$

be a deleted injective resolution of B.

Define a bigraded module M by

$$M_{-p,-q} = \text{Hom}(P_p, E^q);$$

define

$$d'_{-p,-q} = (\Delta'_{p+1})^* : \text{Hom}(P_p, E^q) \to \text{Hom}(P_{p+1}, E^q)$$

and

$$d''_{-p,-q} = (-1)^{p+q+1}(\Delta''_q)_* : \text{Hom}(P_p, E^q) \to \text{Hom}(P_p, E^{q+1}).$$

First of all, the signs of the indices have been chosen to ensure that d' has bidegree $(-1, 0)$ and d'' has bidegree $(0, -1)$. Second, the sign does give anticommutativity so that M is, indeed, a bicomplex (the reason for the more complicated exponent will be seen soon).

Since $M_{p,q} = 0$ for $p > 0$ or $q > 0$, M is an example of a **third quadrant bicomplex**.

As usual, we eliminate negative indices by raising them. Thus, write $M^{p,q}$ instead of $M_{-p,-q}$. In visualizing M, replace points $(-p, -q)$ by points (p, q), so the original picture (third quadrant) is replaced by a new picture (first quadrant) by reflection through the origin. In particular, all arrows retain their slope but reverse their direction.

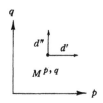

Example: 5. If $(\mathbf{A}, \Delta')$ and $(\mathbf{C}, \Delta'')$ are arbitrary complexes, define a bicomplex M having modules

$$M_{p,q} = \text{Hom}(A_p, C_q),$$

and differentiations

$$d'_{p,q} = (\Delta'_{-p+1})^* : \text{Hom}(A_{-p}, C_q) \to \text{Hom}(A_{-p+1}, C_q)$$

and

$$d''_{p,q} = (-1)^{p+q+1}(\Delta''_q)_* : \text{Hom}(A_{-p}, C_q) \to \text{Hom}(A_{-p}, C_{q-1}).$$

Because interesting maps between complexes usually involve infinitely many indices (see Exercise 11.2), $\text{Tot}(M)$ is not the reasonable complex to associate to M (but see Exercise 11.14). Define

$$\mathbf{Hom}\,(\mathbf{A},\mathbf{C})_n = \prod_{p+q=n} \mathrm{Hom}(A_{-p}, C_q).$$

Thus, $f \in \mathbf{Hom}(\mathbf{A},\mathbf{C})_n$ is a family of maps $f = \{f_{p,q}\}$, where $f_{p,q}: A_{-p} \to C_q$ and $p + q = n$. Observe that a harmless notational change yields

$$\mathbf{Hom}(\mathbf{A},\mathbf{C})_n = \prod_{q-p=n} \mathrm{Hom}(A_p, C_q) = \prod_p \mathrm{Hom}(A_p, C_{p+n}),$$

which is precisely the maps $\mathbf{A} \to \mathbf{C}$ of degree n as usually defined.

Definition: If $(\mathbf{A},\Delta')$ and $(\mathbf{C},\Delta'')$ are complexes, then $\mathbf{Hom}\,(\mathbf{A},\mathbf{C})$ is the complex defined by

$$\mathbf{Hom}(\mathbf{A},\mathbf{C})_n = \prod_{p+q=n} \mathrm{Hom}(A_{-p}, C_q)$$

and

$$D_n : \mathbf{Hom}(\mathbf{A},\mathbf{C})_n \to \mathbf{Hom}(\mathbf{A},\mathbf{C})_{n-1}$$

is given by

$$f = \{f_{p,q}\} \mapsto D_n f = \{g_{p,q}\},$$

where $g_{p,q}: A_{-p} \to C_q$ (with $p + q = n - 1$) is defined by

$$g_{p,q} = (-1)^{p+q} f_{p+1,\,q}\Delta'_{-p} + \Delta''_{q+1} f_{p,\,q+1}.$$

It is a mechanical check that $\mathbf{Hom}(\mathbf{A},\mathbf{C})$ is a complex. Moreover, the definition of D_n is just

$$\begin{aligned}
D_n &= \prod_{p+q=n-1} [(-1)^{p+q}(\Delta'_{-p})^* + (\Delta''_{q+1})_*] \\
&= \prod_{p+q=n-1} [(-1)^{p+q} d'_{p+1,\,q} + (-1)^{-p-(q+1)-1} d''_{p,\,q+1}] \\
&= (-1)^{n-1} \prod_{p+q=n} (d'_{p,q} + d''_{p,q}).
\end{aligned}$$

Exercises: 11.14. If $\mathbf{A}$ is a positive complex (e.g., a deleted projective resolution) and $\mathbf{C}$ is a negative complex (e.g., a deleted injective resolution), then $\mathbf{Hom}(\mathbf{A},\mathbf{C})_n = (\mathrm{Tot}(M))_n$ (where M is the bicomplex with $M_{p,q} = \mathrm{Hom}(A_{-p}, C_q)$) and the differentials agree up to sign. Conclude that, in this case,

$$H_n(\mathbf{Hom}(\mathbf{A},\mathbf{C})) \cong H_n(\mathrm{Tot}(M)).$$

(Hint: For each n, there are only finitely many (p, q) with $p + q = n$.)

11.15.   If **A**, **B**, and **C** are complexes, there is an isomorphism of complexes

$$\mathbf{Hom}\,(\mathbf{A} \otimes \mathbf{B}, \mathbf{C}) \cong \mathbf{Hom}(\mathbf{A}, \mathbf{Hom}(\mathbf{B}, \mathbf{C})).$$

Here is the strategy. Given a bicomplex M, consider the complex $\text{Tot}(M)$. There are two rather natural filtrations of $\text{Tot}(M)$, each of which gives rise to a spectral sequence. If M is either first or third quadrant, then we shall see both spectral sequences converge to $H_*(\text{Tot}(M))$. Moreover, the E^2 terms of these spectral sequences can be computed explicitly, and, in each case, $E_{p,q}^{\infty}$ is $E_{p,q}^{r}$ for large r.

We now give the filtrations $^{\mathrm{I}}F$ and $^{\mathrm{II}}F$ of $\text{Tot}(M)$, where M is a bicomplex. Recall that $\text{Tot}(M)_n = \coprod_{p+q=n} M_{p,q}$.

Definition: The **first filtration** $^{\mathrm{I}}F$ is defined by

$$(^{\mathrm{I}}F^p \, \text{Tot}(M))_n = \coprod_{i \leq p} M_{i, \, n-i}.$$

A picture will make this clear.

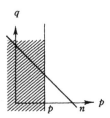

The degree n term of $^{\mathrm{I}}F^p \, \text{Tot}(M)$ consists of the sum of all $M_{p,q}$ with (p,q) on the 45° line $p + q = n$ and to the left of the vertical line. To check that $^{\mathrm{I}}F^p$ is a subcomplex of $\text{Tot}(M)$, recall that $d = \sum (d' + d'')$. Thus,

$$d(M_{i, \, n-i}) = (d' + d'')M_{i, \, n-i} = d'M_{i, \, n-i} + d''M_{i, \, n-i}$$
$$\subset M_{i-1, \, n-i} \oplus M_{i, \, n-i-1} \subset (^{\mathrm{I}}F^p \, \text{Tot}(M))_{n-1},$$

for $i - 1 < i \leq p$.

Definition: The **second filtration** $^{\mathrm{II}}F$ is defined by

$$(^{\mathrm{II}}F^p \, \text{Tot}(M))_n = \coprod_{j \leq p} M_{n-j, \, j}.$$

A picture will make this clear.

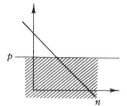

The degree n term of ${}^{II}F^p \operatorname{Tot}(M)$ consists of the sum of all $M_{p,q}$ with (p,q) on the $45°$ line $p + q = n$ and below the horizontal line. One checks, as above, that ${}^{II}F^p$ is a subcomplex of $\operatorname{Tot}(M)$. (Be alert to a possible danger that will be overcome: p signals the first index, but we are using it to restrict the second index.)

Lemma 11.16: (i) *If M is a bicomplex, then both filtrations of $\operatorname{Tot}(M)$ are bounded if and only if each $45°$ line $p + q = n$ meets M in a finite number of points (i.e., there are only finitely many such (p,q) with $M_{p,q} \neq 0$).*

(ii) *If M is either a first or third quadrant bicomplex, then both filtrations of $\operatorname{Tot}(M)$ are bounded.*

Proof: (i) Fix a $45°$ line $p + q = n$. The first filtration is defined by moving a vertical line to the right, and collecting all the terms in back of it. If the $45°$ line meets M in a finite number of points, a vertical line can be drawn to left of all of them (giving $s(n)$) and a vertical line can be drawn to the right of all of them (giving $t(n)$). Similarly, the second filtration is defined by moving a horizontal line upwards, and collecting all the terms below it.

(ii) This is obvious from the first part. However, we can give explicit formulas for $s(n)$ and $t(n)$. If M is first quadrant, then $s(n) = -1$ and $t(n) = n$ works for both filtrations; if M is third quadrant, then $s(-n) = -n - 1$ and $t(-n) = 0$ works for both filtrations. ∎

Theorem 11.17: *Let M be a first or third quadrant bicomplex, and let $\{{}^IE^r\}$ and $\{{}^{II}E^r\}$ be the spectral sequences determined by the first and second filtrations of $\operatorname{Tot}(M)$, respectively.*

(i) *For each p, q, we have ${}^IE_{p,q}^\infty = {}^IE_{p,q}^r$ for large r; similarly for ${}^{II}E_{p,q}^\infty$.*

(ii) *${}^IE_{p,q}^2 \underset{p}{\Rightarrow} H_n(\operatorname{Tot}(M))$ and ${}^{II}E_{p,q}^2 \underset{p}{\Rightarrow} H_n(\operatorname{Tot}(M))$.*

Proof: Lemma 11.16 and Theorem 11.13. ∎

We have created notational monsters, ${}^IE_{p,q}^r$ and ${}^{II}E_{p,q}^r$ (though they could be worse; one corner is still undecorated), but the next two theorems will replace them by something less ugly.

Having two spectral sequences converging to $H_*(\text{Tot}(M))$ is, of course, an advantage. With Theorem 11.17, we may contemplate calculating factor modules of $H_*(\text{Tot}(M))$ as certain subquotients of E^2; obviously, we should at least be able to compute E^2. Let us consider the first filtration (we omit the prescript I). Fix n and recall $(F^p \text{Tot}(M))_n = \coprod_{i \le p} M_{i,\,n-i}$; the differential d is, of course, the restriction of the differential $d = \sum(d' + d'')$ of $\text{Tot}(M)$. Now $(F^p/F^{p-1})_n$ is just $M_{p,q}$, where $p + q = n$, for F_n^{p-1} consists of all

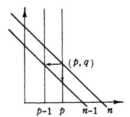

terms on $p + q = n$ to the left of the vertical line at $p - 1$, while F_n^p consists of all such points to the left of the vertical line at p. Also, $d_{p,q}(M_{p,q}) \subset M_{p-1,\,q} \oplus M_{p,\,q-1}$ (since only $d'_{p,q} + d''_{p,q}$ are relevant). However, $M_{p-1,\,q} \subset F^{p-1}$, so that only d'' survives in F^p/F^{p-1}. We conclude that F^p/F^{p-1} is the pth column with differential d''. Therefore

$$E_{p,q} = H_n(F^p/F^{p-1}, d)$$
$$= \ker(d''_{p,q}: M_{p,q} \to M_{p,\,q-1})/\text{im}(d''_{p,\,q+1}: M_{p,\,q+1} \to M_{p,q}) = H_q(M_{p,*}, d'').$$

A better notation for $E_{p,q}$ is $H''_{p,q}(M)$.

For each fixed q, $H''_{p,q}(M)$ is a complex: the differential d' induces maps $\bar{d}': H''_{p,q}(M) \to H''_{p-1,\,q}(M)$ by

$$\bar{d}'_{p,q}: [z_{p,q}] \mapsto [d'_{p,q} z_{p,q}],$$

where $[\ ]$ denotes homology class. It is a routine exercise (left to the reader) that $\bar{d}'$ is well defined. Taking homology of $H''_{p,q}(M)$ with respect to $\bar{d}'$ gives a bigraded module $\{H'_p H''_{p,q}(M)\}$, the **first iterated homology** of M with respect to the first filtration. In short, first take homology of the pth column, then take homology of its rows.

Theorem 11.18:   *If M is a first or third quadrant bicomplex, then*

$$^{1}E^2_{p,q} = H'_p H''_{p,q}(M) \underset{p}{\Rightarrow} H_n(\text{Tot}(M)),$$

where $\{^{1}E^r\}$ is the spectral sequence arising from the first filtration.

Proof: In light of Theorem 11.17, it suffices to compute E^2. We have already verified that $E_{p,q} = H''_{p,q}(M)$. Now invoke Exercise 11.13 which identifies $d^1 = \beta\gamma$ with the connecting homomorphism ∂ arising from the short exact sequence of complexes

$$0 \to F^{p-1}/F^{p-2} \xrightarrow{i} F^p/F^{p-2} \xrightarrow{\pi} F^p/F^{p-1} \to 0$$

(where $F = {}^{\mathrm{I}}F$). Consider this sequence in degrees $n = p + q$ and $n - 1$:

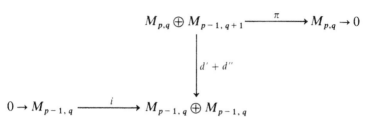

(actually, there should be a third term, $M_{p-2,q+1}$, in $\mathrm{im}(d' + d'')$, but it becomes 0 upon dividing out F^{p-2}). Let $z \in M_{p,q}$ be such that $x = [z] \in H''_{p,q}(M)$; thus $d''z = 0$. Choose $\pi^{-1}z = (z,0)$. The formula $\partial = i^{-1}(d' + d'')\pi^{-1}$ now gives

$$\partial x = [i^{-1} d'z],$$

since $d''z = 0$. This says $d^1 = \partial$ is induced by d', for i is merely an inclusion map whose presence guarantees the element lies in the correct place. Thus $E^2 = H(E, d^1) = H'H''$, the first iterated homology. ∎

Let us evaluate the E^2 term arising from the second filtration of $\mathrm{Tot}(M)$. It is at this point that a little trick enters, necessitated by the "danger alert" mentioned when the second filtration was defined.

Definition: If $M = \{M_{p,q}, d'_{p,q}, d''_{p,q}\}$ is a bicomplex, its **transpose** ${}^{\mathrm{t}}M$ is the bicomplex $\{{}^{\mathrm{t}}M_{p,q}, \Delta'_{p,q}, \Delta''_{p,q}\}$, where

$${}^{\mathrm{t}}M_{p,q} = M_{q,p}, \qquad \Delta'_{p,q} = d''_{q,p}, \qquad \text{and} \qquad \Delta''_{p,q} = d'_{q,p}.$$

It is a simple matter to check that ${}^{\mathrm{t}}M$ is, indeed, a bicomplex; obviously, ${}^{\mathrm{t}}M$ is just M reflected about the line $p = q$. There are two elementary, but important, observations. First, $\mathrm{Tot}({}^{\mathrm{t}}M) = \mathrm{Tot}(M)$ as complexes, i.e., they have the same nth terms and the same differential $d = \sum(d' + d'') = \sum(\Delta'' + \Delta')$. Also, the second filtration of M is the first filtration of ${}^{\mathrm{t}}M$.

Theorem 11.19: *If M is a first or third quadrant bicomplex, then*

$$^{\mathrm{II}}E^2_{p,q} = H''_p H'_{q,p}(M) \underset{p}{\Rightarrow} H_n(\mathrm{Tot}(M)),$$

where $\{{}^{\mathrm{II}}E^r\}$ is the spectral sequence arising from the second filtration.

Proof: Since the second filtration of M is the first filtration of ^1M, Theorem 11.18 gives $H'_p H''_{p,q}(^1M) \underset{p}{\Rightarrow} H_n(\text{Tot}(^1M)) = H_n(\text{Tot}(M))$ (our first observation above).

Now $H''_{p,q}(^1M) = \ker \Delta''_{p,q}/\text{im } \Delta''_{p,\,q+1} = \ker d'_{q,p}/\text{im } d'_{q+1,\,p} = H'_{q,p}(M)$. Finally, $\Delta' = d''$ yields

$$H'_p H''_{p,q}(^1M) = H''_p H'_{q,p}(M). \qquad \blacksquare$$

The **second iterated homology** $H''_p H'_{q,p}(M)$ first takes homology of the pth row, and then takes homology of its columns. It is important that E^2 terms can be computed, so let us review the procedure. First filtration: fix p, so that $H''_{p,q} = H_q(M_{p,*})$; now regard this as a complex and take homology $H'_p H''_{p,q}(M)$. Second filtration: relabel the modules as $M_{q,p}$; fix p, so that $H'_{q,p}(M) = H_q(M_{*,p})$; now compute $H''_p H'_{q,p}(M)$.

Although both spectral sequences converge to $H_*(\text{Tot}(M))$, the corresponding filtrations and factor modules need not be the same, i.e., it is possible that $^1E^\infty_{p,q} \not\cong {}^{II}E^\infty_{p,q}$.

Remark: If M is a third quadrant bicomplex, we have already mentioned that one gets rid of negative indices by raising them: $M_{-p,-q} = M^{p,q}$. The same convention applies to the bigraded modules E^r, except that r does not change sign: thus

$$E^r_{-p,-q} = E^{p,q}_r.$$

The differential $d_r : E_r \to E_r$ now has bidegree $(r, 1 - r)$ and filtrations decrease: $F_p \supset F_{p+1}$ for all p. The reader is requested to restate the definition of convergence $E^{p,q}_r \underset{p}{\Rightarrow} H^n(\text{Tot}(M))$ for the spectral sequences arising from either filtration of $\text{Tot}(M)$ when M is a third quadrant bicomplex. The proof of the next lemma will illustrate this sign change and its notational consequences.

Definition: Let M be a bicomplex; $\{E^r\}$ is a **usual spectral sequence** of M if it is the spectral sequence arising from either the first or second filtration of $\text{Tot}(M)$.

The following simple situation arises often enough to deserve a name.

Definition: A usual spectral sequence $\{E^r\}$ of a first or third quadrant bicomplex **collapses** if $E^2_{p,q} = 0$ for $q \neq 0$.

Thus, collapsing says the bigraded module E^2 can have nonzero terms only on the p-axis.

Lemma 11.20:   *Let $\{E^r\}$ be a usual spectral sequence of a bicomplex M,* *where M is first or third quadrant. If $\{E^r\}$ collapses, then*

$$E_{p,q}^\infty = E_{p,q}^2, \qquad all \quad p,q, \qquad and \qquad H_n(\mathrm{Tot}(M)) \cong E_{n,0}^2.$$

Proof: The differential d^r on E^r has bidegree $(-r, r-1)$, so that the arrow in the p-q plane representing $d_{p,q}^r$ (connecting (p,q) to $(p-r, q+r-1)$) has slope $(r-1)/-r$. Since the only nonzero terms lie on the p-axis, $d^r = 0$ for $r \geq 2$, for every $d_{p,q}^r$ either starts or ends at a 0 module. Therefore, $E^r = \ker d^r = \ker d^r/\mathrm{im}\, d^r = H(E^r) = E^{r+1}$ for all $r \geq 2$, and so $E^2 = E^\infty$.

If M is first quadrant, there is a filtration Φ of $H_n(\mathrm{Tot}(M))$ with

$$0 = \Phi^{-1}H_n \subset \cdots \subset \Phi^n H_n = H_n \qquad and \qquad \Phi^p H_n/\Phi^{p-1}H_n \cong E_{p,q}^\infty = E_{p,q}^2,$$

where $n = p + q$. As there is only one such nonzero term, namely, $E_{n,0}^2$, the filtration Φ has only one nonzero factor module and $H_n(\mathrm{Tot}(M)) = H_n \cong E_{n,0}^2$.

If M is third quadrant, there is a filtration Ψ of $H_{-n}(\mathrm{Tot}(M))$ (Lemma 11.16(ii) gives $s(-n) = -n-1$ and $t(-n) = 0$) with

$$0 = \Psi^{-n-1}H_{-n} \subset \Psi^{-n}H_{-n} \subset \cdots \subset \Psi^{-1}H_{-n} \subset \Psi^0 H_{-n} = H_{-n}$$

and

$$\Psi^p H_{-n}/\Psi^{p-1}H_{-n} \cong E_{p,q}^\infty = E_{p,q}^2.$$

Raise indices and rename the filtration Φ, so that $\Phi^p = \Psi^{-p}$. We now have

$$H^n = \Phi^0 H^n \supset \Phi^1 H^n \supset \cdots \supset \Phi^n H^n \supset \Phi^{n+1}H^n = 0$$

and

$$\Phi^p H^n/\Phi^{p+1}H^n \cong E_{-p,-q}^\infty = E_{-p,-q}^2 = E_2^{p,q}.$$

Again, there is only one nonzero factor, namely, $E_2^{n,0} \cong \Phi^n H^n$; that all previous factors are 0 says $H^n = \Phi^0 H^n = \Phi^1 H^n = \cdots = \Phi^n H^n$, and so $H^n(\mathrm{Tot}(M)) \cong E_2^{n,0}$. ∎

It is necessary to see the index raising in boring detail one time; we shall not be so tiresome again!

Remark: There is a similar result to Lemma 11.20 if a spectral sequence collapses on the q-axis.

Let us use spectral sequences to give another proof of Theorem 7.9: the definition of $\mathrm{Tor}_n^R(A, B)$ is independent of the variable resolved (we shall also obtain a bit more information). What is more natural than to resolve both variables simultaneously?

Theorem 11.21: *Let $\mathbf{P}_A$ and $\mathbf{Q}_B$ be deleted projective resolutions of a right*

R-module A and a left R-module B, respectively. Then

$$H_n(\mathbf{P}_A \otimes B) \cong H_n(\mathbf{P}_A \otimes \mathbf{Q}_B) \cong H_n(A \otimes \mathbf{Q}_B).$$

Proof: Consider the first quadrant bicomplex M of Example 1:

$$\{M_{p,q}\} = \{P_p \otimes Q_q\},$$

so that $\text{Tot}(M) = \mathbf{P}_A \otimes \mathbf{Q}_B$.

The first filtration has E^2 term the iterated homology $H'_p H''_{p,q}(M)$. Now $H''_{p,q}(M) = \ker d''_{p,q}/\text{im } d''_{p,\,q+1}$, and this is just the qth homology of the pth column $P_p \otimes \mathbf{Q}_B$. Since P_p is projective, hence flat,

$$H''_{p,q}(M) = \begin{cases} 0 & \text{if } q > 0, \\ P_p \otimes B & \text{if } q = 0, \end{cases}$$

for $H''_{p,0} = \text{coker}(P_p \otimes Q_1 \to P_p \otimes Q_0) = P_p \otimes B$. It follows that

$$H'_p H''_{p,q}(M) = H_p(\mathbf{P}_A \otimes B)$$

and

$${}^{\text{I}}E^2_{p,q} = \begin{cases} 0 & \text{if } q > 0, \\ H_p(\mathbf{P}_A \otimes B) & \text{if } q = 0; \end{cases}$$

thus, this spectral sequence collapses, and

$$H_n(\mathbf{P}_A \otimes \mathbf{Q}_B) = H_n(\text{Tot}(M)) \cong {}^{\text{I}}E^2_{n,0} \cong H_n(\mathbf{P}_A \otimes B).$$

A similar argument works for the second filtration, now using the fact that each Q_q is projective, hence flat. Thus

$${}^{\text{II}}E^2_{p,q} = \begin{cases} 0 & \text{if } q > 0, \\ H_p(A \otimes \mathbf{Q}_B) & \text{if } q = 0, \end{cases}$$

this spectral sequence also collapses, and

$$H_n(\mathbf{P}_A \otimes \mathbf{Q}_B) \cong H_n(A \otimes \mathbf{Q}_B). \quad \blacksquare$$

We have actually proved more (namely, Corollary 7.10), for we only used projectivity of the modules P_p and Q_q to guarantee they are flat.

Corollary 11.22:   If $\mathbf{P}_A$ and $\mathbf{Q}_B$ are deleted flat resolutions of A_R and $_R B$, respectively, then

$$H_n(\mathbf{P}_A \otimes B) \cong H_n(\mathbf{P}_A \otimes \mathbf{Q}_B) \cong H_n(A \otimes \mathbf{Q}_B).$$

Exercises: 11.16.   Let $\mathbf{P}_A$ be a deleted projective resolution of A, and $\mathbf{E}_B$

be a deleted injective resolution of B. Prove that

$$H^n(\mathrm{Hom}(\mathbf{P}_A, B)) \cong H^n(\mathbf{Hom}(\mathbf{P}_A, \mathbf{E}_B)) \cong H^n(\mathrm{Hom}(A, \mathbf{E}_B))$$

using the third quadrant bicomplex of Example 4 and Exercise 11.14.

11.17.   Let M be a first or third quadrant bicomplex all of whose rows (or columns) are exact. Prove that $\mathrm{Tot}(M)$ is acyclic. (One can prove this without the spectral sequence machine.)

11.18.   Let $G = G' \times G''$ be a direct product of groups. If $\mathbf{P}'$ is a deleted G'-projective resolution of $\mathbf{Z}$, and $\mathbf{P}''$ is a deleted G''-projective resolution of $\mathbf{Z}$, then $\mathbf{P}' \otimes \mathbf{P}''$ is a deleted G-projective resolution of $\mathbf{Z} \otimes \mathbf{Z} = \mathbf{Z}$. (Hint: Use Exercise 5.5: there is a ring isomorphism $\mathbf{Z}G \cong \mathbf{Z}G' \otimes_\mathbf{Z} \mathbf{Z}G''$.)

11.19.   If G is a free abelian group of finite rank r, prove that $\mathrm{cd}(G) \leq r$. (Hint: Use induction on r and Exercise 11.18 to construct a G-projective resolution of $\mathbf{Z}$ of length $\leq r$.)

11.20.   Let $\varphi : R \to R^*$ be a ring map and let A be an R^*-module. Prove

$$\mathrm{pd}_R(A) \leq \mathrm{pd}_{R_*}(A) + \mathrm{pd}_R(R^*).$$

(Hint: Take an R^*-resolution of A, then take an R-resolution of each term, and construct a bicomplex.)

The added information obtained in Theorem 11.21 allows us to complete the discussion of anticommutativity of iterated connecting homomorphisms; we had reached the point, in Theorem 7.26, that a 3×3 commutative diagram of complexes having exact rows and columns yields a certain anti-commutative diagram involving the homology of its four outside corners.

Theorem 11.23:   *Let $0 \to A' \to A \to A'' \to 0$ be an exact sequence of right R-modules, and let $0 \to B' \to B \to B'' \to 0$ be an exact sequence of left R-modules. For each $n \geq 0$, there is an anticommutative diagram*

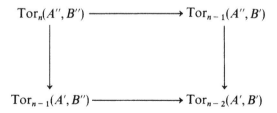

where the maps are connecting homomorphisms.

Proof: By Lemma 6.20, there are deleted projective resolutions $\mathbf{P}'$, $\mathbf{P}$, $\mathbf{P}''$ of A', A, A'', respectively, such that $0 \to \mathbf{P}' \to \mathbf{P} \to \mathbf{P}'' \to 0$ is an exact sequence of complexes. Similarly, there are deleted projective resolutions $\mathbf{Q'}, \mathbf{Q}, \mathbf{Q''}$ of B' B, B'', respectively, such that $0 \to \mathbf{Q'} \to \mathbf{Q} \to \mathbf{Q''} \to 0$ is an exact sequence of complexes. There results a commutative diagram of complexes having exact rows and columns

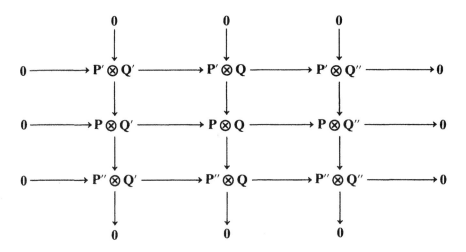

(one has such a diagram in each degree n, for tensor is a bifunctor and each P_n', P_n, P_n'', Q_n', etc., is projective, hence flat). This is exactly the situation to which Theorem 7.26 applies: we obtain the anticommutative diagram with connecting homomorphisms

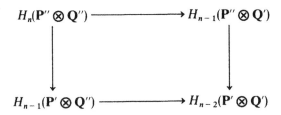

Theorem 11.21 allows us to identify $H_n(\mathbf{P''} \otimes \mathbf{Q''})$ with $\operatorname{Tor}_n(A'', B'')$, etc., and the result follows. ∎

The dual result holds for Ext, using Exercise 11.16 in place of Theorem 11.21.

Theorem 11.24:   *Let* $0 \to A' \to A \to A'' \to 0$ *and* $0 \to C' \to C \to C'' \to 0$ *be exact sequences of left R-modules. For each $n \geq 0$, there is an anticommutative*

diagram

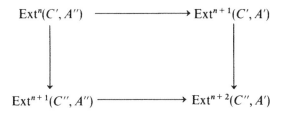

where the maps are connecting homomorphisms.

There is a nice application of this, but it is better if we first present a general lemma.

Lemma 11.25:   (i)   *A commutative diagram of modules*

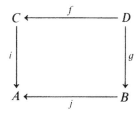

gives rise to a complex $\mathbf{X}$, *where*

$$\mathbf{X} = 0 \to D \xrightarrow{(f, -g)} C \oplus B \xrightarrow{i+j} A \to 0.$$

(ii)   *If i is monic, then*

$$H_1(\mathbf{X}) \cong \ker(\bar{j}: \operatorname{coker} g \to \operatorname{coker} i).$$

(iii)   *If i and j are monic, then*

$$H_1(\mathbf{X}) \cong (\operatorname{im} i \cap \operatorname{im} j)/\operatorname{im} if.$$

Proof: (i)   If we replace g by $-g$, we obtain an anticommutative diagram which may be regarded as a first quadrant bicomplex M (the displayed terms are at positions $(0,0)$, $(0,1)$, $(1,0)$, and $(1,1)$; all other terms are 0), and $\operatorname{Tot}(M)$ is precisely $\mathbf{X}$.

(ii)   Let us compute the iterated homology $H'_p H''_{p,q}(M)$, which may be displayed as a 2×2 matrix. Since i is monic,

$$H''_{p,q}(M) = \begin{bmatrix} 0 & \ker g \\ \operatorname{coker} i & \operatorname{coker} g \end{bmatrix}$$

and

$$E^2_{p,q} = H'_p H''_{p,q}(M) = \begin{bmatrix} 0 & * \\ * & \ker(\bar{j}:\text{coker } g \to \text{coker } i) \end{bmatrix}.$$

Now $E^2_{0,1} = 0$ implies $E^r_{0,1} = 0$ for all $r \geq 2$, since $E^r_{0,1}$ is a subquotient of $E^2_{0,1}$; therefore, $E^\infty_{0,1} = 0$. We know that $E^2_{1,0} = \ker \bar{j}$; what is $E^\infty_{1,0}$? The differential $d^r : E^r \to E^r$ has bidegree $(-r, r-1)$, so that

$$E^{r+1}_{1,0} = \ker d^r_{1,0}/\text{im } d^r_{r+1, 1-r}.$$

If $r \geq 1$, both maps $d^r_{1,0}$ and $d^r_{r+1, 1-r}$ are 0 (the first ends at 0, the second begins at 0, for it is only a question of being outside the little 2×2 square). Therefore $E^2_{1,0} = E^3_{1,0} = \cdots = E^\infty_{1,0}$ (Exercise 11.12).

Further, $E^2_{p,q} \Rightarrow H_n(\mathbf{X})$. In particular, there is a two-step filtration of $H_1(\mathbf{X})$ with factor modules $E^\infty_{0,1} = 0$ and $E^\infty_{1,0} = \ker \bar{j}$. Therefore, $H_1(\mathbf{X}) \cong \ker \bar{j}$.

(iii) Recall that

$$\mathbf{X} = 0 \to D \xrightarrow{(f, -g)} C \oplus B \xrightarrow{i+j} A \to 0,$$

so that

$$H_1(\mathbf{X}) = \ker(i+j)/\text{im}(f, -g).$$

Define $\varphi : \ker(i+j) \to \text{im } i \cap \text{im } j$ by

$$\varphi : (c, b) \mapsto ic.$$

Since $(c, b) \in \ker(i+j)$, $ic + jb = 0$ and $ic = -jb \in \text{im } i \cap \text{im } j$. It is clear that φ is epic and $\text{im}(f, -g) \subset \ker \varphi$. Therefore, φ induces an epimorphism

$$\bar{\varphi} : \ker(i+j)/\text{im}(f, -g) \to (\text{im } i \cap \text{im } j)/\text{im } if$$

(just compose the epimorphism onto $\text{im } i \cap \text{im } j$ with the natural map mod if). A simple calculation, using i and j monic, shows that $\bar{\varphi}$ is monic, hence is an isomorphism. ∎

Theorem 11.26 (Cartan–Eilenberg): *Let A_R and $_RB$ be modules, and let $0 \to K \xrightarrow{\alpha} P \to A \to 0$ and $0 \to L \xrightarrow{\beta} Q \to B \to 0$ be exact, where P and Q are flat. Then*

(i) $\text{Tor}_n(A, B) \cong \text{Tor}_{n-2}(K, L)$ *for $n > 2$;*

(ii) $\text{Tor}_2(A, B) \cong \ker(\alpha \otimes \beta)$;

(iii) $\text{Tor}_1(A, B) \cong [\text{im}(\alpha \otimes 1_Q) \cap \text{im}(1_P \otimes \beta)]/\text{im}(\alpha \otimes \beta)$.

Corollary 11.27: *Let R be a ring with right ideal I and left ideal J. Then*

(i) $\text{Tor}_n(R/I, R/J) \cong \text{Tor}_{n-2}(I, J)$ *for $n > 2$;*

(ii) $\mathrm{Tor}_2(R/I, R/J) \cong \ker(I \otimes J \to IJ)$;

(iii) $\mathrm{Tor}_1(R/I, R/J) \cong (I \cap J)/IJ$.

Proof: This follows immediately from the Theorem applied to the exact sequences $0 \to I \to R \to R/I \to 0$ and $0 \to J \to R \to R/J \to 0$. ∎

As an instance of this, let R be a commutative local ring with maximal ideal $\mathfrak{m}$ and residue field $k = R/\mathfrak{m}$. Then $\mathrm{Tor}_1^R(k, k) \cong \mathfrak{m}/\mathfrak{m}^2$. Of course, one may prove this more simply.

Proof of Theorem 11.26:   (i)   Theorem 11.23 gives an anticommutative diagram of connecting homomorphisms:

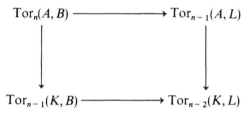

If $n > 2$, these maps are isomorphisms, for all the bordering terms are "honest" Tor's (i.e., not Tor_0) which vanish because a variable is flat. Therefore, for $n > 2$

$$\mathrm{Tor}_n(A, B) \cong \mathrm{Tor}_{n-2}(K, L).$$

(ii)   There is a diagram with exact rows and columns that is almost commutative (the upper left square is only anticommutative):

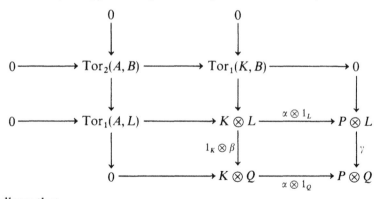

It follows that

$$\mathrm{Tor}_2(A, B) \cong \mathrm{Tor}_1(A, L) \cong \ker(\alpha \otimes 1_L)$$
$$= \ker(\gamma \circ (\alpha \otimes 1_L)), \qquad \text{since } \gamma \text{ is monic,}$$
$$= \ker(\alpha \otimes \beta).$$

(iii)   Apply Lemma 11.25 to the commutative diagram

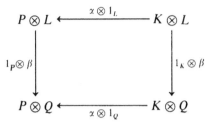

Since P and Q are flat, the maps $i = 1_P \otimes \beta$ and $j = \alpha \otimes 1_Q$ are monic. But the map $\bar{j}: \operatorname{coker}(1_K \otimes \beta) \to \operatorname{coker}(1_P \otimes \beta)$ is just $\alpha \otimes 1_B$, whose kernel is $\operatorname{Tor}_1(A, B)$. The final part of the Lemma gives the desired isomorphism. ∎

Exercise: 11.21.   Let $0 \to K \xrightarrow{\alpha} P \to A \to 0$ and $0 \to B \to E \xrightarrow{\beta} L \to 0$ be exact sequences with P projective and E injective. Then

   (i) $\operatorname{Ext}^n(A, B) \cong \operatorname{Ext}^{n-2}(K, L)$ for $n > 2$;
   (ii) $\operatorname{Ext}^2(A, B) \cong \operatorname{coker} \operatorname{Hom}(\alpha, \beta)$;
   (iii) $\operatorname{Ext}^1(A, B) \cong \ker \operatorname{Hom}(\alpha, \beta)/(\ker \alpha^* + \ker \beta_*)$.

KÜNNETH THEOREMS

The Eilenberg–Zilber Theorem (Example 3) poses the problem: given complexes $\mathbf{A}$ and $\mathbf{C}$ whose homology is known, find $H(\mathbf{A} \otimes \mathbf{C})$. Even in the special case that $\mathbf{C}$ consists of a module C concentrated in degree 0 (i.e., $C_q = C$ if $q = 0$ and $C_q = 0$ otherwise), the problem is not simple. If one further assumes R is a hereditary ring and each term A_p of $\mathbf{A}$ is a projective R-module, then the solution is the Universal Coefficient Theorem 8.22. To indicate why the general problem involving two complexes is difficult, even with the extra assumptions just mentioned, let us see what goes wrong with one's first idea of considering the bicomplex M of Example 2 having $M_{p,q} = A_p \otimes C_q$; after all, $\operatorname{Tot}(M) = \mathbf{A} \otimes \mathbf{C}$. We compute ${}^1E^2_{p,q} = H'_p H''_{p,q}(M)$. The pth column of M is $A_p \otimes \mathbf{C}$; since A_p is projective, hence flat, we know that $H_q(A_p \otimes \mathbf{C}) \cong A_p \otimes H_q(\mathbf{C})$ (Exercise 6.4: homology commutes with exact functors). Hence $H''_{p,q}(M) = A_p \otimes H_q(\mathbf{C})$. The Universal Coefficient Theorem now gives

$$E^2_{p,q} = H_p(\mathbf{A} \otimes H_q(\mathbf{C})) = H_p(\mathbf{A}) \otimes H_q(\mathbf{C}) \oplus \operatorname{Tor}^R_1(H_{p-1}(\mathbf{A}), H_q(\mathbf{C})).$$

Though it is true that $E^2 \Rightarrow H(\mathbf{A} \otimes \mathbf{C})$, we are essentially helpless because we cannot compute E^∞ (the other spectral sequence, arising from the second filtration of $A \otimes C$, is worse: the Universal Coefficient Theorem may not apply, and one may not even compute E^2). Indeed, there seems to be just one meager result that can be salvaged: if A is projective, i.e., each A_p is projective, and if either A or C is acyclic, then $A \otimes C$ is acyclic (either observe that $E^2_{p,q} = 0$, or use the much more elementary Exercise 11.17).

We give two solutions to the problem. The first, strong enough for use of the Eilenberg–Zilber Theorem, does not use spectral sequences (the proof has been deferred to this chapter since it involves bicomplexes and graded modules). The second, more general, solution uses spectral sequences, as well as a special case of the first (Corollary 11.29). It also illustrates a second technique in using spectral sequences. Our first illustration treated a case in which both spectral sequences collapse; this proof is an instance in which only one of them collapses.

Here is a collection of easy exercises which will be used in the forthcoming proof.

Exercises: 11.22. Consider the diagram of modules

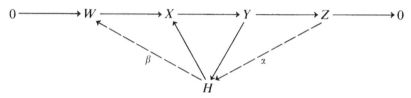

in which the row and the triangle are exact. Prove that there exist unique maps $\alpha: Z \to H$ and $\beta: H \to W$ making the outer triangles commute, and that there is an exact sequence

$$0 \to Z \xrightarrow{\alpha} H \xrightarrow{\beta} W \to 0.$$

The result remains true if one replaces "modules" by "graded modules".

11.23. For fixed $k \geq 0$ and complexes A and C, show there is a bicomplex M with $M_{p,q} = \mathrm{Tor}_k(A_p, C_q)$. (Use the facts that Tor_k is a bifunctor that is additive in each variable, together with the Sign Lemma 11.15.) Denote the complex $\mathrm{Tot}(M)$ by

$$\mathbf{Tor}_k(\mathbf{A}, \mathbf{C}).$$

11.24. If $0 \to A' \to A \to A'' \to 0$ is an exact sequence of complexes, and if C is a complex, then there is an exact sequence of complexes

$$\cdots \to \mathbf{Tor}_1(\mathbf{A}, \mathbf{C}) \to \mathbf{Tor}_1(\mathbf{A}'', \mathbf{C}) \to \mathbf{A}' \otimes \mathbf{C} \to \mathbf{A} \otimes \mathbf{C} \to \mathbf{A}'' \otimes \mathbf{C} \to \mathbf{0}.$$

(Use naturality of connecting homomorphisms and the fact that each Tor_k, $k \geq 0$, is a bifunctor.)

11.25.   If **A** is a complex, then its cycles $\mathbf{Z} = \{Z_n\}$ and boundaries $\mathbf{B} = \{B_n\}$ are subcomplexes having zero differentiation (i.e., each $d_n = 0$).

11.26.   Let $(\mathbf{A}, \Delta)$ and $(\mathbf{C}, d)$ be complexes. If **A** has zero differentiation, then the differentiation in $\mathbf{A} \otimes \mathbf{C}$ is $\{\pm 1 \otimes d_n\}$.

11.27.   Let **A** be a **flat complex** (i.e., A_n is flat for each n) having zero differentiation. Then, for any complex **C**,

$$H_n(\mathbf{A} \otimes \mathbf{C}) = (\mathbf{A} \otimes H_*(\mathbf{C}))_n,$$

where $H_*(\mathbf{C})$ is construed as a complex with zero differentiation.

Theorem 11.28   (Künneth):   *Let **A** be a complex of right R-modules whose subcomplexes of cycles **Z** and boundaries **B** are flat. Given a complex **C**, there is an exact sequence for each n*

$$0 \to \coprod_{p+q=n} H_p(\mathbf{A}) \otimes_R H_q(\mathbf{C}) \xrightarrow{\alpha} H_n(\mathbf{A} \otimes_R \mathbf{C}) \xrightarrow{\beta} \coprod_{p+q=n-1} \mathrm{Tor}_1^R(H_p(\mathbf{A}), H_q(\mathbf{C})) \to 0,$$

where $\alpha : [a] \otimes [c] \mapsto [a \otimes c]$, $a \in Z_p(\mathbf{A})$, $c \in Z_q(\mathbf{C})$. *Moreover, α and β are natural in **A** and **C***.

Proof: (A. Heller):   The usual exact sequences

$$0 \to Z_n \to A_n \to B_{n-1} \to 0$$

give rise to an exact sequence of complexes

$$0 \to \mathbf{Z} \xrightarrow{i} \mathbf{A} \xrightarrow{\pi} \mathbf{B}^+ \to 0,$$

where $(\mathbf{B}^+)_n = B_{n-1}$. Obviously, **B** flat implies $\mathbf{B}^+$ flat, so Exercise 11.24 gives an exact sequence of complexes

$$0 \to \mathbf{Z} \otimes \mathbf{C} \xrightarrow{i \otimes 1} \mathbf{A} \otimes \mathbf{C} \xrightarrow{\pi \otimes 1} \mathbf{B}^+ \otimes \mathbf{C} \to 0.$$

There results an exact triangle of graded modules in which ∂ has degree -1 and the other maps have degree 0:

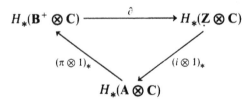

Note that $H_n(\mathbf{B}^+ \otimes \mathbf{C}) = H_{n-1}(\mathbf{B} \otimes \mathbf{C})$. Also, Exercise 11.27 permits us to write

$$H_*(\mathbf{B}^+ \otimes \mathbf{C}) = \mathbf{B}^+ \otimes H_*(\mathbf{C}) \qquad \text{and} \qquad H_*(\mathbf{Z} \otimes \mathbf{C}) = \mathbf{Z} \otimes H_*(\mathbf{C}),$$

where $H_*(C)$ is construed as a complex having zero differentiation. Rewriting with these two observations in mind gives an exact triangle of graded modules

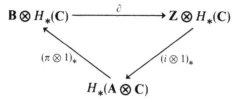

where we now regard $(\pi \otimes 1)_*$ as having degree -1 and ∂ as having degree 0.

The next claim is that the connecting homomorphism ∂ is the map induced by the inclusion $\mathbf{B} \hookrightarrow \mathbf{Z}$. Let Δ be the differentiation in $\mathbf{A}$, let d be the differentiation in $\mathbf{C}$, and consider the defining diagram for the connecting homomorphism (having replaced $\mathbf{B}^+$ by $\mathbf{B}$)

$$
\begin{array}{ccc}
(\mathbf{A} \otimes \mathbf{C})_n & \xrightarrow{\ \Delta \otimes 1\ } & (\mathbf{B} \otimes \mathbf{C})_{n-1} \\
\Big\downarrow{\scriptstyle D} & & \\
(\mathbf{Z} \otimes \mathbf{C})_{n-1} & \xrightarrow[\ j\ ]{} & (\mathbf{A} \otimes \mathbf{C})_{n-1}
\end{array}
$$

Ignoring needed summation, a typical cycle in $(\mathbf{B} \otimes \mathbf{C})_{n-1}$ has the form $b \otimes c$, where $b \in B_{p-1}$, $c \in C_q$, $p + q = n$. If $\Delta a = b$, then

$$
\begin{aligned}
\partial(b \otimes c) &= j^{-1}D(a \otimes c) \\
&= j^{-1}(\Delta a \otimes c + (-1)^p a \otimes dc) \\
&= j^{-1}(b \otimes c + (-1)^p(1 \otimes d)(a \otimes c)).
\end{aligned}
$$

But j^{-1} tells us to regard $b \otimes c + (-1)^p(1 \otimes d)(a \otimes c)$ in $(\mathbf{Z} \otimes \mathbf{C})_{n-1}$, and Exercise 11.26 tells us this element is homologous to $b \otimes c$. The claim is established.

Finally, consider the exact sequence of complexes

$$0 \to \mathbf{B} \to \mathbf{Z} \to H_*(\mathbf{A}) \to 0,$$

where $H_*(\mathbf{A})$ is regarded as a complex with zero differentiation. Since $\mathbf{Z}$ is flat, by hypothesis, there is an exact sequence of graded modules to which we affix our exact triangle (the maps $\mathbf{B} \otimes H_*(\mathbf{C}) \to \mathbf{Z} \otimes H_*(\mathbf{C})$ being the same):

$$0 \to \operatorname{Tor}_1^R(H_*(\mathbf{A}), H_*(\mathbf{C})) \longrightarrow \mathbf{B} \otimes H_*(\mathbf{C}) \longrightarrow \mathbf{Z} \otimes H_*(\mathbf{C}) \longrightarrow H_*(\mathbf{A}) \otimes_R H_*(\mathbf{C}) \to 0$$

$$(\pi \otimes 1)_* \searrow \qquad \swarrow$$

$$H_*(\mathbf{A} \otimes \mathbf{C})$$

Exercise 11.22 applies at once to give an exact sequence of graded modules

$$0 \to H_*(A) \otimes H_*(C) \xrightarrow{\alpha} H_*(A \otimes C) \xrightarrow{\beta} \mathrm{Tor}_1^R(H_*(A), H_*(C)) \to 0$$

in which α has degree 0 and β has degree -1 (our substitution of $\mathbf{B}$ for $\mathbf{B}^+$ forces us to regard $(\pi \otimes 1)_*$ as having degree -1). Translating to ordinary modules, for each n there is an exact sequence

$$0 \to \coprod_{p+q=n} H_p(A) \otimes_R H_q(C) \xrightarrow{\alpha} H_n(A \otimes_R C) \xrightarrow{\beta} \coprod_{p+q=n-1} \mathrm{Tor}_1^R(H_p(A), H_q(C)) \to 0.$$

We let the reader verify the formula for α and the naturality of α and β, using his solution of Exercise 11.22. ∎

Corollary 11.29:   *Let $\mathbf{A}$ be a complex such that $\mathbf{Z} = \mathbf{Z}(\mathbf{A})$ and $H_*(\mathbf{A})$ are projective. For all n, and every complex $\mathbf{C}$,*

$$\coprod_{p+q=n} H_p(A) \otimes H_q(C) \cong H_n(A \otimes C),$$

the isomorphism being the map α of Lemma 11.28.

Proof: Since $H_n(\mathbf{A})$ is projective, the exact sequence

$$0 \to B_n \to Z_n \to H_n(A) \to 0$$

splits. Thus B_n is projective, being a summand of Z_n, and so the complexes $\mathbf{Z}$ and $\mathbf{B}$ are projective, hence flat. The hypothesis of Theorem 11.28 holds, and the Tor term in its conclusion vanishes because $H_p(\mathbf{A})$ is projective. ∎

The next result will be used in proving that the Künneth exact sequence splits when R is hereditary.

Lemma 11.30:   *If R is hereditary and $(\mathbf{A}, \Delta)$ is a complex of R-modules, then there exists a projective complex $\mathbf{P}$ and a chain map $f : \mathbf{P} \to \mathbf{A}$ such that $f_* : H_n(\mathbf{P}) \to H_n(\mathbf{A})$ is an isomorphism for every n.*

Proof: Fix n, and write $Z_n \subset A_n$ as a quotient of a projective V, say, $\varphi : V \to Z_n$. Define $W = \varphi^{-1}(B_n) \subset V$, and consider the diagram

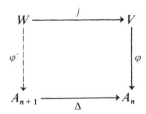

where j is inclusion. Since R is hereditary, Corollary 4.18 shows W is projective, and there thus exists φ' making the diagram commute. Regard the top row as a complex $\mathbf{X}^{(n)}$, where

$$(\mathbf{X}^{(n)})_p = \begin{cases} V & \text{if} \quad p = n, \\ W & \text{if} \quad p = n + 1, \\ 0 & \text{otherwise.} \end{cases}$$

The maps (φ', φ) define a chain map $f^{(n)} : \mathbf{X}^{(n)} \to \mathbf{A}$; moreover, $H_p(\mathbf{X}^{(n)}) = 0$ if $p \neq n$, while

$$H_n(\mathbf{X}^{(n)}) = \operatorname{coker} j = V/W \cong Z_n/B_n = H_n(\mathbf{A}).$$

The lemma is proved by defining $\mathbf{P} = \coprod_n \mathbf{X}^{(n)}$ and $f = \coprod_n f^{(n)}$, and applying Exercise 6.7. ∎

Theorem 11.31 (Künneth formula): *Let R be a hereditary ring, and let $\mathbf{A}$ and $\mathbf{C}$ be complexes with $\mathbf{A}$ flat. There are natural exact sequences*

$$0 \to \coprod_{p+q=n} H_p(\mathbf{A}) \otimes H_q(\mathbf{C}) \xrightarrow{\alpha} H_n(\mathbf{A} \otimes \mathbf{C}) \to \coprod_{p+q=n-1} \operatorname{Tor}_1^R(H_p(\mathbf{A}), H_q(\mathbf{C})) \to 0$$

that split.

Proof: Since R is hereditary (hence semihereditary), Theorem 9.25 says that every submodule of a flat module is flat. The hypothesis of Theorem 11.28 holds, and we have the desired exact sequence.

To prove the sequence splits, let us first consider the special case $\mathbf{A}$ and $\mathbf{C}$ projective. For each p, the exact sequence

$$0 \to Z_p \to A_p \to B_{p-1} \to 0$$

splits, for B_{p-1} is a submodule of the projective module A_{p-1}, hence is projective (Corollary 4.18). Therefore

$$A_p = Z_p \oplus Y_p, \qquad Y_p \cong B_{p-1}.$$

The natural map $Z_p \to H_p(\mathbf{A})$ defined by $a \mapsto [a]$ thus extends (by annihilating Y_p) to a map $\varphi_p : A_p \to H_p(\mathbf{A})$, so $\varphi_p(a) = [a]$ for every cycle $a \in Z_p$. There are similar maps $\psi_q : C_q \to H_q(\mathbf{C})$. These maps induce

$$\varphi \otimes \psi : (\mathbf{A} \otimes \mathbf{C})_n \to \coprod_{p+q=n} H_p(\mathbf{A}) \otimes H_q(\mathbf{C});$$

furthermore, $\varphi \otimes \psi$ annihilates boundaries because φ_p and ψ_q do. We conclude that $\varphi \otimes \psi$ induces a map

$$(\varphi \otimes \psi)_* : H_n(\mathbf{A} \otimes \mathbf{C}) \to \coprod_{p+q=n} H_p(\mathbf{A}) \otimes H_q(\mathbf{C})$$

by

$$[a_p \otimes c_q] \mapsto \varphi_p a_p \otimes \psi_q c_q = [a_p] \otimes [c_q].$$

If $u \in Z_p(\mathbf{A})$ and $v \in Z_q(\mathbf{C})$, then

$$(\varphi \otimes \psi)_* \alpha([u] \otimes [v]) = (\varphi \otimes \psi)_*[u \otimes v] = \varphi_p u \otimes \psi_q v = [u] \otimes [v].$$

Therefore $(\varphi \otimes \psi)_* \alpha = 1$ and the sequence splits.

For the general case, Lemma 11.30 provides projective complexes $\mathbf{P}$ and $\mathbf{Q}$ and chain maps $f : \mathbf{P} \to \mathbf{A}$ and $g : \mathbf{Q} \to \mathbf{C}$ such that

$$f_* : H_p(\mathbf{P}) \xrightarrow{\sim} H_p(\mathbf{A}) \qquad \text{and} \qquad g_* : H_q(\mathbf{Q}) \xrightarrow{\sim} H_q(\mathbf{C})$$

for all p, q. Naturality of α and β gives a commutative diagram with exact rows

$$
\begin{array}{ccccccccc}
0 \to & \coprod H(\mathbf{P}) \otimes H(\mathbf{Q}) & \xrightarrow{\alpha} & H(\mathbf{P} \otimes \mathbf{Q}) & \xrightarrow{\beta} & \coprod \mathrm{Tor}(H(\mathbf{P}), H(\mathbf{Q})) & \longrightarrow & 0 \\
& \downarrow{\scriptstyle f_* \otimes g_*} & & \downarrow{\scriptstyle (f \otimes g)_*} & & \downarrow{\scriptstyle \mathrm{Tor}(f_*, g_*)} & & \\
0 \to & \coprod H(\mathbf{A}) \otimes H(\mathbf{C}) & \xrightarrow{\alpha} & H(\mathbf{A} \otimes \mathbf{C}) & \xrightarrow{\beta} & \coprod \mathrm{Tor}(H(\mathbf{A}), H(\mathbf{C})) & \longrightarrow & 0
\end{array}
$$

Since f_* and g_* are isomorphisms, the outer maps are isomorphisms and hence, by the Five Lemma, the middle map is, too. Splitting of the top sequence now implies splitting of the bottom. ∎

Remarks: 1. If X and Y are topological spaces, their singular complexes $\mathbf{S}(X)$ and $\mathbf{S}(Y)$ are free complexes over $\mathbf{Z}$, so the Künneth formula gives an expression for $H_*(X \times Y)$ in terms of $H_*(X)$ and $H_*(Y)$. In particular, if X and Y are compact triangulable subspaces of euclidean space, each of their homology groups is f.g., and so one may compute the homology groups of $X \times Y$ explicitly (for $G \otimes_{\mathbf{Z}} H$ and $\mathrm{Tor}_1^{\mathbf{Z}}(G, H)$ are computable when G and H are f.g.).

2. Theorem 11.31 generalizes the Universal Coefficient Theorem 8.22 by assuming one complex is concentrated in degree 0. Notice also that one only needs the complex flat, and not projective as in Theorem 8.22.

3. The splitting of the exact sequence need not be natural.

4. Although all the preceding lemmas dualize, we only state the dual of the last result.

Theorem 11.32 (Künneth formula): *Let R be a hereditary ring, and let $\mathbf{A}$*

and **C** *be complexes with* **A** *projective. Then there is a natural exact sequence*

$$0 \to \prod_{q-p=n+1} \mathrm{Ext}_R^1(H_p(\mathbf{A}), H_q(\mathbf{C})) \to H^n(\mathrm{Hom}(\mathbf{A}, \mathbf{C}))$$

$$\to \prod_{q-p=n} \mathrm{Hom}(H_p(\mathbf{A}), H_q(\mathbf{C})) \to 0$$

that splits.

Exercises: 11.28. Assume R is commutative hereditary and A, B, C are R-modules. Writing Tor for Tor_1^R, prove that

$$\mathrm{Tor}(A, B) \otimes C \oplus \mathrm{Tor}(A \otimes B, C) \cong A \otimes \mathrm{Tor}(B, C) \oplus \mathrm{Tor}(A, B \otimes C)$$

and

$$\mathrm{Tor}(\mathrm{Tor}(A, B), C) \cong \mathrm{Tor}(A, \mathrm{Tor}(B, C)).$$

(Hint: Use the Künneth formula 11.31 and the associative law for tensor product applied to deleted projective resolutions of A, B, C, respectively.)

11.29. Assume R is commutative hereditary and A, B, C are R-modules. Writing Ext for Ext_R^1 and Tor for Tor_1^R, prove that

$$\mathrm{Hom}(\mathrm{Tor}(A, B), C) \oplus \mathrm{Ext}(A \otimes B, C) \cong \mathrm{Hom}(A, \mathrm{Ext}(B, C))$$

$$\oplus \mathrm{Ext}(A, \mathrm{Hom}(B, C))$$

and

$$\mathrm{Ext}(\mathrm{Tor}(A, B), C) \cong \mathrm{Ext}(A, \mathrm{Ext}(B, C)).$$

(Hint: Use the Kunneth Formula 11.32 and the adjoint isomorphism, Exercise 11.15, applied to suitable deleted resolutions.)

11.30. Let G be an abelian group with $\mathrm{Hom}(\mathbf{Q}, G) = 0$. Such a group is called **cotorsion** if $G \cong \mathrm{Ext}_\mathbf{Z}^1(A, B)$ for some abelian groups A, B. Prove that G is cotorsion if and only if $\mathrm{Ext}_\mathbf{Z}^1(\mathbf{Q}, G) = 0$. (Hint: For necessity, use Exercise 11.29; for sufficiency, show $G \cong \mathrm{Ext}_\mathbf{Z}^1(\mathbf{Q}/\mathbf{Z}, G)$.)

We seek a generalization of the Künneth theorem, but we have already seen that the "obvious" bicomplex M having $M_{p,q} = A_p \otimes C_q$ is too complicated. A constant theme in homological algebra is to replace a module by a resolution of it. It is thus quite natural to replace a complex by a resolution of it, and then use this resolution as an ingredient of a bicomplex.

Definition: A **projective resolution** of a complex **C** is an exact sequence of complexes

$$0 \leftarrow \mathbf{C} \leftarrow \mathbf{M}_0 \xleftarrow{d''} \mathbf{M}_1 \xleftarrow{d''} \mathbf{M}_2 \leftarrow \cdots$$

in which each complex $\mathbf{M}_q$ is projective.

A projective resolution of $\mathbf{C}$ is thus a large commutative diagram of projective modules $M_{p,q}$ whose rows $\mathbf{M}_q$ are complexes and whose columns are exact:

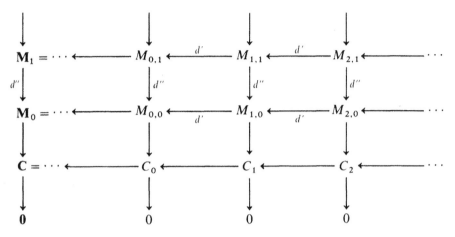

If we delete $\mathbf{C}$ and adjust signs via the Sign Lemma 11.15, then $M = \{M_{p,q}, d', d''\}$ is a bicomplex.

We want to make the bicomplex M as nice as possible. Let $\mathbf{Z}'_q$, $\mathbf{B}'_q$, $\mathbf{H}'_q$ denote the complexes (with zero differentiation) consisting of the cycles, boundaries, and homology, respectively, of the qth row $\mathbf{M}_q$ (the single prime reminds one of rows). Each of these determines a bicomplex. For example,

$$0 \leftarrow \mathbf{Z}(\mathbf{C}) \leftarrow \mathbf{Z}'_0 \leftarrow \mathbf{Z}'_1 \leftarrow \mathbf{Z}'_2 \leftarrow \cdots$$

has the following picture:

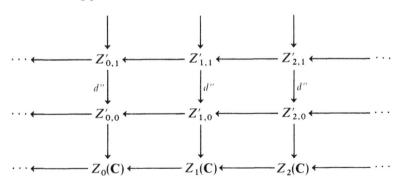

The vertical maps d'' are just restrictions of the vertical maps in M. It is foolish to worry about the horizontal maps, for they are all zero. Thus, it is only the columns

$$Z_p(\mathbf{C}) \leftarrow Z'_{p,0} \leftarrow Z'_{p,1} \leftarrow Z'_{p,2} \leftarrow \cdots$$

that are of interest.

Definition: A **proper projective resolution** of a complex $\mathbf{C}$ is a projective resolution

$$0 \leftarrow \mathbf{C} \leftarrow \mathbf{M}_0 \leftarrow \mathbf{M}_1 \leftarrow \mathbf{M}_2 \leftarrow \cdots$$

whose corresponding bicomplex M has the following property: for each p,

    (i) $0 \leftarrow C_p \leftarrow M_{p,0} \leftarrow M_{p,1} \leftarrow \cdots$,
   (ii) $0 \leftarrow Z_p(\mathbf{C}) \leftarrow Z'_{p,0} \leftarrow Z'_{p,1} \leftarrow \cdots$,
  (iii) $0 \leftarrow B_p(\mathbf{C}) \leftarrow B'_{p,0} \leftarrow B'_{p,1} \leftarrow \cdots$,

and

  (iv) $0 \leftarrow H_p(\mathbf{C}) \leftarrow H'_{p,0} \leftarrow H'_{p,1} \leftarrow \cdots$

are projective resolutions.

Of course, condition (i) is part of the definition of projective resolution of $\mathbf{C}$.

Lemma 11.33: *Every complex $\mathbf{C}$ has a proper projective resolution.*

Proof: As usual, the complex $\mathbf{C}$ determines two families of short exact sequences:

$$0 \leftarrow H_p \leftarrow Z_p \leftarrow B_p \leftarrow 0 \qquad \text{and} \qquad 0 \leftarrow B_{p-1} \leftarrow C_p \leftarrow Z_p \leftarrow 0.$$

For each p, choose projective resolutions $H'_{p,*}$ and $B'_{p,*}$ of $H_p(\mathbf{C})$ and $B_p(\mathbf{C})$, respectively. The Horseshoe Lemma 6.20 provides a projective resolution $Z'_{p,*}$ of $Z_p(\mathbf{C})$ that fits in the middle. Now use the second short exact sequences, these resolutions of $B_{p-1}(\mathbf{C})$ and $Z_p(\mathbf{C})$, and the Horseshoe Lemma to obtain projective resolutions $M_{p,*}$ of C_p that fit in the middle. Defining $M_{p,q} \to M_{p-1,q}$ as the composite

$$M_{p,q} \to B'_{p-1,q} \to Z'_{p-1,q} \to M_{p-1,q},$$

we have constructed a proper projective resolution of $\mathbf{C}$. ∎

Of course, there is a dual notion of proper injective resolutions, and the dual proof shows they always exist.

Theorem 11.34 (Künneth Spectral Sequence): *Let $\mathbf{A}$ and $\mathbf{C}$ be positive complexes with $\mathbf{A}$ flat. There is a first quadrant spectral sequence*

$$E^2_{p,q} = \coprod_{s+t=q} \mathrm{Tor}_p(H_s(\mathbf{A}), H_t(\mathbf{C})) \underset{p}{\Rightarrow} H_n(\mathbf{A} \otimes \mathbf{C}).$$

Proof: Take a proper projective resolution of **C** with corresponding bicomplex M (which is first quadrant since **C** is positive). Form a new bicomplex K with

$$K_{p,q} = \coprod_{s+t=p} A_s \otimes M_{q,t}$$

and obvious differentiations: d'' is built from the differentiation in **A** and the vertical differentiation in M; d' is built from the horizontal differentiation in M.

Consider the first filtration: $H''_{p,q}(K)$ is the homology of the pth column. For fixed s, t with $s + t = p$, the pth column is

$$\cdots \to A_s \otimes M_{2,t} \to A_s \otimes M_{1,t} \to A_s \otimes M_{0,t} \to 0.$$

Now this is just $A_s \otimes$ applied to

$$\cdots \to M_{2,t} \to M_{1,t} \to M_{0,t} \to 0,$$

a deleted projective resolution of C_t. Since A_s is flat, the pth column is a deleted flat resolution of $A_s \otimes C_t$, and so $H''_{p,q}(K) = \coprod_{s+t=p} \mathrm{Tor}_q(A_s, C_t)$. Using flatness of A_s once again, we see all such terms vanish for $q > 0$. We conclude that

$${}^{1}E^2_{p,q} = H'_p H''_{p,q}(K) = \begin{cases} H_p\left(\coprod_{s+t=p} A_s \otimes C_t \right) & \text{if } q = 0, \\ 0 & \text{if } q > 0. \end{cases}$$

The first spectral sequence thus collapses, and Lemma 11.20 gives

$$H_n(\mathrm{Tot}(K)) = {}^{1}E^2_{n,0} = H_n(\mathbf{A} \otimes \mathbf{C}).$$

Since the second spectral sequence also converges to $H_n(\mathrm{Tot}(K))$, it is only necessary to show its E^2 term $H''_p H'_{q,p}(K)$ is as stated in the Theorem. For fixed p (noting the transposing of p and q), we must first consider the homology of the complex with qth term $\coprod_{s+t=q} A_s \otimes M_{p,t}$; this complex is just $\mathrm{Tot}(\mathbf{A} \otimes \mathbf{M}_p)$. Since M arose from a proper projective resolution of **C**, the cycles and homology of $\mathbf{M}_p$ are projective; Corollary 11.29 applies to give

$$H'_{q,p}(K) = H_q(\mathrm{Tot}(\mathbf{A} \otimes \mathbf{M}_p)) = \coprod_{s+t=q} H_s(\mathbf{A}) \otimes H_t(\mathbf{M}_p).$$

We also know that

$$\cdots \to H_t(\mathbf{M}_2) \to H_t(\mathbf{M}_1) \to H_t(\mathbf{M}_0) \to H_t(\mathbf{C}) \to 0$$

is a projective resolution. The homology of this complex tensored by $H_s(\mathbf{A})$

is precisely the definition of Tor:

$$^{II}E^2_{p,q} = H'_p H''_{q,p}(K) = H_p\left(\coprod_{s+t=q} H_s(\mathbf{A}) \otimes H_t(\mathbf{M}_p)\right)$$

$$= \coprod_{s+t=q} H_p(H_s(\mathbf{A}) \otimes H_t(\mathbf{M}_p))$$

$$= \coprod_{s+t=q} \mathrm{Tor}_p(H_s(\mathbf{A}), H_t(\mathbf{C})). \qquad \blacksquare$$

We merely state the dual result for cohomology.

Theorem 11.34   (Künneth Spectral Sequence):   *Let* $\mathbf{A}$ *be a positive complex and* $\mathbf{C}$ *a negative complex. If either* $\mathbf{A}$ *is projective or* $\mathbf{C}$ *is injective, there is a third quadrant spectral sequence*

$$E^{p,q}_2 = \coprod_{s+t=q} \mathrm{Ext}^p(H_s(\mathbf{A}), H_t(\mathbf{C})) \Rightarrow H^n(\mathbf{Hom}(\mathbf{A}, \mathbf{C})).$$

We give a second proof of the Künneth Theorem 11.28 using spectral sequences. As the proof will yield an E^2 consisting of only two columns, we first dispose of this situation.

Lemma 11.35:   *Let* M *be a first quadrant bicomplex for which* E^2 ($=^1E^2$) *consists of two columns: there is* $t > 0$ *with* $E^2_{p,q} = 0$ *for* $p \neq 0$ *and* $p \neq t$. *Then*

(i) $E^2 = E^3 = \cdots = E^t$;

(ii) $E^{t+1} = E^\infty$;

(iii)   *there are exact sequences*

$$0 \to E^{t+1}_{0,n} \to H_n(\mathrm{Tot}(M)) \to E^{t+1}_{t, n-t} \to 0;$$

(iv)   *there is an exact sequence*

$$\cdots \to H_{n+1}(\mathrm{Tot}(M)) \to E^2_{t, n+1-t} \xrightarrow{d^t} E^2_{0,n} \to H_n(\mathrm{Tot}(M)) \to E^2_{t, n-t} \xrightarrow{d^t} E^2_{0, n-1} \to \cdots.$$

Proof: The differential $d^r: E^r \to E^r$ has bidegree $(-r, r-1)$: d^r goes r steps left and $r-1$ steps up. It follows that $d^r = 0$ for $r \neq t$, since it either begins or ends at 0. Exercise 11.12 now applies to prove (i) and (ii).

Consider the picture of the E^2 plane below.

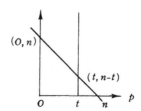

Since M is first quadrant, we know $E^2_{p,q} \underset{p}{\Rightarrow} H_n(\mathrm{Tot}(M))$, i.e.,

$$0 = \Phi^{-1}H_n \subset \Phi^0 H_n \subset \cdots \subset \Phi^n H_n = H_n \qquad \text{and} \qquad \Phi^p H_n / \Phi^{p-1} H_n \cong E^\infty_{p,q}.$$

Thus

$$\Phi^0 H_n = E^\infty_{0,n} \subset H_n.$$

Also, $\Phi^1 H_n / \Phi^0 H_n = E^\infty_{1,n-1} = 0$, so that $\Phi^0 H_n = \Phi^1 H_n$. Indeed, $\Phi^0 H_n = \Phi^1 H_n = \cdots = \Phi^{t-1} H_n$. When we reach t,

$$\Phi^t H_n / \Phi^{t-1} H_n \cong E^\infty_{t,n-t},$$

and then all remains constant again: $\Phi^t H_n = \Phi^{t+1} H_n = \cdots = \Phi^n H_n = H_n$. We conclude there is an exact sequence

$$0 \to E^\infty_{0,n} \to H_n \to E^\infty_{t,n-t} \to 0.$$

In light of (ii), this may be rewritten as

$$(*) \qquad\qquad 0 \to E^{t+1}_{0,n} \to H_n \to E^{t+1}_{t,n-t} \to 0,$$

which is item (iii).

There is an exact sequence obtained from $d^t = d^t_{t,\,n+1-t}$:

$$0 \to \ker d^t \to E^t_{t,\,n+1-t} \xrightarrow{d^t} E^t_{0,n} \to \mathrm{coker}\, d^t \to 0.$$

By definition,

$$E^{t+1}_{t,\,n+1-t} = \ker d^t_{t,\,n+1-t} / \mathrm{im}\, d^t_{2t,\,n+2-2t} = \ker d^t_{t,\,n+1-t}$$

and

$$E^{t+1}_{0,n} = \ker d^t_{0,n} / \mathrm{im}\, d^t_{t,\,n+1-t} = E^t_{0,n} / \mathrm{im}\, d^t_{t,\,n+1-t} = \mathrm{coker}\, d^t_{t,\,n+1-t}.$$

The sequence may thus be rewritten

$$(**) \qquad\qquad 0 \to E^{t+1}_{t,\,n+1-t} \to E^t_{t,\,n+1-t} \xrightarrow{d^t} E^t_{0,n} \to E^{t+1}_{0,n} \to 0.$$

Splicing $(*)$ and $(**)$ together and remembering that $E^t = E^2$ yields the desired exact sequence

$$\cdots \to H_{n+1} \to E^2_{t,\,n+1-t} \xrightarrow{d^t} E^2_{0,n} \to H_n \to E^2_{t,\,n-t} \to \cdots \qquad \blacksquare$$

Observe that it is possible that the terms E^r of a spectral sequence may remain constant for a while before moving. Thus, it is not true in general that $E^r = E^{r+1}$ implies $E^r = E^\infty$.

Exercise: 11.31.   Let M be a first quadrant bicomplex for which E^2 consists of two rows: $E^2_{p,q} = 0$ for $q \neq 0$ and $q \neq t$. Show there is an exact sequence

$$\cdots \to H_{n+1}(\text{Tot}(M)) \to E_{n,0}^2 \xrightarrow{d^{t+1}} E_{n-t-1,\,t}^2 \to H_n(\text{Tot}(M)) \to \cdots .$$

(Hint: d^r vanishes unless $r = t + 1$.)

We really need only a special case of this lemma.

Lemma 11.36: *Let M be a first quadrant bicomplex for which E^2 ($={}^{I}E^2$) consists of two adjacent columns: $E_{p,q}^2 = 0$ for $p \neq 0, 1$. Then there are exact sequences*

$$0 \to E_{0,n}^2 \to H_n(\text{Tot}(M)) \to E_{1,\,n-1}^2 \to 0.$$

Remark: If M is third quadrant, arrows are reversed, and there is an exact sequence under the adjacent column hypothesis:

$$0 \to E_2^{1,\,n-1} \to H^n(\text{Tot}(M)) \to E_2^{0,n} \to 0.$$

Here is the fancy proof of the Künneth theorem when **A** and **C** are positive complexes.

Theorem 11.37 (=Theorem 11.28): *Let **A** be a positive complex whose subcomplexes **Z** and **B** are flat. Given a positive complex **C**, there is an exact sequence for each n*

$$0 \to \coprod_{p+q=n} H_p(\mathbf{A}) \otimes H_q(\mathbf{C}) \to H_n(\mathbf{A} \otimes \mathbf{C}) \to \coprod_{p+q=n-1} \text{Tor}_1(H_p(\mathbf{A}), H_q(\mathbf{C})) \to 0.$$

Proof: First, let us show **A** is flat. For each p, exactness of $0 \to Z_p \to A_p \to B_{p-1} \to 0$ gives exactness, for any module X, of

$$\text{Tor}_1(Z_p, X) \to \text{Tor}_1(A_p, X) \to \text{Tor}_1(B_{p-1}, X).$$

The outer terms vanish because Z_p and B_{p-1} are flat; hence $\text{Tor}_1(A_p, X) = 0$ for every X, and so A_p is flat.

Since **A** and **C** are positive, Theorem 11.34 applies to give

$$E_{p,q}^2 = \coprod_{s+t=q} \text{Tor}_p(H_s(\mathbf{A}), H_t(\mathbf{C})) \underset{p}{\Rightarrow} H_n(\mathbf{A} \otimes \mathbf{C}).$$

Now exactness of $0 \to B_s \to Z_s \to H_s(\mathbf{A}) \to 0$ gives exactness of

$$\text{Tor}_p(Z_s, X) \to \text{Tor}_p(H_s(\mathbf{A}), X) \to \text{Tor}_{p-1}(B_s, X).$$

If $p - 1 \geq 1$, flatness of Z_s and B_s force the middle term to vanish:

$$\text{Tor}_p(H_s(\mathbf{A}), X) = 0 \qquad \text{for} \quad p \neq 0, 1.$$

Thus $E^2_{p,q} = 0$ for $p \neq 0, 1$, which is precisely the adjacent column hypothesis of Lemma 11.36. There are thus exact sequences

$$0 \to E^2_{0,n} \to H_n(\mathbf{A} \otimes \mathbf{C}) \to E^2_{1,n-1} \to 0,$$

and this is what is sought, once one replaces the E^2 terms by their values.   ∎

GROTHENDIECK SPECTRAL SEQUENCES

The main theorem of this section shows that a composite of functors often leads to a spectral sequence. For a change, we will prove the third quadrant (cohomology) version and merely state the dual first quadrant (homology) version. As usual, all functors are between module categories.

Theorem 11.38   (Grothendieck):   *Let* $G:\mathfrak{A} \to \mathfrak{B}$ *and* $F:\mathfrak{B} \to \mathfrak{C}$ *be functors with F left exact such that E injective in* $\mathfrak{A}$ *implies GE is right F-acyclic. For each module A in* $\mathfrak{A}$*, there exists a third quadrant spectral sequence with*

$$E^{p,q}_2 = R^p F(R^q G(A)) \underset{p}{\Rightarrow} R^n(FG)(A).$$

Proof: Choose an injective resolution $0 \to A \to E^0 \to E^1 \to \cdots$, and apply G to its deletion to obtain a complex

$$GE_A = 0 \to GE^0 \to GE^1 \to GE^2 \to \cdots.$$

By the dual of Lemma 11.33, there is a proper injective resolution of GE_A. The corresponding bicomplex M may be pictured, after indices are raised, as a diagram of injective modules $M^{p,q}$:

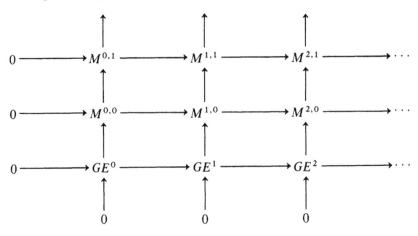

Consider the bicomplex FM and its total complex $\text{Tot}(FM)$. It is very easy to compute the spectral sequence arising from the first filtration. For fixed p, the pth column $M^{p,*}$ is a deleted injective resolution of GE^p. Therefore, $FM^{p,*}$ is a complex whose homology is R^qF: we have

$$H''^{p,q}(FM) = H^q(FM^{p,*}) = (R^qF)(GE^p).$$

But E^p injective implies GE^p is right F-acyclic, so $(R^qF)(GE^p) = 0$ for $q > 0$. Also, F left exact implies $R^0F = F$. We conclude that

$$H''^{p,q}(FM) = \begin{cases} FG(E^p) & \text{if} \quad q = 0, \\ 0 & \text{if} \quad q > 0. \end{cases}$$

All that survives is on the p-axis:

$$0 \to FG(E^0) \to FG(E^1) \to FG(E^2) \to \cdots.$$

Applying H'^p, we see that

$$E_2^{p,q} = \begin{cases} R^p(FG)(A) & \text{if} \quad q = 0, \\ 0 & \text{if} \quad q > 0. \end{cases}$$

Therefore, the first spectral sequence collapses, and we have

$$H^n(\text{Tot}(FM)) \cong R^n(FG)(A).$$

Now consider the spectral sequence arising from the second filtration. Let us recall that our choice of a proper injective resolution of GE_A means the following. If the qth row is $\mathbf{M}^q$, and $\mathbf{Z}'^q$, $\mathbf{B}'^q$, and $\mathbf{H}'^q$ denote the complexes with zero differentiation of cocycles, coboundaries, and cohomology, respectively, then for each p there are injective resolutions

$$0 \to Z^p(GE_A) \to Z'^{p,0} \to Z'^{p,1} \to \cdots,$$
$$0 \to B^p(GE_A) \to B'^{p,0} \to B'^{p,1} \to \cdots,$$

and

$$0 \to H^p(GE_A) \to H'^{p,0} \to H'^{p,1} \to \cdots.$$

For every p, q, we transpose indices and consider the usual exact sequence

$$0 \to Z'^{q,p} \to M^{q,p} \xrightarrow{d'} B'^{q+1,p} \to 0;$$

it splits because all terms are injective. The sequence thus remains split after applying F, and

$$\ker(Fd'^{q,p}) = FZ'^{q,p} \quad \text{and} \quad \text{im}(Fd'^{q-1,p}) = FB'^{q,p}.$$

Therefore,

$$H'^{q,p}(FM) = H^q(FM^p) = \ker(Fd'^{q,p})/\text{im}(Fd'^{q-1,p}) = FZ'^{q,p}/FB'^{q,p}.$$

Since

$$0 \to B'^{q,p} \to Z'^{q,p} \to H'^{q,p} \to 0$$

is also split exact, $FZ'^{q,p}/FB'^{q,p} = FH'^{q,p}$. We conclude that

$$H'^{q,p}(FM) = FH'^{q,p}.$$

Now $^{II}E_2^{p,q} = H''^p H'^{q,p}(FM) = H^p(FH'^{q,p})$. But the modules $H'^{q,p}$ form an injective resolution of $H^q(GE_A) = (R^qG)(A)$; by definition of right derived functor, $H^p(FH'^{q,p}) = R^pF(R^qG(A))$. Hence

$$(R^pF)(R^qG(A)) \underset{p}{\Rightarrow} R^n(FG)(A). \quad \blacksquare$$

Here are three more versions; the proofs are the same.

Theorem 11.39:   *Let* $G:\mathfrak{A} \to \mathfrak{B}$ *and* $F:\mathfrak{B} \to \mathfrak{C}$ *be functors with F right exact and P projective in* $\mathfrak{A}$ *implies GP left F-acyclic. For each module A in* $\mathfrak{A}$, *there is a first quadrant spectral sequence*

$$E_{p,q}^2 = L_pF(L_qG(A)) \underset{p}{\Rightarrow} L_n(FG)(A).$$

Theorem 11.40:   *Let* $F:\mathfrak{B} \to \mathfrak{C}$ *be a contravariant left exact functor, and let* $G:\mathfrak{A} \to \mathfrak{B}$ *be a functor such that P projective in* $\mathfrak{A}$ *implies GP right F-acyclic. For each module A in* $\mathfrak{A}$, *there is a third quadrant spectral sequence*

$$E_2^{p,q} = R^pF(L_qG(A)) \underset{p}{\Rightarrow} R^n(FG)(A).$$

Theorem 11.41:   *Let* $F:\mathfrak{B} \to \mathfrak{C}$ *be a contravariant left exact functor, and let* $G:\mathfrak{A} \to \mathfrak{B}$ *be a contravariant functor such that P projective in* $\mathfrak{A}$ *implies GP right F-acyclic. For each module A in* $\mathfrak{A}$, *there is a first quadrant spectral sequence*

$$E_{p,q}^2 = R^pF(R^qG(A)) \underset{p}{\Rightarrow} L_n(FG)A.$$

There are other Grothendieck spectral sequences for other combinations of variances whose statements are left to the reader.

Before giving applications of the Grothendieck spectral sequence, let us give the standard proof of the five-term exact sequence. We could have given the proof some time ago, but the advantage of waiting is that we may now give another proof of Theorem 11.2.

Theorem 11.42:   *Assume* $E_{p,q}^2 \underset{p}{\Rightarrow} H_n(M)$, *where* $\{E^r\}$ *is a first quadrant spectral sequence.*

(i)   *For each n, there is an epimorphism $E_{0,n}^2 \to E_{0,n}^\infty$;*
(ii) *for each n, there is a monomorphism $E_{n,0}^\infty \to E_{n,0}^2$;*
(iii) *for each n, there is a monomorphism $E_{0,n}^\infty \to H_n(M)$;*
(iv) *for each n, there is an epimorphism $H_n(M) \to E_{n,0}^\infty$;*
(v)   *there is an exact sequence*

$$H_2(M) \to E_{2,0}^2 \xrightarrow{d^2} E_{0,1}^2 \to H_1(M) \to E_{1,0}^2 \to 0.$$

Remark: The maps in (i) and (ii) are called **edge homomorphisms**.

Proof: (i)   By definition, $E_{0,n}^3 = \ker d_{0,n}^2 / \operatorname{im} d_{2,n-1}^2$. But $\ker d_{0,n}^2 : E_{0,n}^2 \to E_{-2,n+1}^2$ is the zero map (since $E_{p,q}^2$ is first quadrant). Therefore $\ker d_{0,n}^2 = E_{0,n}^2$ and there is an epimorphism $E_{0,n}^2 \to E_{0,n}^3$. The argument can be repeated for each $d^r : E^r \to E^r$, using only the fact that d^r has bidegree $(-r, r-1)$. There is thus a chain of epimorphisms $E_{0,n}^2 \to E_{0,n}^3 \to E_{0,n}^4 \to \cdots \to E_{0,n}^r$. This completes the argument, for we know $E_{0,n}^\infty = E_{0,n}^r$ for large r.
(ii)   This argument is the dual of that just given: $E_{n,0}^3 \hookrightarrow E_{n,0}^2$ because $E_{n,0}^3 = \ker d_{n,0}^2 / \operatorname{im} d_{n+2,-1}^2 = \ker d_{n,0}^2 \subset E_{n,0}^2$. Now iterate, noting that all $\operatorname{im} d^r = 0$.
(iii)   The definition of convergence yields

$$E_{0,n}^\infty = \Phi^0 H_n / \Phi^{-1} H_n = \Phi^0 H_n \subset H_n.$$

(iv)   The definition of convergence yields

$$E_{n,0}^\infty = \Phi^n H_n / \Phi^{n-1} H_n = H_n / \Phi^{n-1} H_n.$$

(v)   There is an exact sequence

$$0 \to \ker d_{2,0}^2 \to E_{2,0}^2 \xrightarrow{d^2} E_{0,1}^2 \to \operatorname{coker} d_{2,0}^2 \to 0.$$

Now $\ker d_{2,0}^2 = \ker d_{2,0}^2 / \operatorname{im} d_{4,-1}^2 = E_{2,0}^3$; iterating this argument gives $\ker d_{2,0}^2 = E_{2,0}^r$ for large r, and hence is $E_{2,0}^\infty$. Dually,

$$\operatorname{coker} d_{2,0}^2 = E_{0,1}^2 / \operatorname{im} d_{2,0}^2 = \ker d_{0,1}^2 / \operatorname{im} d_{2,0}^2 = E_{0,1}^3;$$

iteration gives $\operatorname{coker} d_{2,0}^2 = E_{0,1}^\infty$. The exact sequence is thus

$$0 \to E_{2,0}^\infty \to E_{2,0}^2 \xrightarrow{d^2} E_{0,1}^2 \to E_{0,1}^\infty \to 0.$$

The epimorphism $H_2(M) \to E_{2,0}^\infty$ of (iv) gives exactness of

$$H_2(M) \to E_{2,0}^2 \xrightarrow{d^2} E_{0,1}^2 \to E_{0,1}^\infty \to 0.$$

Finally, the equations

$$H_1 = \Phi^1 H_1, \qquad E_{1,0}^\infty = \Phi^1 H_1 / \Phi^0 H_1,$$

As every finite l-group has a nontrivial center, there is a (necessarily normal) subgroup N of order l with $N \subset Z(\Pi)$. Consider the corresponding Lyndon–Hochschild–Serre spectral sequence with

$$E_{p,q}^2 = H_p(\Pi/N, H_q(N, \mathbf{Z})) \underset{p}{\Rightarrow} H_n(\Pi, \mathbf{Z}).$$

There is thus a filtration

$$0 = \Phi^{-1} H_2 \subset \Phi^0 H_2 \subset \Phi^1 H_2 \subset \Phi^2 H_2 = H_2$$

with

$$\Phi^p H_2 / \Phi^{p-1} H_2 \cong E_{p,q}^\infty, \qquad p + q = 2.$$

Writing $|X|$ for card(X) when X is a set,

$$|H_2(\Pi, \mathbf{Z})| = |E_{2,0}^\infty| |E_{1,1}^\infty| |E_{0,2}^\infty| \leq |E_{2,0}^2| |E_{1,1}^2| |E_{0,2}^2|$$

because $E_{p,q}^\infty$ is a subquotient of $E_{p,q}^2$.

In computing $E_{p,q}^2$, we use the remark that $H_q(N, \mathbf{Z})$ is Π/N-trivial. Now $E_{2,0}^2 = H_2(\Pi/N, H_0(N, \mathbf{Z})) \cong H_2(\Pi/N, \mathbf{Z})$, so induction gives $|E_{2,0}^2| \leq l^{(n-1)(n-2)/2}$. The term $E_{1,1}^2 = H_1(\Pi/N, H_1(N, \mathbf{Z})) \cong H_1(\Pi/N, \mathbf{Z}/l\mathbf{Z})$, for $H_1(N, \mathbf{Z}) \cong N/N' = N$. Exercise 10.5 applies: if we write G for Π/N, then $H_1(G, \mathbf{Z}/l\mathbf{Z}) \cong G/G'G^l$, a quotient of G. It follows that $|E_{1,1}^2| \leq |\Pi/N| = l^{n-1}$. Finally, $E_{0,2}^2 = H_0(\Pi/N, H_2(N, \mathbf{Z})) = 0$, for $H_2(N, \mathbf{Z}) = 0$ (case $n = 1$); hence $|E_{0,2}^2| = 1$. We conclude that

$$|H_2(\Pi, \mathbf{Z})| \leq l^{(n-1)(n-2)/2} \cdot l^{n-1} = l^{n(n-1)/2}. \qquad \blacksquare$$

The next result shows Green's inequality is best possible. Remember our labors with the 4-group, when $l = 2$ and $n = 2$.

Theorem 11.48:   *Let l be a prime and let Π be an elementary abelian group of order l^n (i.e., Π is the direct product of n cyclic groups of order l). Then $H_2(\Pi, \mathbf{Z})$ is elementary abelian of order $l^{n(n-1)/2}$.*

Proof: First of all, let us show $l H_2(\Pi, \mathbf{Z}) = 0$. The group Π has the presentation

$$\Pi = (x_1, \ldots, x_n \,|\, x_1^l, \ldots, x_n^l, [x_i, x_j], i < j).$$

Let F be free with basis $\{x_1, \ldots, x_n\}$, and let R be the normal subgroup of F generated by the relators. We proceed as we did when computing the multiplier of the 4-group. Define $K = F/[R, F]$, so that $K' = F'/[R, F] \cong H_2(\Pi, \mathbf{Z})$ (Hopf's formula, for $F' \subset R$). Define $a_i = x_i[R, F]$ and observe (as in the 4-group discussion) that

$$K' = \langle [a_i, a_j], \quad i < j \rangle,$$
$$[a_i, a_j] \in Z(K), \quad \text{and} \quad a_j^l \in Z(K).$$

Hence

$$a_j^l = a_i a_j^l a_i^{-1} = (a_i a_j a_i^{-1})^l = ([a_i, a_j] a_j)^l = [a_i, a_j]^l a_j^l.$$

Therefore $[a_i, a_j]^l = 1$ for all $i < j$, as desired.

The theorem is proved by induction on n, the case $n = 1$ being covered by the previous theorem. Choose a subgroup N of Π with Π/N cyclic of order l (this choice of N differs from that of the previous theorem). The corresponding Lyndon–Hochschild–Serre spectral sequence satisfies

$$E_{p,q}^2 = H_p(\Pi/N, H_q(N, \mathbf{Z})) \underset{p}{\Rightarrow} H_n(\Pi, \mathbf{Z}).$$

Since Π/N is cyclic, $E_{2,0}^2 = 0$ and hence $E_{2,0}^\infty = 0$; the filtration of $H_2(\Pi, \mathbf{Z})$ thus has only two steps. The usual bidegree argument shows $E_{0,2}^\infty = E_{0,2}^2$ and $E_{1,1}^\infty = E_{1,1}^2$. There is thus an exact sequence

$$0 \to E_{0,2}^2 \to H_2(\Pi, \mathbf{Z}) \to E_{1,1}^2 \to 0.$$

As $lH_2(\Pi, \mathbf{Z}) = 0$, the middle term $H_2(\Pi, \mathbf{Z})$ is a vector space over $\mathbf{Z}/l\mathbf{Z}$; thus the outside terms are also vector spaces and the sequence splits. It only remains to compute dimensions. Now

$$H_1(N, \mathbf{Z}) \cong N/N' = N = \coprod_{n-1} \mathbf{Z}/l\mathbf{Z},$$

so Exercise 10.5 gives

$$E_{1,1}^2 = H_1(\Pi/N, H_1(N, \mathbf{Z})) \cong \coprod_{n-1} H_1(\Pi/N, \mathbf{Z}/l\mathbf{Z}) \cong \coprod_{n-1} \mathbf{Z}/l\mathbf{Z},$$

for N is elementary abelian of order l^{n-1}. The term

$$E_{0,2}^2 = H_0(\Pi/N, H_2(N, \mathbf{Z})) \cong H_2(N, \mathbf{Z});$$

by induction, $E_{0,2}^2$ is elementary abelian of order $l^{(n-1)(n-2)/2}$. Therefore $H_2(\Pi, \mathbf{Z})$ has dimension $(n-1) + \frac{1}{2}(n-1)(n-2) = \frac{1}{2}n(n-1)$ and order $l^{n(n-1)/2}$. ∎

Theorem 11.49: *If Π is free abelian of finite rank k, then $H^n(\Pi, \mathbf{Z})$ is free abelian of rank $\binom{k}{n}$, the binomial coefficient. (If $n > k$, we interpret $\binom{k}{n}$ as 0.)*

Proof: First of all, Exercise 11.19 shows $H^n(\Pi, \mathbf{Z}) = 0$ for $n > k$. We proceed by induction on k.

If $k = 1$, then $\Pi \cong \mathbf{Z}$. Now $H^0(\Pi, \mathbf{Z}) = \mathbf{Z}^\Pi = \mathbf{Z}$ and $H^1(\Pi, \mathbf{Z}) = \text{Hom}(\Pi, \mathbf{Z}) \cong \mathbf{Z}$, so the induction begins.

For the inductive step, choose a subgroup N with $\Pi/N \cong \mathbf{Z}$ (N is necessarily free abelian of rank $k - 1$). The Lyndon–Hochschild–Serre spectral sequence satisfies

$$E_2^{p,q} = H^p(\Pi/N, H^q(N, \mathbf{Z})) \underset{p}{\Rightarrow} H^n(\Pi, \mathbf{Z}).$$

Since $\Pi/N \cong \mathbf{Z}$, we know $E_2^{p,q} = 0$ for $p \neq 0, 1$. Therefore, the "two adjacent column" hypothesis of Lemma 11.36 (rather, the third quadrant version of that Lemma) holds, so there are exact sequences

(∗) $0 \to E_2^{1,\,n-1} \to H^n(\Pi, \mathbf{Z}) \to E_2^{0,n} \to 0.$

Now $E_2^{1,\,n-1} = H^1(\Pi/N, H^{n-1}(N, \mathbf{Z}))$. By the remark before Theorem 11.5, $H^{n-1}(N, \mathbf{Z})$ is Π/N-trivial, and so

$$H^1(\Pi/N, H^{n-1}(N, \mathbf{Z})) \cong \operatorname{Hom}(\mathbf{Z}, H^{n-1}(N, \mathbf{Z})) \cong H^{n-1}(N, \mathbf{Z});$$

by induction, this group is free abelian of rank $\binom{k-1}{n-1}$. The term $E_2^{0,n} = H^0(\Pi/N, H^n(N, \mathbf{Z})) \cong H^n(N, \mathbf{Z})$, using the remark once again. By induction, this group is free abelian of rank $\binom{k}{n-1}$. It follows that the exact sequence (∗) of abelian groups splits, whence $H^n(\Pi, \mathbf{Z})$ is free abelian of rank $\binom{k-1}{n-1} + \binom{k-1}{n} = \binom{k}{n}$. ∎

Corollary 11.50: *If Π is free abelian of finite rank k, then*

$$\operatorname{cd}(\Pi) = k.$$

Proof: Exercise 11.19 shows $\operatorname{cd}(\Pi) \leq k$, while Theorem 11.49 shows this inequality cannot be strict. ∎

Lyndon [1948] computes $H^n(\Pi, \mathbf{Z})$ for any f.g. abelian group Π.

MORE MODULES

We now examine composites of suitable pairs of functors and the corresponding Grothendieck spectral sequences. Let R and S be rings, and consider the situation $(A_R, {}_R B_S, {}_S C)$; the associative law for tensor product gives a natural isomorphism

$$A \otimes_R (B \otimes_S C) \cong (A \otimes_R B) \otimes_S C.$$

Thus, if $G = B \otimes_S : {}_S\mathfrak{M} \to {}_R\mathfrak{M}$ and $F = A \otimes_R : {}_R\mathfrak{M} \to \mathbf{Ab}$, then $FG = (A \otimes_R B) \otimes_S$. Clearly F is right exact, so Theorem 11.39 applies when there is acyclicity.

Theorem 11.51: *Assume A_R and ${}_R B_S$ satisfy $\operatorname{Tor}_i^R(A, B \otimes_S P) = 0$, all $i > 0$, whenever ${}_S P$ is projective. Then there is a first quadrant spectral sequence for every ${}_S C$:*

$$\operatorname{Tor}_p^R(A, \operatorname{Tor}_q^S(B, C)) \underset{p}{\Rightarrow} \operatorname{Tor}_n^S(A \otimes_R B, C).$$

In this case, let us write down the five-term exact sequence, so the reader may see it is not something he would have invented otherwise:

$$\operatorname{Tor}_2^S(A \otimes_R B, C) \to \operatorname{Tor}_2^R(A, B \otimes_S C) \to A \otimes_R \operatorname{Tor}_1^S(B, C)$$
$$\to \operatorname{Tor}_1^S(A \otimes_R B, C) \to \operatorname{Tor}_1^R(A, B \otimes_S C) \to 0.$$

We can do something more interesting. Assume A_R is flat (which ensures the hypothesis of Theorem 11.51). The spectral sequence collapses on the q-axis and Lemma 11.20 gives

$$A \otimes_R \operatorname{Tor}_n^S(B, C) \cong \operatorname{Tor}_n^S(A \otimes_R B, C),$$

which is the isomorphism of Theorem 9.48.

Observe that there is a second way to look at the associative law. If $\Gamma = \otimes_R B : \mathfrak{M}_R \to \mathfrak{M}_S$ and $\Phi = \otimes_S C : \mathfrak{M}_S \to \mathbf{Ab}$, then $\Phi\Gamma = \otimes_R (B \otimes_S C)$. Here is the corresponding Grothendieck spectral sequence.

Theorem 11.52: *Assume $_RB_S$ and $_SC$ satisfy $\operatorname{Tor}_i^S(Q \otimes_R B, C) = 0$, all $i > 0$, whenever Q_R is projective. Then there is a first quadrant spectral sequence for each A_R:*

$$\operatorname{Tor}_p^S(\operatorname{Tor}_q^R(A, B), C) \underset{p}{\Rightarrow} \operatorname{Tor}_n^R(A, B \otimes_S C).$$

There is a second way to force these spectral sequences to collapse.

Theorem 11.53: *If $_RB_S$ is flat on either side, then*

$$\operatorname{Tor}_n^S(A \otimes_R B, C) \cong \operatorname{Tor}_n^R(A, B \otimes_S C).$$

Proof: If $_RB$ is flat and Q_R is projective (hence flat), then $Q \otimes_R B$ is a flat right S-module, and the hypothesis of Theorem 11.52 holds. Moreover, the spectral sequence collapses and we obtain the desired isomorphism. If B_S is flat, a similar argument works using Theorem 11.51. ∎

The same idea applies for other "associativity" laws. For example, consider the adjoint isomorphism in the situation $(_RA, _SB_R, _SC)$:

$$\operatorname{Hom}_S(B \otimes_R A, C) \cong \operatorname{Hom}_S(A, \operatorname{Hom}_R(B, C)).$$

If $G = B \otimes_S$ and $F = \operatorname{Hom}_R(, C)$, then $FG = \operatorname{Hom}_S(, \operatorname{Hom}_R(B, C))$. Obviously F is left exact, being a Hom, and Theorem 11.40 applies.

Theorem 11.54: *Assume $_SB_R$ and $_SC$ satisfy $\operatorname{Ext}_S^i(B \otimes_R P, C) = 0$, all $i > 0$, whenever $_RP$ is projective. Then there is a third quadrant spectral sequence for every $_RA$:*

$$\operatorname{Ext}_R^p(\operatorname{Tor}_q^S(B, A), C) \underset{p}{\Rightarrow} \operatorname{Ext}_S^n(A, \operatorname{Hom}_R(B, C)).$$

Looking at the adjoint isomorphism another way gives $\Gamma = \operatorname{Hom}_R(B, \ )$, $\Phi = \operatorname{Hom}_S(A, \ )$, and $\Phi\Gamma = \operatorname{Hom}_R(B \otimes_R A, \ )$.

Theorem 11.55:   *Assume $_RA$ and $_SB_R$ satisfy $\operatorname{Ext}_S^i(A, \operatorname{Hom}_R(B, Q)) = 0$, all $i > 0$, whenever $_SQ$ is injective. Then there is a third quadrant spectral sequence for every $_SC$:*

$$\operatorname{Ext}_S^p(A, \operatorname{Ext}_R^q(B, C)) \underset{p}{\Rightarrow} \operatorname{Ext}_R^n(B \otimes_S A, C).$$

Theorem 11.56:   *If $_SB_R$ is projective on either side, then*

$$\operatorname{Ext}_S^n(A, \operatorname{Hom}_R(B, C)) \cong \operatorname{Ext}_S^n(B \otimes_R A, C).$$

Proof: If B is projective on either side, then the hypothesis of either Theorem 11.54 or 11.55 holds, and the corresponding spectral sequence collapses.   ∎

This is the isomorphism of Exercise 9.23. If, instead, one assumes $_RA$ projective, he obtains the isomorphism of Exercise 9.22:

$$\operatorname{Ext}_S^n(B \otimes_R A, C) \cong \operatorname{Hom}_R(A, \operatorname{Ext}_S^n(B, C)).$$

Another isomorphism results if one chooses $_SC$ injective. The hypothesis of Theorem 11.54 holds and the spectral sequence collapses (on the q-axis): there are isomorphisms

$$\operatorname{Hom}_R(\operatorname{Tor}_n^S(B, A), C) \cong \operatorname{Ext}_S^n(A, \operatorname{Hom}_R(B, C)).$$

Theorem 11.57 ($=$ Theorem 9.51):   *Let R and S be rings with R left noetherian, and consider the situation $(_RA, \ _RB_S, C_S)$ in which A is f.g. and C is injective. Then there are isomorphisms for all $n \geq 0$*

$$\operatorname{Hom}_S(\operatorname{Ext}_R^n(A, B), C) \cong \operatorname{Tor}_n^R(\operatorname{Hom}_S(B, C), A).$$

Proof: Since R is left noetherian, A f.g. implies A finitely related. The hypotheses of Lemma 3.60 are satisfied, and there is thus a natural isomorphism

$$\operatorname{Hom}_S(B, C) \otimes_R A \cong \operatorname{Hom}_S(\operatorname{Hom}_R(A, B), C).$$

Let $G = \operatorname{Hom}_R(\ , B)$ and $F = \operatorname{Hom}_S(\ , C)$, so that $FG = \operatorname{Hom}_S(B, C) \otimes_R$. Now $G: \mathfrak{A} \to {}_S\mathfrak{M}$, where $\mathfrak{A}$ is the category of all f.g. left R-modules (this is the reason for insisting R be left noetherian: otherwise, there might not exist a projective resolution of A within $\mathfrak{A}$, and we would not be able to apply the appropriate Grothendieck spectral sequence). If P is a f.g. R-projective, then $\operatorname{Ext}_S^i(\operatorname{Hom}_R(P, B), C) = 0$ for $i > 0$ because C is injective. Having verified acyclicity, Theorem 11.41 gives

$$\operatorname{Ext}^p_S(\operatorname{Ext}^q_R(A, B), C) \underset{p}{\Rightarrow} \operatorname{Tor}^R_n(\operatorname{Hom}_S(B, C), A).$$

Since C is injective, this spectral sequence collapses on the q-axis, yielding isomorphisms

$$\operatorname{Hom}_S(\operatorname{Ext}^n_R(A, B), C) \cong \operatorname{Tor}^R_n(\operatorname{Hom}_S(B, C), A). \quad \blacksquare$$

There is a variation of this technique which may be applied when the composite functor is neither Hom nor tensor (in which case its derived functors may be unknown). We illustrate the technique in the next proof; the result generalizes Theorem 9.50.

Theorem 11.58: *Let R be a commutative noetherian ring with subset S, and let N be finitely generated. Then for every R-module M, there are isomorphisms*

$$S^{-1} \operatorname{Ext}^n_R(N, M) \cong \operatorname{Ext}^n_{S^{-1}R}(S^{-1}N, S^{-1}M).$$

Proof: The hypothesis allows us to assert Theorem 3.84: for every R-module M, there is a natural isomorphism

$$S^{-1} \operatorname{Hom}_R(N, M) \cong \operatorname{Hom}_{S^{-1}R}(S^{-1}N, S^{-1}M).$$

Let $G = \operatorname{Hom}_R(N, \ )$ and $F = S^{-1} = S^{-1}R \otimes_R$. Now F is left exact (even exact) and the acyclicity condition holds: if E is injective, then

$$\operatorname{Tor}^R_i(S^{-1}R, \operatorname{Hom}_R(N, E)) = 0$$

for $i > 0$ (for $S^{-1}R$ is flat). There is thus a spectral sequence

$$\operatorname{Tor}^R_p(S^{-1}R, \operatorname{Ext}^q_R(N, M)) \underset{p}{\Rightarrow} R^n(FG)M.$$

Now let $\Gamma = S^{-1}$ and $\Phi = \operatorname{Hom}_{S^{-1}R}(S^{-1}N, \ )$. Clearly Φ is left exact and the acyclicity condition holds: if E is R-injective, then

$$\operatorname{Ext}^i_{S^{-1}R}(S^{-1}N, S^{-1}E) = 0$$

for $i > 0$ (for $S^{-1}E$ is $S^{-1}R$-injective, by Theorem 3.85). There is thus a spectral sequence

$$\operatorname{Ext}^p_{S^{-1}R}(S^{-1}N, \operatorname{Tor}^R_q(S^{-1}R, M)) \underset{p}{\Rightarrow} R^n(\Phi\Gamma)M.$$

Since $S^{-1}R$ is R-flat, both spectral sequences collapse, yielding isomorphisms:

$$S^{-1}R \otimes_R \operatorname{Ext}^n_R(N, M) \cong R^n(FG)M;$$
$$\operatorname{Ext}^n_{S^{-1}R}(S^{-1}N, S^{-1}R \otimes_R M) \cong R^n(\Phi\Gamma)M.$$

But Theorem 3.84 asserts the natural equivalence of FG and $\Phi\Gamma$, hence the natural equivalence of their derived functors: $R^n(FG)M \cong R^n(\Phi\Gamma)M$. By definition, $S^{-1} = S^{-1}R \otimes_R$, so

$$S^{-1}\mathrm{Ext}^n_R(N, M) \cong \mathrm{Ext}^n_{S^{-1}R}(S^{-1}N, S^{-1}M). \qquad \blacksquare$$

Our final aim is a discussion of "change of rings". Recall that a ring map

$$\varphi : R \to T$$

defines an exact functor $U : {}_T\mathfrak{M} \to {}_R\mathfrak{M}$ (in essence, every T-module and T-map may be viewed as an R-module and R-map, respectively). If M is a T-module and $r \in R$, define

$$r \cdot m = \varphi(r) \cdot m, \qquad m \in M.$$

At our pleasure, M may be regarded as either a T-module or an R-module (and we shall not provide different notations).

The ring T plays a special role: it is an $(R\text{-}T)$ bimodule, for the associative law in T gives

$$r \cdot (t_1 t_2) = \varphi(r)(t_1 t_2) = (\varphi(r)t_1)t_2 = (r \cdot t_1)t_2.$$

Similarly, one may equip T with a right R-module structure, namely, $t \cdot r = t\varphi(r)$, and T becomes a $(T\text{-}R)$ bimodule.

Since a T-module may also be regarded as an R-module, one may ask about the relation between "homological properties" of a module over T and over R.

Although it is not necessary to be fancy, let us put things in proper context.

Lemma 11.59:   *Let $\varphi : R \to T$ be a ring map, and let $U : {}_T\mathfrak{M} \to {}_R\mathfrak{M}$ be the corresponding change of rings functor. Then $(U, \mathrm{Hom}_R(T, \ ))$ and $(T \otimes_R \ , U)$ are adjoint pairs.*

Proof: A routine exercise. $\blacksquare$

There is also a change of rings functor $\mathfrak{M}_T \to \mathfrak{M}_R$, with a similar lemma.

Nature thus tells us to look at $\mathrm{Hom}_R(T, \ )$ and $T \otimes_R$; of course, we have already had reason to deal a bit with these functors. Actually, there are more cases to consider because of the left-right possibilities due to the noncommutativity of the rings R and T. Let us assume $A = A_R$ and $B = {}_RB$. Denote

$$_\varphi B = T \otimes_R B \in {}_T\mathfrak{M}, \qquad T = {}_T T_R$$

and

$$A_\varphi = A \otimes_R T \in \mathfrak{M}_T, \qquad T = {}_R T_T.$$

There are also left and right modules resulting from $\operatorname{Hom}_R(T, \ )$:

$$B^\varphi = \operatorname{Hom}_R(T, B) \in \mathfrak{M}_T, \qquad T = {}_RT_T,$$

and

$$\,^\varphi A = \operatorname{Hom}_R(T, A) \in {}_T\mathfrak{M}, \qquad T = {}_TT_R.$$

In short, the symbol φ is a subscript tensor, is a superscript for Hom, and is set left or right depending on whether the resulting T-module is left or right.

Lemma 11.60: *Let* $\varphi : R \to T$ *be a ring map,* $A = A_R$, $B = {}_RB$.

(i) *If A is R-projective, then A_φ is T-projective.*
(ii) *If B is R-projective, then ${}_\varphi B$ is T-projective.*
(iii) *If A is R-injective, then $\,^\varphi A$ is T-injective.*
(iv) *If B is R-injective, then B^φ is T-injective.*

Proof: (i) and (ii): Exercise 3.10.
(iii) and (iv): Exercise 3.22. ∎

Exercise: 11.33. Use Theorem 11.8 to state assumptions on T (projective or flat R-module, on the appropriate side) to ensure the change of rings functors preserve projectivity or injectivity.

Remember that a left T-module L may be viewed as a left R-module, and a right T-module M may be viewed as a right R-module.

Lemma 11.61: *If $\varphi : R \to T$ is a ring map, then there are natural isomorphisms in the following cases:*

(i) $(A_R, {}_TL)$: $A \otimes_R L \cong A_\varphi \otimes_T L$;
(ii) $(M_T, {}_RB)$: $M \otimes_R B \cong M \otimes_T ({}_\varphi B)$;
(iii) $({}_RB, {}_TL)$: $\operatorname{Hom}_R(B, L) \cong \operatorname{Hom}_T({}_\varphi B, L)$;
(iv) $({}_TL, {}_RB)$: $\operatorname{Hom}_R(L, B) \cong \operatorname{Hom}_T(L, \,^\varphi B)$.

Proof: (i) and (ii) follow from the associativity of tensor product. For example,

$$A_\varphi \otimes_T L = (A \otimes_R T) \otimes_T L \cong A \otimes_R (T \otimes_T L) \cong A \otimes_R L.$$

(iii) and (iv) follow from the adjoint isomorphisms:

$$\operatorname{Hom}_T({}_\varphi B, L) = \operatorname{Hom}_T(T \otimes_R B, L)$$
$$\cong \operatorname{Hom}_R(B, \operatorname{Hom}_T(T, L)) \cong \operatorname{Hom}_R(B, L);$$
$$\operatorname{Hom}_T(L, \,^\varphi B) = \operatorname{Hom}_T(L, \operatorname{Hom}_R(T, B))$$
$$\cong \operatorname{Hom}_R(T \otimes_T L, B) \cong \operatorname{Hom}_R(L, B). \quad ∎$$

Theorem 11.62   (Change of Rings):   *Let $\varphi: R \to T$ be a ring map, and let $A = A_R$ and $L = {}_T L$. There is a spectral sequence*

$$\operatorname{Tor}_p^T(\operatorname{Tor}_q^R(A, T), L) \underset{p}{\Rightarrow} \operatorname{Tor}_n^R(A, L).$$

Proof: Let $G = (\ )_\varphi = \otimes_R T$, and let $F = \otimes_T L$; by Lemma 11.61, $FG = \otimes_R L$. Now F is right exact and left acyclicity holds: if P_R is projective, then P_φ is T-projective and $\operatorname{Tor}_i^T(P_\varphi, L) = 0$ for $i > 0$. Now apply Theorem 11.39.   ∎

Corollary 11.63:   *Let $\varphi: R \to T$ be a ring map; if ${}_R T$ is flat, then there are isomorphisms for all $n \geq 0$,*

$$\operatorname{Tor}_n^T(A_\varphi, L) \cong \operatorname{Tor}_n^R(A, L)$$

for $A = A_R$ and $L = {}_T L$.

Proof: Flatness of T forces the spectral sequence to collapse.   ∎

Here is an elementary proof of this corollary. If $\mathbf{P}$ is an R-projective resolution of A, there is an isomorphism of complexes:

$$(\mathbf{P} \otimes_R T) \otimes_T L \cong \mathbf{P} \otimes_R L.$$

Since $\mathbf{P} \otimes_R T$ is a flat resolution of $A \otimes_R T = A_\varphi$ (because ${}_R T$ is flat), taking homology of both sides yields the desired isomorphisms.

Compare this result with Theorem 9.49 when R is commutative and $T = S^{-1}R$.

The other isomorphisms of Lemma 11.61 also produce spectral sequences; it suffices to state the theorems.

Theorem 11.64   (Change of Rings):   *If $\varphi: R \to T$ is a ring map, there is a spectral sequence*

$$\operatorname{Tor}_p^T(M, \operatorname{Tor}_q^R(T, B)) \underset{p}{\Rightarrow} \operatorname{Tor}_n^R(M, B)$$

for $M = M_T$ and $B = {}_R B$. Moreover, if T_R is flat, there are isomorphisms

$$\operatorname{Tor}_n^T(M, {}_\varphi B) \cong \operatorname{Tor}_n^R(M, B).$$

Theorem 11.65   (Change of Rings):   *If $\varphi: R \to T$ is a ring map, there is a spectral sequence*

$$\operatorname{Ext}_T^p(\operatorname{Tor}_q^R(T, B), L) \underset{p}{\Rightarrow} \operatorname{Ext}_R^n(B, L)$$

for $B = {}_RB$ and $L = {}_TL$. Moreover, if T_R is flat, there are isomorphisms

$$\text{Ext}_T^n({}_\varphi B, L) \cong \text{Ext}_R^n(B, L).$$

Theorem 11.66   (Change of Rings):   *If $\varphi: R \to T$ is a ring map, there is a spectral sequence*

$$\text{Ext}_T^p(L, \text{Ext}_R^q(T, B)) \underset{p}{\Rightarrow} \text{Ext}_R^n(L, B)$$

for $B = {}_RB$ and $L = {}_TL$. Moreover, if ${}_RT$ is projective, there are isomorphisms

$$\text{Ext}_T^n(L, {}^\varphi B) \cong \text{Ext}_R^n(L, B).$$

We give only three applications of these theorems in order to show how they may be used. The first is another proof of Shapiro's Lemma 10.32.

Corollary 11.67:   *If S is a subgroup of a group Π and B is a Π-module, then*

$$H^n(S, B) \cong H^n(\Pi, \text{Hom}_S(\mathbf{Z}\Pi, B)).$$

Proof: First of all, the inclusion $S \hookrightarrow \Pi$ induces a ring map $\varphi: \mathbf{Z}S \to \mathbf{Z}\Pi$. Since $H^n(\Pi, \ ) = \text{Ext}_{\mathbf{Z}\Pi}^n(\mathbf{Z}, \ )$, we are in the situation of Theorem 11.66 with $L = \mathbf{Z}$. As $\mathbf{Z}\Pi$ is a projective S-module (even free, by Exercise 10.9), and ${}^\varphi B = \text{Hom}_S(\mathbf{Z}\Pi, B)$, one merely asserts the isomorphism of Theorem 11.66.   ∎

Here is the spectral sequence proof of Theorem 9.37 of Rees.

Corollary 11.68:   *Assume R is a ring and $x \in R$ is a central element that is neither a unit nor a zero divisor; let $T = R/xR$, and assume $B = {}_RB$ is such that multiplication $B \xrightarrow{x} B$ is monic. Then there are isomorphisms for all $n \geq 0$*

$$\text{Ext}_T^n(L, B/xB) \cong \text{Ext}_R^{n+1}(L, B)$$

for every left T-module L.

Proof: Since x is central, multiplication by x is an R-map; since x is not a zero divisor, there is an R-exact sequence

$$0 \to R \xrightarrow{x} R \to T \to 0.$$

Applying $\text{Hom}_R(\ , B)$ yields exactness of

$$0 \to \text{Hom}_R(T, B) \to \text{Hom}_R(R, B) \xrightarrow{x} \text{Hom}_R(R, B) \to \text{Ext}_R^1(T, B) \to \text{Ext}_R^1(R, B)$$

and, also, exactness of the sequences

$$\text{Ext}_R^q(R, B) \to \text{Ext}_R^{q+1}(T, B) \to \text{Ext}_R^{q+1}(R, B)$$

for all $q \geq 1$. Since $\mathrm{Hom}_R(R, B) = B$ and multiplication by x on B is monic, we have

$$\mathrm{Hom}_R(T, B) = 0.$$

Since $\mathrm{Ext}_R^1(R, B) = 0$, we also obtain

$$\mathrm{Ext}_R^1(T, B) \cong B/xB.$$

Finally, the other exact sequences give

$$\mathrm{Ext}_R^q(T, B) = 0, \qquad q \geq 2.$$

Consider the change of rings spectral sequence 11.66

$$E_2^{p,q} = \mathrm{Ext}_T^p(L, \mathrm{Ext}_R^q(T, B)) \underset{p}{\Rightarrow} \mathrm{Ext}_R^n(L, B).$$

We have just seen that $E_2^{p,q} = 0$ for all $q \neq 1$, so there is collapsing on the line $q = 1$. There are thus isomorphisms

$$\mathrm{Ext}_T^{n-1}(L, \mathrm{Ext}_R^1(T, B)) \cong \mathrm{Ext}_R^n(L, B);$$

replacing n by $n + 1$ yields

$$\mathrm{Ext}_T^n(L, B/xB) \cong \mathrm{Ext}_R^{n+1}(L, B). \quad \blacksquare$$

Our final application is a solution to Exercise 11.20 = Theorem 9.32.

Corollary 11.69: *If $\varphi : R \to T$ is a ring map and L is a T-module, then*

$$\mathrm{pd}_R(L) \leq \mathrm{pd}_T(L) + \mathrm{pd}_R(T).$$

Proof: Clearly we may assume both $\mathrm{pd}_T(L) = l$ and $\mathrm{pd}_R(T) = t$ are finite. By Theorem 11.66, there is a spectral sequence

$$E_2^{p,q} = \mathrm{Ext}_T^p(L, \mathrm{Ext}_R^q(T, B)) \underset{p}{\Rightarrow} \mathrm{Ext}_R^n(L, B)$$

for every R-module B. If $n = p + q > l + t$, then either $p > l$ or $q > t$. If $p > l$, then $E_2^{p,q} = 0$ because we have exceeded $\mathrm{pd}_T(L)$; if $q > t$, then $E_2^{p,q} = 0$ because we have exceeded $\mathrm{pd}_R(T)$. It follows that $E_\infty^{p,q} = 0$ for all p, q with $p + q = n$, whence $\mathrm{Ext}_R^n(L, B) = 0$. As B is arbitrary, $\mathrm{pd}_R(L) \leq l + t$. $\quad \blacksquare$

The reader should now be convinced that virtually every purely homological result may be proved with spectral sequences. Even though "elementary" proofs may exist for many of these results, spectral sequences offers a systematic approach in place of sporadic success.

References

Albrecht, F., On projective modules over a semihereditary ring, *Proc. Amer. Math. Soc.* **12** (1961), 638–639.

Anderson, F., and Fuller, K., *Rings and Categories of Modules*. Springer-Verlag, Berlin and New York, 1974.

Auslander, M., and Buchsbaum, D. A., Unique factorization in regular local rings, *Proc. Nat. Acad. Sci. U.S.A.* **45** (1959), 733–734.

Bass, H., Big projective modules are free, *Illinois J. Math.* **7** (1963), 24–31.

Berstein, I., On the dimension of modules and algebras IX, direct limits, *Nagoya Math. J.* **13** (1958), 83–84.

Brewer, J. W., and Costa, D. L., Projective modules over some non-Noetherian polynomial rings, *J. Pure Appl. Algebra* **13** (1978), 157–163.

Cartan, H., and Eilenberg, S., *Homological Algebra*. Princeton, Univ. Press, Princeton, New Jersey, 1956.

Chase, S. U., Direct products of modules, *Trans. Amer. Math. Soc.* **97** (1960), 457–473.

Chase, S. U., A generalization of the ring of triangular matrices, *Nagoya Math. J.* **18** (1961), 13–25.

Cohen, D. E., *Groups of Cohomological Dimension One*. Lecture Notes 245, Springer-Verlag, Berlin and New York, 1972.

Cohn, P. M., Some remarks on the invariant basis property, *Topology* **5** (1966), 215–228.

Cohn, P. M., *Free Rings and Their Relations*. Academic Press, New York, 1971.

Curtis, C., and Reiner, I., *Representation Theory of Finite Groups and Associative Algebras*. Wiley (Interscience), New York, 1962.

Dade, E. C., Localization of injective modules, *J. Algebra* **69** (1981), 415–425.

DeMeyer, F., and Ingraham, E., *Separable Algebras over Commutative Rings*. Lecture Notes 181, Springer-Verlag, Berlin and New York, 1971.

Endo, S., On semihereditary rings, *J. Math. Soc. Japan* **13** (1961), 109–119.

Endo, S., On flat modules over commutative rings, *J. Math . Soc. Japan* **14** (1962), 284–291.

Faith, C., *Algebra: Rings, Modules, and Categories*, Vol. I. Springer-Verlag, Berlin and New York, 1973.

Fossum, R., Griffith, P., and Reiten, I., *Trivial Extensions of Abelian Categories.* Lecture Notes 456, Springer-Verlag, Berlin and New York, 1975.

Fuchs, L., *Infinite Abelian Groups,* Vol. I. Academic Press, New York, 1970.

Fuchs, L., *Infinite Abelian Groups,* Vol. II. Academic Press, New York, 1973.

Goodearl, K. R., Global dimension of differential operator rings, *Proc. Amer. Math. Soc.* **45** (1974), 315–322.

Griffith, P., On the problem of intrinsically characterizing the functor Ext, *Symp. Math. XIII,* p. 207–232. Academic Press, New York, 1974.

Gruenberg, K., *Cohomological Topics in Group Theory.* Lecture Notes 143, Springer-Verlag, Berlin and New York, 1970.

Helmer, O., Divisibility properties of integral functions, *Duke Math. J.* **6** (1940), 345–356.

Herstein, I. N., *Noncommutative Rings.* Carus Mathematical Monograph, Math. Assoc. Amer., Providence, Rhode Island, 1968.

Isaacs, I. M., *Character Theory of Finite Groups.* Academic Press, New York, 1976.

Jans, J., *Rings and Homology.* Holt, New York, 1964.

Jategaonkar, A. V., A counter-example in ring theory and homological algebra, *J. Algebra* **12** (1969), 418–440.

Jensen, C. U., Homological dimensions of rings with countably generated ideals, *Math. Scand.* **18** (1966), 97–105.

Kaplansky, I., Projective modules, *Ann. Math.* **68** (1958a), 372–377.

Kaplansky, I., On the dimension of rings and modules, X, *Nagoya Math. J.* **13** (1958b), 85–88.

Kaplansky, I., *Commutative Rings.* Allyn and Bacon, Boston, Massachusetts, 1970.

Kong, M., Euler classes of inner product modules, *J. Algebra* **49** (1977), 276–303.

Lam, T. Y., *Serre's Conjecture.* Lecture Notes 635, Springer-Verlag, Berlin and New York, 1978.

Lambek, J., *Lectures on Rings and Modules.* Ginn (Blaisdell), Boston, Massachusetts, 1966.

Lang, S., *Algebra.* Addison-Wesley, Reading, Massachusetts, 1965.

Lyndon, R. C., The cohomology theory of group extensions, *Duke Math. J.* **15** (1948), 271–292.

MacLane, S., *Homology.* Springer-Verlag, Berlin and New York, 1963.

MacLane, S., *Category Theory for the Working Mathematician.* Springer-Verlag, Berlin and New York, 1971.

Matsumura, H., *Commutative Algebra.* Benjamin (Addison-Wesley), Reading, Massachusetts, 1970.

Milnor, J., *Introduction to Algebraic K-Theory.* Princeton, Univ. Press, Princeton, New Jersey, 1971.

Nagata, M., A general theory of algebraic geometry over Dedekind rings, II, *Amer. J. Math.* **80** (1958), 382–420.

Northcott, D. G., *An Introduction to Homological Algebra.* Cambridge, Univ. Press, London and New York, 1960.

Nunke, R. J., and Rotman, J. J., Singular cohomology groups, *J. London Math. Soc.* **37** (1962), 301–306.

Ojanguren, M., and Sridharan, R., Cancellation of Azumaya algebras, *J. Algebra* **18** (1971), 501–505.

Ol'šanskii, A. Y. On the Novikov–Adyan Theorem, *Mat. Sb. (N.S.)* **118** (160), No. 2 (1982). 203–235 (in Russian). [*Engl. transl.: Math USSR-Sb.* **46**, No. 2 (1983), 203–236.]

Osofsky, B. L., Upper bounds of homological dimension, *Nagoya Math. J.* **32** (1968), 315–322.

Osofsky, B. L., *Homological Dimensions of Modules, CBMS Regional Conf. Ser. Math.* 12. Amer, Math. Soc., Providence, Rhode Island, 1973.

Roitman, M., On projective modules over polynomial rings, *J. Algebra* **58** (1979), 51–63.

Rotman, J., *The Theory of Groups: An Introduction,* 2nd ed. Allyn and Bacon, Boston, Massachusetts, 1973.

Shelah, S., Infinite abelian groups, Whitehead problem, and some constructions, *Israel Math. J.* **18** (1974), 243–256.

Small, L., Hereditary rings, *Proc. Nat. Acad. Sci. U.S.A.* **55** (1966), 25–27.

Stallings, J., On torsion-free groups with infinitely many ends, *Ann. Math.* **88** (1968), 312–334.

Stammbach, U., *Homology in Group Theory*. Lecture Notes 359, Springer-Verlag, Berlin and New York, 1973.

Swan, R. G., Vector bundles and projective modules, *Trans. Amer. Math. Soc.* **105** (1962), 264–277.

Swan, R. G., Groups of cohomological dimension one, *J. Algebra* **12** (1969), 585–610.

Weiss, E., *Cohomology of Groups*, Academic Press, New York, 1969.

Zariski, O., and Samuel, P., *Commutative Algebra*, Vol. I. Van Nostrand Reinhold, Princeton, New Jersey, 1958 (2nd ed., Springer-Verlag, Berlin and New York, 1975).

Index

Printed and bound by CPI Group (UK) Ltd, Croydon, CR0 4YY

03/10/2024

01040425-0011